LEARNING RESOURCES SERVICE

Taunton Campus
Wellington Road, Taunton TA1 5AX
01823 366 372
tauntonlrc@btc.ac.uk

SOMERSET COLLEGE

T0144173

The book is due for return on or before the last date shown

ENVIRONMENTAL SCIENCES

A STUDENT'S COMPANION

Kenneth J. Gregory
Ian G. Simmons
Anthony J. Brazel
John W. Day
Edward A. Keller
Arthur G. Sylvester and
Alejandro Yáñez-Arancibia

Los Angeles • London • New Delhi • Singapore • Washington DC

SAGE Publications Ltd
1 Oliver's Yard
55 City Road
London EC1Y 1SP

SAGE Publications Inc.
2455 Teller Road
Thousand Oaks, California 91320

SAGE Publications India Pvt Ltd
B 1/I 1 Mohan Cooperative Industrial Area
Mathura Road
New Delhi 110 044

SAGE Publications Asia-Pacific Pte Ltd
33 Pekin Street #02-01
Far East Square
Singapore 048763

Library of Congress Control Number: 2008933203

British Library Cataloguing in Publication data

A catalogue record for this book is available from the British Library

ISBN 978-1-4129-4704-6
ISBN 978-1-4129-4705-3 (pbk)

Typeset by Cepha Imaging Pvt Ltd, Bangalore, India
Printed in Great Britain by The Cromwell Press Ltd, Trowbridge,
Wiltshire
Printed on paper from sustainable resources

CONTENTS

PREFACE

At least three major developments have affected the environmental sciences in the 21st century. First, there has been a more holistic approach as it was appreciated that the increasingly specialized reductionist trends of the 20th century had to be complemented by a more integrated understanding, by more multidisciplinary research and studies, and by interdisciplinary endeavours. Knowing more and more about less and less was no longer sufficient to progress the environmental sciences. Second, there has been the greater acceptance of global climate change and greater awareness of the implications that it may bring. Therefore the environmental sciences have assumed greater familiarity in the minds of the general public. Third, the communication revolution, with availability of material through the internet, has altered the way in which students of the environmental sciences have access to data.

The idea for this volume was stimulated by the UK benchmark statement on Environmental Sciences and Environmental Studies at first degree level prompting a suggestion from Robert Rojek. We envisaged that there could be a volume which introduced the field of environmental sciences in such a way that it would appeal to students in English-speaking countries, giving much relevant material for first-year students following courses in subjects which make up the range of environmental sciences and environmental studies. It has been compiled so that it would be of most relevance to first year higher education (HE) students but would also serve as a reference during more advanced courses, have entries that are informative but concise, which are presented in a thought-provoking way, and use a standard level of reference citation with emphasis upon recent key references that can be used to follow up specific topics. Although an introductory text with some basic references this work could be consulted throughout the degree course, so it includes 'scaled' references pertinent to different levels of HE programmes.

Introductory material is already provided in two ways, in basic texts and in dictionaries or encyclopaedias. There are already a number of basic texts available and a range of dictionaries has already been published covering environmental sciences, earth sciences, ecology and geography, and other volumes are in preparation. However many existing volumes are so substantial that they would not be purchased by individual students. This publication therefore attempts a third type of provision, as a short benchmark volume introducing major concepts and contemporary issues that need to be appreciated and understood. This 'third way' benchmark volume is called a Companion, intended to be intermediate between a dictionary (expensive and extensive) and a textbook (several needed to cover the field of environmental sciences). It has entries, arranged alphabetically in sections, each aiming to give a current and up-to-date summary of critical issues and concepts in the field of environmental science. The book should be a starting point but is compiled bearing in mind the needs of students in the US and North America and in the southern hemisphere as well as in the UK and Europe. It is therefore a volume that first-year students can continue to use throughout their undergraduate course. Each article has recent key references and diagrams as necessary, introduces key words that are included in the index, cites websites that are pertinent and active in the summer of 2007, and reflects the current state of thinking. The index is extensive, containing many more terms than are elaborated in the text articles.

Entries are arranged alphabetically in six major sections which are entitled:

» Environmental Sciences;
» Environments;
» Paradigms/Concepts;
» Processes and Dynamics;
» Scales and Techniques;
» Environmental Issues.

The entry in the first section, on environmental sciences, provides an introduction to the volume, showing how a number of disciplines constitute the field of environmental sciences.

In *The Revenge of Gaia* James Lovelock suggests that we need 'a book of knowledge .... for anyone interested in the state of the Earth and of us – a manual for living well and for survival .....it would be a primer of philosophy and science' (Lovelock, 2006, p. 158). This Companion cannot aspire to meet that suggestion but it may perhaps contribute towards it.

Ken Gregory acknowledges the assistance of his wife in preparing the index together with her unfailing support for this venture. Tony Brazel gratefully acknowledges help by Sophia Tsong.

Every attempt has been made to obtain permission to reproduce copyright material, but if proper acknowledgement has not been made we invite copyright holders to please inform us of the oversight.

Kenneth J. Gregory
Ian G. Simmons

Further reading

Lovelock, James, 2007. *The Revenge of Gaia*. Allen Lane, London.

PART I

ENVIRONMENTAL SCIENCES

INTRODUCTION

A landscape induces various reactions in humans. To some, it is a picture in the sense that it is an assemblage of elements such as slopes, soil and sky which make a scene that evokes a basically emotional reaction: 'isn't that pretty/magnificent/dreary?' It may then produce the desire to capture that landscape in pencil, paint or pixels. Equally, some people's reaction is to want to write about it as a travelogue or part of their blog and they may even be moved to capture certain characteristics in poetic form. There is another category of enquirers – those who are interested primarily in the land as reflecting the human place in it. The types of buildings; the use of the land, whether agriculture, forestry or suburbia, for instance; the ways in which rivers have been bridged and shorelines reinforced, are central to their envisioning of what is before them. Anything they then communicate to others about it is likely to be seen through the prism of human usefulness, that is the landscape is a set of resources and shows the ways in which societies past and present have perceived and made use of those resources. The time element is important since nearly all landscapes derive features from the past as well as the present. The time depth may be immense and measured in millions of years, as with some elements of the geological history of a region, or very recent as with, for example, the spread of silt over part of a flood plain from the previous night's overspill of a swollen river.

There is yet another category of approach to a landscape, one which is central to this book, namely that of the natural sciences. The approach of science is not as uniform as is sometimes portrayed and our environment provides good examples of the diversity of scientific endeavour. Nevertheless, science has certain organizing principles which define it as different from the formulations described in the first paragraph. In the first place, a scientific investigation tries to eliminate the difference between individual human observers. This is impossible in its entirety, of course, but science tries to express its findings in a common language that is independent of the individual and of the multitude of natural languages that have arisen. Hence in its early development in early modern Europe, science was written in Latin, as was law, theology and political discourse. Nowadays, the preferred language of science is English, though this is mostly a post-1950 development. Latin is most obviously seen now in the names of living organisms, where international practice still uses the scheme developed by Linnaeus in the 18th century. Transcending Latin, however, is mathematics. The use of the equation to describe a process in nature, such as (the orbit of a planet, or) the movement of particles down a slope is a major tool of the environmental sciences and examples are seen throughout this book. Mathematics is free of any connotations except those given by the user and agreed by all the users: this finds expression in simple forms such as measurements of length, where for example

'1 metre' is referable to a standard metre in a National Physical Laboratory, or the emission of radiation by a radioactive element such as uranium is a standard quantity agreed by all those who work with it. There is not much debate over the speed of light, for example. Now, not all measurements and descriptions in science are easily agreed, and many pages of scientific journals are given over to debate and disagreement; this is a major characteristic of this way of thought. The aim is to eliminate the differences due to individual investigations and so by following agreed methods (known as protocols) make it possible to replicate findings. The measurement of CO_2 concentration at 10,000 m above Singapore on January 3rd 2006 at noon local time should yield the same number whether done by a team of scientists in Singapore, Osaka or Oxford. But as many examples in this section of the book show, reduction of knowledge to numerical form is not always possible, and the use of words is always necessary and always need careful scrutiny as to meaning. Many scientific concepts are expressed, for example, as metaphors, which can be misleading either because the relationship is not adequately conveyed by the metaphor or because the metaphor was an early attempt at explanation which is now superseded by newer knowledge but which has lodged in the popular mind. Most of us think of an atom as having a dense nucleus around which whirl other particles like the runners on an athletics track; the reality seems to need far greater amounts of space and higher levels of unpredictability. When James Lovelock (see p. 207) wanted to conceptualize his Gaia hypothesis for the 'physiology' of the whole of planet Earth, he used the idea of a single living organism in all its complexity as a metaphor and so called the Earth a 'superorganism'.

The demand that the results of science should be replicable has led to some common features of practice. The general rule is that the first approach to a feature of the world is that of observation. Whether of atoms, molecules, cells, organs, organisms, physical environments such as air and water, or combinations of any or all of these, a description must be made of size, number, distribution and behaviour. It may be that this is enough: if only an inventory is required since the place is going to be left alone (as in a National Park, perhaps) then it is sufficient to know that there are six individual specimens of sloths per 10 square kilometres and leave it at that. But a classic next step is to formulate a hypothesis which looks at the possible explanation for the observed patterns. We ask not only, 'what' but 'why'? In this case the hypothesis might be along the lines of 'we think that the density of sloths is related to presence of a particular tree species, which is their main food'. Science then demands that this hypothesis be tested. Now, in a piece of laboratory chemistry or physics, this can be done indoors in controlled conditions. In much of environmental science this is not the case, for the objects of investigation are embedded in the outside world and wrenching them free of it for investigation may well rob the project of any reality whatsoever. So testing sloth density might mean plotting the distribution of the likely food tree and comparing the sloth numbers in areas of different tree density. Unless a connection between sloths and trees is made, however, fallacies could arise. So more observation is needed: a student is hired to sit in the forest for a few days so as to watch and record sloth activities: do they eat the leaves of that particular type of tree? Do they eat leaves from other tree-types also? In a few cases, the equivalent of the laboratory experiment is possible: in order to study the rates of colonization of bare islands, some small islets in the Caribbean were fumigated, so that insect influx could be measured from a zero datum line.

To see how clays and silts behave during glaciations, samples can be put into cold chambers and repeatedly frozen and unfrozen to mimic natural conditions. So in the end, the combination of observation and tested hypothesis allows a scientific law to be stated: 'sloths of this species require trees of that species for their existence'. Such a law, as with many laws in physics, has a predictive quality. Here the prediction would be that if we fell all the trees of the food species, then the sloths will die out, just as the knowledge of planetary orbits allows the prediction of eclipses.

In the environmental context the aims of scientific endeavour may sometimes have to depart from the purer protocols of e.g., physics and chemistry since the accessibility of the process to be investigated may be problematic. However, as the ingenuity and skill of investigators is cumulative, improvements are constant: the concentration of contaminants like pesticides in the water of the oceans was in the 1950s not measurable at all, then became possible in parts per million (ppm) and now is often expressible in parts per billion (bbp). The concentration of trace gases in the upper atmosphere has had a comparable history. Looked at overall, the sciences of the environment fall into a number of categories, none of which is exclusive. There are first of all the historical sciences which by definition examine the past. Parts of geology (such as palaeontology) are good examples, as is palaeoecology. Obviously enough, these are immune to experiment in the true sense: the past is not available for experiment. However, some attempts at analogy may be made since restoration projects are in themselves a form of experiment. When rivers have their training walls, dams and straightened sections removed in favour of a more 'natural' condition then this is akin to experimenting in the laboratory. A river flume in a lab may also be rigged to mimic conditions in the past; a period of exceptionally heavy rainfall, for instance. It is uncommon for historic sciences to be predictive since the future is rarely totally imitative of the past but findings like those of the atmospheric sciences that carbon dioxide levels have never been so high as now will carry some predictive value even if they are inexact.

A second and major category deals with the major divisions of the study of the environment. Here, the major natural component 'spheres' of the planet are described, classified and where possible laws of behaviour set out. Where measurements are not possible, nor prediction certain, the models based on an explicit set of assumptions have to be made. Thus the connection between rising levels of 'greenhouse gases' and planetary warming is subject to uncertainties, not least because the relationship may not be linear, i.e., there may be sudden flips. So the science is expressed as a series of models, most of which have a number of states: if gases increase by x parts per million (ppm), then temperature goes up by y degrees, is the general form and different outcomes are predicted. So, there are environmental sciences of the lithosphere (rocks), the biosphere, atmosphere, the oceans and freshwater, and ways of portraying the science such as cartography. In each of these, the scientists attempt to make inventories of the materials concerned (how much of the planet's water at any one time is in the oceans?), their physics and chemistry, and their dynamics: how do they move around and at what speeds, how much of them are affected by other processes?

Much of the work done might be said to be analytical, in the sense that smaller and smaller pieces are investigated, with the aim of making more and more precise measurements and other descriptions of the phenomena involved. 'People who know more and more about less and less' is the common jibe. But there is a third category of science which aims at synthesis, that is of looking at the interfaces between the different sciences mentioned in the previous paragraph, so that important transfers of energy and materials (as at the atmosphere–ocean interface) are not overlooked and indeed so that the landscape in its entirety can be seen as single entity and described as such: this is the role of studies like Physical Geography, Earth System Science and Environmental Sciences. Since human activity can now reach into all the natural systems of the earth and affect them in ways not dreamed of before the 20th century, these synthetic appraisals gain an ever greater importance. However, they rest upon tested and replicated analytical work which is continually being evaluated by a sceptical audience and so science always claims that its findings are provisional. The way in which the knowledge is acquired and care taken to make it independent of the prejudices of its discoverers gives the findings of science an authority in the world not held by any other form of knowledge.

Fountain Hills east of Phoenix Arizona (Fig. 1) has developed since 1971, increasing to a population of 17,000 by 1999 and nearly 27,000 in 2007. This completely new urban area has modified the physical environment and affected the natural system of wash channels that flow east to the Verde River, between the urban area and the mountains in the distance. This interaction between human activity and the environment is relatively well planned and includes a fountain constructed in 1971 in the centre of the town, which operates from 9am to 9pm daily sending a stream of water to a height of 170 m and is recorded in the Guinness Book of Records as the tallest fountain in the world. (Photo K.J. Gregory (facing page top) and Anne Chin (facing page bottom).)

Figure 1 *Fountain Hills east of Phoenix Arizona © Ken J. Gregory and Anne Chin*

ANTHROPOLOGY

Anthropology emerged as a social science in the 19th century as an attempt to develop scientific methods to address social phenomena and provide a universal basis for social knowledge. Using the methods of the natural sciences and developing new techniques involving not only structured interviews, but unstructured participant observation, and drawing on the new theory of evolution through natural selection, the branches of anthropology proposed the scientific study of a new object: humankind, conceived as a whole. Crucial to this study is the concept of culture, which anthropologists saw as containing adaptations to local conditions, which takes the form of beliefs and practices about their natural environments, among many other elements. Thus, anthropology eliminates the divisions between the natural sciences, social sciences, and humanities to explore the biological, linguistic, material, and symbolic dimensions of humankind in all forms. Its relevance to environmental matters is mostly in the domain of material use of environmental resources: how and why do people carry out their agriculture, forest and soil management, how and why do they dispose of wastes? These are the concerns of a branch called cultural ecology, which has much in common with other environmental social sciences such as sociology and politics. Physical anthropology is more purely scientific and is much concerned with the material evidence for human evolution and spread. Anthropology has however, much to offer in the study of the non-material, as in the case of environmental mythologies. Though these are often focused on relatively simple cultures and relate to e.g., the treatment of hunted animals, they could also be seen to apply to today's urban myths such as the belief that technology is the solution to all environmental anxieties. Anthropology is also well placed to comment on the question of human emotion in relation to the environment: though much derided by 'pure' scientists and technologists, there is no denying its power and motivational effects, especially at local scales.

See also: CULTURE, RESOURCES, SOIL

Further reading

Jurmain, R., Kilgore, L. and Ciochon, R., 2007. *Introduction to physical anthropology.* Wadsworth, Belmont CA.

Moran, E., 2006. *People and nature: an introduction to human ecological relations.* Blackwell, Oxford.

Peoples, J.G. and Bailey, G., 2006. *Humanity: an introduction to cultural anthropology.* Thomson, Belmont CA.

ARCHAEOLOGY

Archaeology is the study of human cultures through the recovery, documentation and analysis of material remains and environmental data, including architecture, material artifacts, biological and human remains, and landscapes. The goals of archaeology are to document and explain the origins and development of human culture and human ecology for both prehistoric and historic societies. The characteristic initial mode of enquiry is 'the dig' though many other techniques, such as the interpretation of aerial and satellite photography, field walking, geographical information systems (GIS) and the whole suite of sciences which constitute geoarchaeology are used. Whereas these scientific accompaniments were once the subject of minor appendices to published work by experts hastily assembled for a day, their work is now regarded as integral to the planning of an archaeological enterprise. The relevance of archaeology to environmental science is considerable and is especially so in the response of human societies to environmental change. Such changes may be exogenous as when rapid climatic change occurs (as in long periods of drought such as afflicted the American South-West during the Medieval Warm Period) or when extreme events like volcanic eruptions force communities to migrate or be killed, as with Roman Pompeii. They may be endogenous as when population growth exhausts resources (Easter Island/Rapanui is an example) or when a different set of cultural attitudes allows or forbids access to a different set of resources, as with the cow and the pig in Asia. All these changes can be deduced from high-quality excavations and careful post-excavation analysis. Included are the many dating techniques which inform environmental sciences. Controversial areas within archaeology are often concerned with the degree to which the non-material culture can be extrapolated from the material remains: do wall paintings or stauettes of domesticated cattle indicate the actuality of the change to a tamed animal or do they indicate the desire to bring in the species from the wild?

See also: CULTURE, CLIMATIC CHANGE, GEOARCHAEOLOGY, HUMAN ECOLOGY

Further reading

Bayley, J. (ed) 1998. *Science in archaeology*. English Heritage, London.

Scarre, C. (ed) 2005. *The human past: world prehistory and the development of human societies*. Thames and Hudson, London.

ATMOSPHERIC SCIENCES

Atmospheric sciences study the atmosphere, its processes, the effects from other systems, and the effects of the atmosphere on these other systems. Meteorology includes atmospheric chemistry and atmospheric physics with a major focus on weather forecasting. Climatology is the spatio-temporal study of atmospheric changes (both long- and short-term) that define average climates, their variability and drivers, and their change over time, which include natural climate and anthropogenic climate processes. Meteorologists study the atmosphere's physical characteristics, motions, and processes, and how these factors affect the environment, with a strong focus on forecasting of the weather. In addition to predicting the weather, atmospheric scientists attempt to identify and interpret climate trends, understand past weather, and engage in scenarios of future changes. Weather information and meteorological research are also applied in air-pollution control, agriculture, forestry, air and sea transportation, defense, and the study of possible trends in the Earth's climate, which include droughts, ozone depletion, and global warming. Meteorologists use sophisticated computer models to make long- and short-term forecasts of the world's atmosphere and tailor forecasts to local areas. Computer models, satellite data, and climate theory are used by meteorologists and climatologists to interpret research results to make better local weather predictions. Forecasts inform the general public and industries that need accurate weather information for both economic and safety reasons, such as in shipping, air transportation, agriculture, fishing, forestry, and utilities. Physical meteorologists study chemical and physical properties of the atmosphere; the transfer of energy in the atmosphere; and the transmission of light, sound, and radio waves. In addition, they examine factors that affect clouds, rain, and snow; air pollutants in urban areas; and weather phenomena such as the severe storms. Synoptic meteorologists constantly work at discovering new ways to forecast weather. Climatologists study climactic variations spanning thousands of years. They collect, interpret, and analyze past records of surface and upper air observations of many variables including temperatures, humidity, wind, rainfall, and radiation. Climate studies are used to design buildings, plan heating and cooling systems, and in planning as well as in agricultural production. Meteorologists study problems in order to evaluate, analyze, and report on air quality for environmental impact assessments. Research meteorologists also examine and present effective ways to control or diminish air pollution. Because atmospheric science is a small field, relatively few universities specifically offer degree programs in meteorology or atmospheric science, although many departments of earth science, geography, geophysics, and similar disciplines offer atmospheric science and related coursework. Entry level operational meteorologists in the US Federal Government usually are placed in intern positions for training and experience. During this period, they learn about the Weather Service's forecasting equipment and procedures and rotate to various offices to learn about weather systems and other operations and programmes. After completing the training period, these interns are assigned to a permanent station with more specific assignments. Farmers, commodity investors, radio and television stations, as well as transportation, utilities, and construction firms greatly benefit from additional weather information more closely meeting their needs than the general information provided by the US National Weather Service. Research on seasonal and other long-range forecasting has yielded positive responses, demonstrating the immediate need for atmospheric scientists to interpret forecasts and advise climate-sensitive industries. Because many customers for private weather services are in industries sensitive to fluctuations in the economy, the growth and development of private weather services in turn depend on the health of the economy.

See also: CLIMATOLOGY, METEOROLOGY, MICROCLIMATE

Further reading

Fleming, J.R., 1990. *Meteorology in America, 1800–1870*. Johns Hopkins University Press, Baltimore.

Hewitt, C.N. and Jackson, A., 2003. *Handbook of atmospheric science principles and applications*. Blackwell, Oxford.

Wallace, J.M. and Hobbs, P., 2006. *Atmospheric science: an introductory survey.*   Elsevier Academic Press, Burlington, MA.

Useful Resources for learning about the Atmospheric Sciences

American Geophysical Union

Find links to information about AGU's meetings, publications, news items, or access its archives which includes journal table of contents, a complete index to all AGU journals, books, and publications, a meetings database, and a FTP dataset archive.

American Meteorological Society

Information about services, programs, publications, meetings, and policies.

Committee for the National Institute for the Environment

Site includes information on the origin of the NIE, its accomplishments, its current activities, as well as sections on membership, fund-raising, and financial and legislative sponsors.

International Global Atmospheric Chemistry

Provides links to IGAC's field efforts, programs, conferences, meetings, workshops, and its structure. http://www.igac.noaa.gov/

Resources of Scholary Societies – Environmental Sciences

Access to websites and gophers maintained by or for scholarly environmental science societies around the world.

Royal Meteorological Society

Includes information on publications, journals and awards.

World Meteorological Association

This U.N. organization site offers a News Centre, official weather forecasts from around the world, and information for members.

Canada – Meterological Service (MSC)

Canada's Source for Meteorological Data.

United Kingdom Meteorological Office

UK's biggest weather information provider; includes links to other weather sites.

World Meteorological Organization – member list of National Weather Services

Lists National Weather Services or National Hydrological Services for almost 187 member states and territories. (Select 'WMO Members' from right side of page.)

BIOGEOGRAPHY

Biogeography is the science of the distribution of plants and animals. Starting from synoptic pictures of those distributions (usually in the form of maps) it investigates the processes which have brought about those patterns. In that sense it has mostly been a historic science but in recent years some engagement with predictability has emerged. An increasing preoccupation with biodiversity is also characteristic of recent years. It starts inevitably with a concern for classification (taxonomy) for if the basic units of the distributions (be they sub-species, species, genera, families or even larger classification units, or taxa) are not defined then the portrayal of distributions is meaningless.

The distribution of biological taxa involves different spatial scales and is also concerned with time: some species' distributions are the result of continental drift many millions of years ago whereas others reflect recent transfer due to human activity. The distributions of species reflects both their evolutionary history and current interactions with other species as well as the environment. Physical forces have a strong impact on species distributions. For example, continental drift has both allowed interchange and movement of species in the past and prevented it as the continents changed positions. There are similar species of organisms on both sides of the Atlantic in Africa and South America. Continental is responsible for the similarity and for the process of speciation that occurred when the continents separated. Continental drift and the presence of other natural boundaries such as deserts and mountains in combination with evolution have resulted in a number of broad distinctive plant and animal associations over the earth.

These are the Palearctic covering much of Europe, north Africa, and northern Asia, the Oriental covering much of India and southern China, the Australian covering Australia and adjacent islands, the Ethiopian covering mid and southern Africa, the Nearctic covering most of North America, and the Neotropical covering much of Mexico, Central America, and South America. Few plants or animals can be said to have a cosmopolitan distribution, i.e., occurring on every continent except Antarctica. One such is the broad-leaved plantain, *Plantago major*. It seems unaffected by soils or climate and to have an extremely effective dispersal mechanism which has enabled to benefit from human activities such as grazing management and trampling. A broad-scale distribution that is less ubiquitous is that of a damselfly *Enallagma cyathigerum* which is found in northern North America and across northern Eurasia, i.e., a circumpolar range. Given that its dispersal abilities are poor, it seems likely that the species evolved at a time when the continents were one land mass. Another kind of distribution is clearly a relict of former conditions: the Norwegian mugwort, *Artemisia norvegica*, is found in the Highlands of Scotland, Norway and the Ural mountains. Fossil pollen evidence shows that it was widespread during the last glaciation of those regions but become restricted as forests spread after the retreat of the ice. Conversion of tropical forest to savanna has isolated populations of the gorilla from each other so that they cannot interbreed. Then there are species found in restricted areas from which they cannot spread due to barriers of topography or habitat type: islands are very good examples. The classificatory label given is endemic: many European mountain ranges house 12–40 of their total flora as endemics which are found nowhere else. The whole of California is rich in endemics since migration is inhibited by the Pacific Ocean on the west and the mountain ranges of the Western Cordillera on the east.

Much of the interest in distributions is not simply in their spatial arrangement but in the combinations of climate, soil, plants and animals which constitute ecosystems. These again can be described on a variety of scales and biogeography looks at the distribution of the larger of these: those found in a worldwide and continental scales in particular. The major world units of climate-soil-biota such as tundra, boreal forest, deciduous forest, deserts and the like are called biomes. There is no doubt that these are in a major way determined by climate and so changes in climate such as the fluctuations of the ice-cover during the Pleistocene have been key periods in changing biome patterns, just as possible future climate shifts may alter the patterns formed in the last 10,000 years. Many maps of biomes, too, fail to indicate the extent to which human activities over that 10,000-year period have altered their species mix, ecology and distribution patterns.

A current focus for much biogeographical study is that of biodiversity. There are probably about 10 milion macroscopic species on the Earth and an unknown number of microscopic species such as fungi, viruses and bacteria. The knowledge of the relations within and between species in being much enhanced by studies based on genetic material. Biogeographers like many other biological scientists are much concerned with the ways in which the human tenure of the planet is reducing this richness at genetic level as well as in species and higher taxonomic groups. The historical approach of much of biogeography was complemented in the 1960s by some mathematical modelling which looked at the biota of islands and which predicted their biotic richness from finding equilibrium points between rates of immigration of new species and the extinction of those already there. The theory was tested on small islands off the coast of Florida which were fumigated so as to kill their insect and arthropod faunas and then the recolonization was monitored. Factors such as the size of the island, the diversity of habitats and its distance from a continent can all be built into increasingly complex and predictive models. The number of species on an island is determined by two processes; the rate at which new species migrate to the island and the rate at which species become extinct. This balance is affected by the size of the island and the distance from the mainland. Since a larger island has more resources and more ecological niches (different habitats for species to inhabit), it can support more species. An extension of the theory took it into terrestrial habitats such as mountain ranges as topographic 'islands' and even into nature reserves acting as 'islands' of a particular management type.

The development of Biogeography as a subfield within the discipline of Geography has been dominated by work in palaeoecology and by the role of wild vegetation in landscapes. However, some theoretical basis was installed in the mid-20th century with the ideas of Pierre Dansereau (1911–) who suggested that since humans were inevitably involved in much alteration of plants, animals and their communities, that the central themes for geographers might be that humans created new genotypes and new ecosystems. The 21st century concerns with the effects of genetically modified organisms and of the role of the earth's land use and land cover in its carbon budgets may well give added impetus to this organizing principle.

Biogeography is thus linked to major themes in the biological sciences, not least into theories of evolution and extinction, and hence is central to many of the concerns currently expressed about human–environmental relations.

See also: EXTINCTION, EVOLUTION, LANDSCAPES, PALAEOECOLOGY, BIODIVERSITY, ECOSYSTEMS, TAXONOMY

Further reading

Briggs, J.C., 1995. *Global biogeography*. Elsevier, Amsterdam.

Cox, B. and Moore, P.D., 2005. *Biogeography: an ecological and evolutionary approach*. Blackwell Scientific, Oxford, 7th edn.

Crosby, A.W., 2003. *The Columbian exchange: biological and cultural consequences of 1492*. Praeger, Westport, CN.

MacArthur, R.H. and Wilson, E.O., 1967. *The theory of island biogeography*. Princeton University Press, Princeton.

Whittaker, R.J., 1998. *Island biogeography: ecology, evolution and conservation*. OUP, Oxford.

BIOLOGY

Biology is the study of life. As such it is a highly diversified science ranging from studies of individual parts of cells to the entire biosphere. But there are some unifying concepts in biology. Evolution is the central core idea in biology. Evolution explains how the grand diversity of life came about. At the molecular level, the replication of DNA explains how inheritance takes place through reproduction. The biosphere is composed of a number of hierarchical levels, each building on the ones below it and being affected by the levels below and above it. Cells are the basic building block of living systems. Within the cell, molecules form the basic chemical building blocks and these are incorporated into organelles in the cell. Groups of cells form tissues, which then form organisms. Organisms are grouped into populations of the same species. These populations interact with each other in a given area to form a community. The interaction of all organisms in a community with each other and with the abiotic (non-living) environment form an ecosystem. At a landscape level, similar ecosystems form a biome (i.e., temperate deciduous forest biome). All life interacting with the environment forms the biosphere. Many believe that the biosphere can be considered as a single global super-entity which is self-regulating and homeostatic. This is called the Gaia hypothesis. The study of biology involves a broad range of sciences, reflecting the nature of biology itself. Major subdivisions of biology include molecular biology (the study of the chemistry of life), cell biology (the study of the basic building blocks of life), genetics (the study of reproduction and mechanisms of inheritance), evolution (the study of changes in biodiversity over time), the study of the structure and function of organisms, and ecology (the interactions of organisms with each other and with the environment and with each other).

See also: ECOLOGY, ECOSYSTEM, EVOLUTION, GENETICS, GAIA HYPOTHESIS

Further reading

Audesirk, T.G., Byers, B.E., 2002. *Biology – life on earth*. Prentice Hall, Upper Saddle River, New Jersey. 6th edn.

Botkin, D.B. and Keller, E.A., 2007. *Environmental science – earth as a living planet*. Wiley, New York.

Campbell, N.A., Reece, J.B. and Mitchell, L.G., 1999. *Biology*. Addison Wesley Longman, Inc., Menlo Park, CA. 5th edn.

CARTOGRAPHY

Cartography (from the Greek *chartis* = map and *graphein* = write) is the art and science of map and chart making including topographical, geological, pedological and thematic maps, was essential from the early days of exploration when maps were required to chart the progress of overseas discovery. A map can fulfil any one of four functions because it can:

> » display information that varies across space;
> » act as an information storage tool;
> » provide a mechanism which links different sources of information;
> » be regarded as an object of art.

Maps have been used for some 8000 years and have recently been radically changed as a result of advances in computing technology and remote sensing which enabled GIS (Geographical Information Systems) and in satellite provision which facilitated GPS (Global Positioning Systems). Thus, whereas cartography was essential for demonstrating the relative position of places, as part of the work of national agencies responsibility for recording national assets, it now uses computational methods and visual displays to increase understanding of the world by employing interactive, learning environments. Significant advances occurred with the development of the needle-pointing magnetic compass (in China in the 8th century AD), together with optical technology involving the telescope, the sextant and the ability to fix latitude and longitude. Originally dependent upon ground topographic survey methods, more rapid production of maps became possible when photogrammetry provided many of the characteristics of topographic maps from aerial photographs. The need to represent a sphere on a flat surface prompted the development of map projections of different types. Further advances occurred with satellite coverage which brought the advantage of frequently repeated remote sensing.

Cartography has emerged as a distinct field of study, the International Cartographic Association was founded in 1960, and the *Cartographic Journal* established in 1964. The superimposition of spatially located variables on to existing maps has opened up many new uses for maps and has instigated new industries to exploit the potential. As maps function as visualization tools for spatial data which can be stored in a database, recent developments have progressed from analogue methods of mapmaking towards creation of increasingly dynamic, interactive maps that can be manipulated digitally. Thus specific places can be identified by a postcode reference enabling ready application and analysis. Satellite navigation systems in cars and vehicles may apparently reduce the need for maps but this can mean that people's awareness of space relationships is reduced.

See also: FIELD STUDIES AND SURVEYING IN GEOLOGY, GIS, GLOBAL POSITIONING SYSTEMS, LANDSCAPE ECOLOGY

Further reading

Kain, R. and Delano-Smith, C., 2003. Geography displayed. In Johnston, R. and Williams, M. (eds), *A century of British geography*. Oxford University Press, Oxford, 371–427.

MacEachren, A.M., 1995. *How maps work: representation, visualization and design*. Guilford Press, New York.

MacEachren, A.M. and Kraak, M.J., 1997. Exploratory cartographic visualization: advancing the agenda. *Computers and Geosciences* 23, 335–343.

Rhind, D., 2003. The geographical underpinning of society and its radical transition. In Johnston, R. and Williams, M. (eds), *A century of British Geography*. Oxford University Press, Oxford, 429–461.

CLIMATOLOGY

Two major fields of the atmospheric sciences are meteorology, which deals with short-term weather phenomena, and climatology, which studies the frequency of weather systems on large scales of both space and time. Thus the weather experienced this week is a topic for the meteorologists, whereas its incidence over the last 50 years, or the difference between the patterns for Buenos Aires and Hong Kong, are the domain of climatologists. The investigation of the dynamics of the atmosphere as they impinge upon the surface of the earth is inevitably complex since the science captures the eventual outcome of many phenomena and processes. The circulation patterns of air and water, the chemical and physical composition of the atmosphere, the interactions between it and both land- and sea-surfaces, are all relevant. A good number of other sciences are thus related to climatology, including chemistry, physics, geophysics, oceanography, geomorphology and volcanology, The figure shows the main interactions that produce climate and are also the sources of its changes. Since the atmosphere is a single global system, the outcome of differences in these interactions is subject to unpredictability and especially to non-linear changes or 'flips'. Thus forecasting the weather for a few days may be problematic in some regions but doing so for the next twenty years is fraught with major difficulties.

Many academic fields have become interested in palaeoclimatology. This tries to reconstruct the climates of the past in order to provide a setting for historical processes such as the migrations of plants and animals. As well, the climatic context for parts of human history may be important. The role of climatic shifts in the eastern Mediterranean at the end of the Pleistocene in the development of early agriculture is one such theme of interest to archaeologists; the storminess of the North Sea in the later 13th century AD has many implications for economies and landscapes of the bordering countries. The resurrection of past climates can rely on instrumental records for only about 200 years and so before that, many types of documents have to be interpreted for their climatic information and 'proxy measures' such as plant and animal remains in fossil and sub-fossil contexts have to be examined for their climatic relevance.

The sub-field with the greatest current profile is the modelling of climatic change. The considerable likelihood that human activities are affecting climatic change at the global scale and in possibly unpredictable ways, has led to a huge amount of scientific effort. This is put into assembling as much information as possible about present and past climates and their rates of change, and then making computer-based predictive models. The uncertainties in these models are made explicit and these often lead to hitherto unexplored processes in the atmosphere, some of which are quite subtle and difficult to measure. The sensitivity of the whole system to small changes is very difficult to forecast, hence the popularity of the 'butterfly effect' in which the beating of the creature's wings in Tokyo causes a storm in New York. Strictly speaking this is meteorology but large-scale versions are easily imagined by alert students.

See also: CLIMATE CHANGE, PLEISTOCENE, PALAEOCLIMATOLOGY, OCEANOGRAPHY, GEOMORPHOLOGY, ECOLOGY, ATMOSPHERIC SCIENCES, METEOROLOGY

Further reading

Barry, R.G. and Chorley. R.J., 2003. *Atmosphere, weather and climate*. London and New York, Routledge, 8th edn.

Strahler, A. and Strahler, A., 2005. *Physical geography: science and systems of the human environment*. Wiley, Chichester, 3rd edn.

Fagan, B., 2005. *The long summer: how climate changed civilization*. Granta, Cambridge.

Flannery, T., 2006. *The weather makers: the history and future impact of climatic change*. Allen Lane, London.

IPCC, 2001 *Third assessment report* @ www.ipcc.ch

Smithson, P., Addison, K. and Atkinson, K., 2002. *Fundamentals of physical geography*. Routledge, London and New York, 3rd edn.

EARTH SYSTEM SCIENCE

Earth system science is the study of earth as a system in terms of its various component systems, such as the atmosphere, hydrosphere, biosphere, and lithosphere. Of particular importance are the linkages between the various systems and how they have been formed, evolved and then maintained through Earth history. There is particular concern with how these systems will evolve and change over periods of time which are of significance to people today. Generally, this is a timescale from a decade or so to perhaps a century or longer. The linkages of the system are important if it is going to be feasible to predict the impacts of a change of one part of a system with another. For example, today in environmental science we are concerned with linkages between the hydrosphere and atmosphere in global warming and how the warming is linked to life on earth. The major challenge is to be able to predict changes that are important to society and use this information to develop management strategies to minimize potential adverse environmental impacts resulting from change. The emergence of earth system science has resulted in a much better understanding of how our planet works. It is now understood that life on earth plays an important role in regulating the global climate through a series of feedback mechanisms that helps maintain the global environment in a more favourable condition for life. For example, single-celled plants floating near the surface of the ocean are a sink for carbon dioxide that helps cool the planet. Some single-celled organisms also release gases such as dimethyl sulphide, which in the atmosphere forms particles important through precipitation in the transfer of sulphur back to the land from the oceans. What is being learned from earth system science is that the various systems on the planet are tightly linked and together allow for the global cycling of important nutrients and other elements that maintain ecosystems on a planetary scale.

See also: ENVIRONMENTAL SCIENCES, GEOLOGY, HOLISM, PARADIGMS

Further reading

Clifford, N. and Richards, K., 2005. Earth system science: an oxymoron? *Earth Surface Processes and Landforms* 30, 379–383.

Earth Systems Science Committee. 1988. *Earth systems science*. National Aeronautics and Space Administration, Washington, DC.

Lawton, J., 2001. Earth system science. *Science* 292, 1965.

ECOLOGY

E cology is a very broad science that spans the entirety of the biosphere. It includes aspects of biological, chemical, geological, and physical factors. Ecology is defined as the study of the relation of organisms or groups of organisms to each other and to their environment. In 1968, Ramón Margalef gave a definition of ecology that stressed the systems level focus of ecology. He stated

> ecology is the study of systems at the level in which individuals or whole organisms can be considered as elements of interaction, either among themselves, or with a loosely organized environmental matrix. Systems at this level are called ecosystems, and ecology is the biology of ecosystems.

Thus, ecology is the environmental matrix of the biosphere, the interactions among specific organisms and the environment, and the structure and functioning of whole ecosystems.

Ecology is normally studied at the different levels of organization of the biosphere. Thus, there is individual or species ecology, population ecology, community ecology, and systems ecology. Species ecology focuses on such topics as the relationship of an organism to the abiotic environment, the energy balance of an individual organism, organism behaviour (mainly animal), and evolutionary consideration of the fitness of and adaptation of species. Energy balance is important to an organism because an organism must maintain a positive energy balance if it is to survive, and organisms have developed many adaptations to do this. Important abiotic factors affecting organisms include such factors as temperature, moisture, soil, nutrients, and fire. The affect of human activities such as pollution is an important aspect of species ecology. Population ecology focuses on all the individuals of the same species. Populations have characteristics different from individuals that make up the population. Individuals are born and die, but populations continue. An individual is generally a male or female, but a population has a sex ratio. Population ecology is concerned with such issues as rates of birth and death, and growth, density, and regulation of populations. Evolutionary considerations are important in population ecology in terms of the relationships of such factors as feeding efficiency, social organization, mating systems, and sex ratios to evolutionary fitness.

Community ecology is the study of the interactions of populations. Trophic or feeding relationships are a fundamental aspect of community ecology. Trophic interactions included a variety of complicated activities. Optimal foraging is concerned with the finding and catching of food. Foraging involves where to look for food, how long to look, and whether to pursue. Herbivory is the consumption of plants. This is the most common type of feeding because herbivores are the link between plants and all other animals. Herbivory involves a number of different types of feeding behaviours such as consumption of the whole plant or only part of it, sucking plant juices or sap, seed predation, and boring into the plant to eat inner parts. Plants have evolved a number of defences against herbivory including production of noxious chemicals and development of tough structural components. Competition is a fundamental aspect of interacting populations, and includes the concepts of the ecological niche and competitive exclusion. Mutualism expresses the way that populations deal in a positive manner with competition including the idea of symbiotic relationships.

At the highest level of organization is ecosystem ecology, which studies the interaction of the community with the abiotic environment. Systems ecology studies the way that the environment affects community structure and diversity. Energy flow in ecosystems is fundamental to this level of ecology. Energy flows into an ecosystem from the sun as it is captured by plants. Other energy sources are also important such as winds, flowing water, and organic matter imported from other ecosystems. The trophic structure of an ecosystem is characterized by food chains and food webs. There is often a pyramid of numbers and biomass, where the high density and biomass of organisms is formed by the plant base of the trophic pyramid. Food can flow through the ecosystem via grazing (where food is consumed alive) and detrital (food is consumed dead) food chains. Biological processes strongly interact with and affect the chemical cycles of nature. This is called biogeochemistry and includes the important cycles of carbon, nitrogen, phosphorus, and sulphur. Many important steps in these cycles are controlled by ecological processes. These include uptake and release of O_2 and CO_2 and biological fixation of nitrogen. Many consider that the entire biosphere is strongly controlled and regulated by ecological processes. This is called the Gaia hypothesis.

Community change and the concepts of ecological succession are aspects of ecosystem ecology. Palaeoecology is the study of past patterns of ecological processes. At the level of the biosphere, ecosystems of the world have been divided into biomes. Biomes are biotic communities of great geographical extent that are characterized by the characteristic climax communities. Climax communities are those that represent the stable stage of an ecosystem for a given set of climatic and environmental conditions. Landscape ecology is the study of ecosystems at a regional level. Examples of important biomes include tropical rain forest, temperate deciduous forest, tundra, desert, savannah, woodland, coniferous forest, and temperate grasslands. There are also aquatic biomes such as streams, lakes, wetlands, and oceans.

Ecology has great practical value in helping us understand human impacts on the biosphere and how to manage these resources in a sustainable manner. This can be generally called applied ecology and is just as important as basic ecology. A number of fields address these issues including conservation biology, restoration ecology, ecological engineering, and eco-technology. The conservation of biodiversity is a central concern of applied ecology. Conserving biodiversity means protecting species, ecosystems, and landscapes. The establishment of parks and refuges is an important element of conservation. But control of pollution is also important because humans are affecting ecology at the level of the biosphere. Global climate change is an important aspect of this. It does no good to protect the Great Barrier Reef in Australia if climate change heats up the ocean so that corals cannot survive. Thus, a great goal of ecology is to help humans manage the biosphere in a sustainable manner.

See also: BIOLOGY, CONSERVATION OF NATURAL RESOURCES, ECOSYSTEM, ECOTECHNOLOGY

Further reading

Allmon, W.D. and Bottjer, D.J. (eds) 2001. *Evolutionary palaeoecology*. Columbia University Press, New York.

Calow, P.P. (ed) 1999. *Blackwell's concise encyclopedia of Ecology*. Blackwell Publishing, Oxford; Malden, Mass.

Brewer, R., 1994. *The science of ecology*. W. B. Saunders, Philadelphia.

Cotgreave, P. and Forseth, I., 2002. *Introductory ecology*. Blackwell Scientific, Oxford.

Kormondy, E., 1995. *Concepts of ecology*. Prentice Hall, New Jersey.

Margalef, R., 1968. *Perspectives in ecological theory*. University of Chicago Press, Chicago.

Margalef, R., 1978. *Perspectives in ecological theory*. The University of Chicago Press, Chicago, IL, Editorial Blume, Spain, 111 pp.

Odum, E.P., 1971. *Fundamentals of ecology*. W.B. Saunders, Philadelphia, 3rd edn. (the most influential ecology text of the mid 20th century).

Odum, H.T., 2007. *Environment, power and society for the twenty-first century: the hierarchy of energy*. Columbia University Press, New York.

Stiling, P., 2001. *Ecology: theory and applications*. Prentice Hall, New Jersey.

Townsend, C.R., Harper, J.L. and Begon, M., 2002. *Essentials of ecology*. Blackwell Science, Oxford.

ECOSYSTEM

An ecosystem is an organized system of land, water, mineral cycles, living organisms and their programmatic behavioural control mechanisms. It includes all the organisms living in a community as well as the abiotic factors with which they interact. Forests, lakes, estuaries, tundra, coral reefs are all different types of ecosystems. Ecosystems have been studied in terms of trophic (feeding) relationships, energy flow, and cycling of chemical elements. Human impacts are often described at the ecosystem level. Trophic or feeding relationships include consideration of primary production which occurs by green plants, algae, and some species of bacteria, consumption by animals and other consumers, and decomposition by microorganisms such as bacteria and fungi. These trophic relationships have a strong impact on energy flow and chemical cycling in ecosystems. Energy flow in ecosystems begins with primary productivity and energy imported into the system. As energy flows through a system, some energy is lost at each step, but materials are conserved. Materials cycle through an ecosystem following trophic and energy pathways. Nutrients, such as nitrogen and phosphorus, and other chemicals, such as metals like iron, are taken up during primary production and incorporated into organic matter. When organic material is consumed and decomposed, these materials are released back into the environment. The availability of these materials often controls the rate of primary production. Human impacts often affect entire ecosystems. Increases in inorganic nutrients, such as runoff from agricultural fields, lead to enrichment in aquatic ecosystems and cause eutrophication (algal blooms, fish kills, and low oxygen). Human activities often physically change ecosystems such as the construction of dams, dredging of canals, cutting of forests, and building roads. Toxic materials are released into the environment leading to both immediate, acute impacts (such as poisoning) and long-term, cumulative impacts (such as concentration of pesticides in the food chain). The effect of DDT on the eggs of hawks, where it thins the eggshells so that they break when incubated is an example of this.

See also: CYCLES, ENERGY, EUTROPHICATION, NITROGEN CYCLE

Further reading

http://www.millenniumassessment.org/en/Index.aspx for a global view

Botkin, D.B. and Keller, E.A., 2007. *Environmental science – earth as a living planet.* Wiley, New York.

Brewer, R., 1994. *The science of ecology.* Saunders College Publishing, Orlando, FA, 2nd edn.

Day, J.W., Hall, C.A.S., Kemp, W.M. and Yáñez-Arancibia, A. 1989. *Estuarine ecology.* Wiley, New York.

Odum, E.P., 1971. *Fundamentals of ecology.* W.B. Saunders, Philadelphia. 3rd edn. (the most influential ecology text of the mid 20th century).

Ruhl, J. B., Kraft, S.E. and Lant, C.L., 2007. *The law and policy of ecosystem services.* Island Press, Washington, DC.

Schmitz, D.J., 2007. *Ecology and ecosystem conservation.* Island Press, Washington, DC.

Valiela, I., 1995. *Marine ecological processes.* Springer-Verlag, New York, 2nd edn.

ENVIRONMENTAL CHEMISTRY

Environmental chemistry is the study of the chemistry of the natural world. This is sometimes related to, but is often separate from chemistry studies carried out in chemical laboratories. Environmental chemistry includes a number of sub-disciplines including atmospheric chemistry, geochemistry (see separate entry), biochemistry, and the related biogeochemistry. Biogeochemistry is concerned with chemical processes that take place as organisms interact with their environment. This might also be called ecological chemistry. Nutrient cycling is the study of the dynamics of the major chemical elements or macronutrients that control the productivity and metabolism of natural systems, including the carbon, nitrogen, phosphorus, sulphur, and silicon cycles. Plants require sufficient quantities of these and other elements for growth to take place and if one element is deficient compared to others, it is said to be limiting. In marine ecosystems, the relative availability of the essential macronutrients is referred to as the Redfield ratio after the oceanographer Alfred Redfield who first described this relationship. In marine plankton, the molar ratio of C:N:Si:P is about 106:16:15:1. If ocean water contains the inorganic forms of these compounds in this ratio, then none will limit nutrient uptake and photosynthesis. If, however, one of the elements is lower than the Redfield ratio, then that element is likely limiting. The study of limiting nutrients is an important area of study in environmental chemistry. Another important area of environmental chemistry is exchanges between the soil, or submerged sediments in aquatic systems, living organisms, and the atmosphere. For example, gaseous nitrogen (N_2) can be taken from the atmosphere by certain plants and microorganisms and converted into nitrogen in living tissues in a process called nitrogen fixation. Inorganic forms of nitrogen (i.e., NO_3, NH_4) are taken up by plants and microorganisms and converted to forms of organic nitrogen in living tissue. When organic matter decomposes, the inorganic forms are released back into the environment again. Similar cycles take place for the other elements. The impact of human activities on environmental chemistry is a major area of study. These include such areas as water quality deterioration and toxic pollution.

See also: CARBON CYCLE, EARTH SYSTEM SCIENCE, GEOCHEMISTRY, NITROGEN CYCLE

Further reading

Bianchi, T., Pennock, J. and Twilley, R., 1999. *Biogeochemistry of Gulf of Mexico estuaries*. Wiley, New York.

Botkin, D.B. and Keller, E.A., 2007. *Environmental science – Earth as a living planet*. Wiley, New York.

Nriagu, J. (ed). 1976. *Environmental biogeochemistry*. Ann Arbor Science. Ann Arbor, Michigan.

ENVIRONMENTAL SCIENCES

Awareness of environmental science is now very high – in the media, in legislation, in politics and even in popular literature and novels. One environmental 'thriller', *State of Fear* by Michael Crighton in 2004, includes the author's conclusions at the end, beginning with 'We know astonishingly little about every aspect of environment ... in every debate all sides overstate the extent of existing knowledge and its degree of certainty' and towards the end includes 'I am certain there is too much certainty in the world'.

There are really at least two ways in which the term 'environmental sciences' are used. Firstly it can be a specific term for a single-science, multidisciplinary field that began to develop in the 1960s and 1970s. In the UK several new universities established Schools of Environmental Science (East Anglia, Lancaster, Stirling, Ulster) and although the founding professors often included physical geographers, the universities did not have separate geography departments; thus it has been suggested that geographers and ecologists were the first to rise to the challenge of environmental science and the opportunities that it presented by developing joint honours courses usually called environmental studies or environmental science. Secondly 'environmental sciences' is employed as a generic term for all those disciplines which contribute to, and illuminate investigation of, the environment. There are therefore inevitably a range of definitions of environmental science (Table 1). In view of these definitions the entries in this Companion are organized in six groups:

- » the sciences and disciplines that are included in the broad spectrum of investigations concerned with the environment;
- » the major types of environments that exist on the earth's surface;
- » the concepts and paradigms that have been necessary for the development of the environmental sciences;
- » the process and dynamics of environmental systems;
- » the subjects associated with scales and techniques of study;
- » the environmental issues which are increasingly evident.

Environment is often synonymous with place, and is usually taken to mean the total range of conditions within which an organism lives. The organism or community of organisms are encompassed by both physical and cultural surroundings. Environmental sciences are therefore concerned with organisms and with where the organisms live, thus embracing the living (biotic) and inanimate (abiotic) components of the earth's surface concentrated in the envelope within 50 km above the surface and a few hundred metres below it. This explains the two types of environmental sciences: the single-science field developed particularly emphasizing environment whereas in addition a spectrum of sciences is necessary for the complete understanding of human environments. Environment therefore includes spatial, cultural, economic and political aspects, as well as the more usually understood subjects such as atmosphere, soil, climate, the land surface, ecosystems, vegetation and land use. The term environment has been

Table 1   *Some definitions of environmental science*

Definition	Source
The definition of 'environmental science' adopted for the review is deliberately wide-ranging, namely 'the sciences concerned with investigating the state and condition of the Earth'. This definition includes the 'traditional environmental science' disciplines (e.g., ecology, atmospheric sciences, marine sciences etc.) as well as pressures on the environment (such as resource usage, pollution, waste, land use etc.) and the interactions of society, individuals and the environment. It also includes those disciplines concerned with managing and improving the environment.	Environmental Research Funders Forum (ERFF) report August 2003
The recently emerging, interdisciplinary field of scientific study examining the complex interactions of human beings with the natural environment in which they live … . Because modern environmental problems cannot be satisfactorily remedied by the application of any one discipline, environmental science is based on a number of scientific disciplines (including chemistry, biology, physics, geography, geology, hydrology, ecology, meteorology and oceanography) and social science disciplines such as economics and social policy.	Mathews, 2001
Environmental science is a multidisciplinary field that includes elements of agronomy, biology, botany, chemistry, climatology, ecology, entomology, geography, geology, geomorphology, hydrology, limnology, meteorology, oceanography, pedology, political science, psychology, remote sensing, zoology, and many other disciplines. It also draws upon a large number of more specialized subjects, such as biogeography, mycology, and toxicology.	Alexander and Fairbridge, 1999

qualified in numerous ways: thus the natural environment is that created before the influence of human activity whereas the built environment refers to the artefacts created by man in the evolution of the cultural landscape. Different environments can embrace a range of types in the way that marine environments can include abyssal, aphotic (and photic), bathyal, littoral, neritic zones; and continental environments include aquatic environments fluvial, lacustrine, limnic, paludal, and paralic. Environment can also denote origin or formation, for example environments of deposition.

Environmentalism has emerged as concerned with the preservation or improvement of the natural environment and therefore of the quality of life involving conservation of natural resources and the control of environmental pollution. In the social sciences environmentalism can be used to denote the importance of environmental factors in the development of culture and society. Environmentalism began with the conservation movement in the nineteenth century which fostered the establishment of state and national parks and forests, wildlife refuges, and national monuments intended to preserve noteworthy natural features. Although the environmental movement exerted significant influence upon national and international developments and saw the inception of nongovernmental organizations, by the 1950s and 1960s, the public was becoming increasingly aware that conservation of wilderness and wildlife was but one aspect of protecting an endangered environment. Books such as Rachel Carson's influential *Silent Spring* in 1962, fostered concerns about a range of matters including air pollution, water pollution, solid waste disposal, dwindling energy resources, radiation, pesticide poisoning, noise pollution, and other environmental problems thus engaging a broadening number of sympathizers and gave rise to what became known as the 'new environmentalism'. The broad objective of this new movement was concerned with preservation of life on the planet and fostered the creation and growth of a number of environmental organizations and subsequently of political groups, some of which believed that further industrial development is incompatible with environmentalism. Some of the new groups, such as Greenpeace, advocated direct action to preserve endangered species, often clashing violently with opponents in highly publicized protestations. Other organizations have been less militant, calling instead for sustainable development and the need to balance environmentalism with economic development. It has thus been suggested by O'Riordan in 1999 that there are three views of the world encompassed by environmentalism. The *technocentric view* sees humanity in a manipulative way, capable of transforming the Earth for the benefit of both people and nature. The *ecocentric view* embraces the costs of altering the natural world and so has fostered the development and acceptance of sustainable development; the precautionary principle; ecological or environmental economics; environmental impact assessment; and ecoauditing. A *deep green view* of the world, embraces deep ecology and steady state economics: deep ecology maintains that all species have an intrinsic right to exist in the natural environment and this view is extended to one which confronts globalism in the economy and in political dependency thus promoting the cause of pacifism, ecofeminsim, consumer rights and animal welfare generally believing in the 'ubiquitousness of the natural world to include humans and human desires … (Crighton, 2004: 614). Regarding the imperative for sustainable development as an opportunity to link social welfare policies, disarmament strategies and peaceful coexistence into the essence of collective survival'.

Study of the environment began in biology including biological classification or taxonomy, and in geology in which at the end of the 18th century uniformitarianism succeeded catastrophism: thence a continuing uniformity of existing processes was regarded as providing the key to understanding of the history of the earth. This uniformitarian viewpoint promulgated the idea that 'the present is the key to the past' and was facilitated with the development of stratigraphy. In the 1860s the concept of evolution, expressing the relatively gradual change in the characteristics of successive generations of a species or race of an organism, was accepted after publication in 1859 of Charles Darwin's *Origin of Species* and evolution was subsequently a concept which can be traced through other disciplines concerned with environment including the approach of the Davisian cycle of erosion to the earth's surface, the zonal soil as the culmination of soil profile development, and of the climax community as the consequence of plant succession. It was particularly the development of the discipline of ecology from 1930 to 1960 that in many respects formed the core of environmental science. Although the word ecology coined in 1869 is usually attributed to Ernst Heinrich Haeckel (1834–1919), it is derived from the Greek *oikos* meaning house or dwelling place, its major development was in the 20th century. It is generally thought of as the scientific study of the interrelationships among organisms and between organisms, and between them and all aspects, living and non living, of their environment. The development of ecology was broadened beyond its purely biological content by the advent of the concept of the ecosystem, a term proposed by the plant ecologist A.G. Tansley in 1935 as a general term for both the biome, which was the whole complex of organisms – both animals and plants – naturally living together as a sociological unit, and for its habitat. The ecosystem involving synergy between organism and habitat thus gave formal expression to a variety of concepts covering habitat and biome which date back to the 19th century. Extension of the ecosystem concept more widely throughout the environmental sciences was enabled by the development of general systems theory, proposed in the 1950s and 1960s by the biologist Ludwig von Bertalanffy as an analytical framework and procedure for all sciences. Although the systems approach proved to be more pervasive and durable in some environmental sciences than in others, its inception at a time when developments in computing allowed mathematical and statistical methods and modelling to become more widely available facilitated the elucidation of system structures and the creation of complex models.

Although initially a unitary holistic approach could be adopted for the study of environment in the 19th century, this was succeeded by more specialized investigations as reductionism affected sciences throughout the 20th century. During the development of the environmental sciences over a period of more than 200 years a framework was necessary as a context for scientific investigations of the environment, for the study of place. Places can be located within the spheres in the envelope of the earth extending from about 200 km above the earth's surface to the centre of the earth. The atmosphere was recognized as early as the late 17th century, and in 1875 the Austrian geologist Suess invented the terms hydrosphere, lithosphere and biosphere, which were later complemented by others including the pedosphere, cryosphere and noosphere. Because no one single discipline could study all the major spheres now identified, specialist disciplines evolved to focus on particular spheres.

By the 1960s there was a feeling that increasingly specialized single disciplines did not fully do justice to the multidisciplinary investigations required for environmental science, a view which was reinforced by the advent of the systems approach, by greater environmental monitoring and analysis as well as by greater environmental awareness, often thought of as beginning with *Silent Spring*, with repercussions for politics, education and lifestyles and in many countries the creation of Green Parties. By the 1980s it was suggested by Frodeman that the 'green movement' had caused a shift in the political landscape because environmental issues, once dismissed as a fringe movement of tree huggers, became the concern of millions of people in many countries. A single-discipline environmental science then evolved. The emergence of a separately identifiable environmental science was because the problems addressed cannot be solved entirely within the bounds of existing single disciplines. Environmental science emerged from public interest in environmental problems with the development of environmentalism and with the creation of a more widely accepted ethic in societies. In the development of disciplines it was realized that there are paradigm phases, each characterized by a dominating school of thought and each separated by a crisis phase which occurs because problems accumulate that cannot be solved by the prevailing paradigm. In this way the emphasis upon the evolutionary paradigm in many disciplines has been succeeded by other emphases upon processes for example. However it is also possible to distinguish between basic sciences, namely physics, chemistry, biology which customarily have concrete problem solutions, from composite sciences including ecology, geology and geomorphology which are too complex to be thus defined.

The single-discipline environmental science came to have a greater involvement with aspects of social sciences. It has been suggested that the history of environmental sciences has shown them to be different from physical sciences in that they have tended to be associated with sounding of alarms about the effects of economic development; stated results or conclusions from them have been contentious with much of the resulting debate between scientists with opposing results or interpretation; have lacked the established rigour of the established sciences; and have tended to group together in pursuit of trans-boundary problems, particularly during the late 1980s with the rise of global environmental challenges. Therefore aspects of the social sciences have been drawn into an increasingly holistic framework, and environmental science was seen by Budd and Young in 1999 as 'an interdisciplinary inquiry that deals primarily with the variety of environmental problems caused by humans as they live their lives: satisfying needs and wants, processing materials, and releasing unwanted products back into the environment' with some general areas of study (e.g., environmental impact assessment, pollution prevention, waste management) being identified closely as environmental science rather than with any specific discipline. Indeed it has been suggested that as environmental science has passed through the descriptive-empirical stage of research evolution to a more physical-deterministic model-based approach, it then needs to integrate the human dimension since global change is a key issue.

Just as single-discipline environmental science developed as the response to a need in the 1960s and 1970s, other developments more recently have included earth system science (ESS). As scientists may underestimate the complexity of interactions between the earth's atmosphere, ocean, geosphere and biosphere then it has been suggested that an earth system science may be necessary, as a structure which can be employed for strategic planning of funding. However although ESS has been described by Pitman, in 2005 as 'the study of the earth as a single, integrated physical and social system', others have suggested that this is not new in view of the antecedents of the Huxley physiographic approach, the ecosystem proposed by Tansley in 1935, the systems approach and general systems theory. Furthermore if ESS does not accept closure, then whenever an ESS question is posed that is researchable, it necessarily ceases to be ESS as it is currently considered (ERFF, 2003, p. 380). In 2005 Clifford and Richards therefore concluded that ESS should not be seen as an alternative to the traditional scientific disciplines, or to environmental science itself, nor regarded as a wholesale replacement for a traditional vision of environmental science, but as an adjunct approach.

An area of study such as that of the environment is necessarily extensive and wide-ranging and therefore requires some form of system closure to limit the fields involved in the investigation of specific aspects of environmental science. What is required for the study of the environment is in fact three types of investigation. First, single-discipline investigations including those by biology including ecology, geography including physical geography and geomorphology, as well as oceanography, hydrology and environmental chemistry. Secondly there is a need for a single-discipline environmental science, a more composite discipline but one which is limited by closure. Thirdly, there is always however a need for multidisciplinary and interdisciplinary investigations because so many problems and issues transcend the way in which closure delimits single disciplines. These multidisciplinary and interdisciplinary investigations change with time as environmental problems rise up the political agenda. Thus global warming was debated in the 1980s, was becoming accepted in the 1990s and in the 21st century is seldom questioned but requires investigations beyond the closure of any one discipline. Such studies involving multiple disciplines are often achieved by establishing government enquiries, both nationally and internationally and UN initiatives are exemplars of this type of development.

Any discipline must be dynamic in character and must adapt and evolve to assimilate developments in technology including those in computing, in information provision and access through the internet, as well as reflecting trends in the philosophy of science. Thus environmental science and the sciences which it requires must evolve and remain dynamic. In the 21st century there is perhaps a more urgent attitude to environmental issues than previously. Thus Gaia was presented by Sir James Lovelock in 1979 as a way of looking at an earth that was resilient. However in his subsequent book *The revenge of Gaia* a vision is given of the earth fighting back. Debate over this as one issue for the future will continue and other current and challenging issues are contained in the final section of this Companion.

See also: ECOLOGY, EARTH SPHERES, ENVIRONMENT, ENVIRONMENTALISM

Further reading

Alexander, D.J., 1999. Preface. In Alexander, D.E. and Fairbridge, R.W. (eds): *Encyclopedia of environmental science*. Kluwer Academic Publishers, Dordrecht, Boston and London, xxix–xxx.

Budd, W.W. and Young, G.L., 1999. Environmental science. In Alexander, D.E. and Fairbridge, R.W. (eds) *Encyclopedia of environmental science*. Kluwer Academic Publishers, Dordrecht, Boston and London, 224.

Clifford, N. and Richards, K., 2005. Earth system science: an oxymoron? *Earth Surface Processes and Landforms* 30, 379–383.

Crighton, M., 2004. *State of fear*. London, Harper Collins.

Environmental Research Funders Forum (ERFF), 2003. *Analysis of Environmental Science in the UK*. http://www.erff.org.uk/

Frodeman, R., 1995: 'Radical environmentalism and the political roots of postmodernism-differences that make a difference'. In Oelschlaeger, M. (ed.), *Post modern environmental ethics*. State University of New York Press, Albany NY, 121–135.

Johnston, R.J., 2003. The institutionalisation of geography as an academic discipline. In Johnston, R. and Williams, M. (eds) *A century of British geography*. Oxford University Press, Oxford, 45–90.

Lane, S.N., 2001. Constructive comments on D. Massey 'Space-time, "science" and the relationship between physical and human geography'. *Transactions of the Institute of British Geographers* 26, 243–256.

Lawton, J., 2001. Earth system science. *Science* 292, 1965.

Lovelock, J.E., 1979. *Gaia: a new look at life on earth*. Oxford University Press, Oxford.

Lovelock, J.E., 2006. *The revenge of Gaia*. Allen Lane, London.

O'Riordan, T., 1999. Environment and environmentalism. In Alexander, D.E. and Fairbridge, R.W. (eds). *Encyclopedia of environmental science*. Kluwer Academic Publishers, Dordrecht, Boston and London, 192–193.

Newson, M.D., 1992. Twenty years of systematic physical geography: issues for a 'New Environmental Age'. *Progress in Physical Geography* 16, 209–221.

Osterkamp, W.R. and Hupp, C.R., 1996. The evolution of geomorphology, ecology and other composite sciences. In Rhoads, B.L. and Thorn, C.E. (eds) *The scientific nature of geomorphology*. Wiley, Chichester, 415–441.

Pitman, A.J., 2005. On the role of geography in earth system science. *Geoforum* 36, 137–148.

Stonehouse, B., 1999. Environmental education. In Alexander, D.E. and Fairbridge, R.W. (eds): *Encyclopedia of environmental science*. Kluwer Academic Publishers, Dordrecht, Boston and London, 203.

GENETICS

Genetics is the study of biologically inherited traits and characteristics. These inherited traits are determined by genes that are transmitted to offspring from parents via reproduction. It has been known for millennia that biological traits are inherited and this knowledge has been used extensively to produce organisms with desirable characteristics. These include agricultural plant species, livestock, horses, and dogs. The basic elements and rules of heredity were discovered by the monk Gregor Mendel (1822–1884) in his famous experiments with peas. In the late 19th century, the role of the cell nucleus in genetics was discovered when it was observed that male and female reproductive cells fused during fertilization. By the beginning of the 20th century, chromosomes and their characteristic splitting were discovered and in the mid 20th century, Watson and Crick demonstrated the double helix structure of DNA, gaining the Nobel Prize in 1962. The double helix structure separates during replication and nucleic acids are added to each separate strand and two new double helixes are formed.

A knowledge of genetics has allowed us to understand the process of biological evolution. When DNA replicates, random mutations occur. Most of these mutations are bad or neutral. But sometimes, mutations make an organism more fit and this mutation is selected for. This is the basic idea of natural selection as developed by Darwin, though he knew little about genetic mechanisms. One of the exciting recent developments in genetics is the elucidation of the genomes of numerous organisms, including the human genome. This is accomplished by large-scale automated DNA sequencing. This has opened up the possibility of genetic engineering for such processes as the production of drugs and to correct undesirable mutations. It has also opened up further the possibilities for the production of hybrid forms within species, which often show enhanced characteristics such as resistance to drought or pests, and even for cloning, which is the exact reproduction of an individual organism. Genetics is currently an area of rapid advance in biological science.

See also: BIOLOGY, EVOLUTION

Further reading

Hartl, D.L. and Jones, E.W., 1998. *Genetics: principles and analysis*. Jones and Bartlett, Sudbury MA. 4th edn.

Johnson, R., 2006. *Genetics*. Twenty-First Century Books, Minneapolis, Minn.

http://gslc.genetics.utah.edu/ has basic and research material

GEOARCHAEOLOGY

This area of the natural sciences brings to bear relevant techniques which allow the reconstruction of the environment of early human communities, where these are being investigated by the methods of archaeology. The main categories are those which deal with sediments and those that look into biological remains. With sediments, analysis of constituents such as particle size, mineral element composition and degree of weathering all help to answer questions about the landforms and soils of an archaeological site and may lead to important discoveries about the role which humans played in the formation of sediments, their transport and their deposition. For example, the human-directed removal of forest from a steep hillside will probably mean the downwashing of soil and weathered rock to form colluvium at the foot of the slope which may well mask earlier human remains such as graves or even settlement structures. Biological remains include animal bones and dried faeces (coprolites) and sometimes skins, the remains of parts of plants including leaves, seeds, pollen, spores and phytoliths and, especially, wood where tree rings often allow precise dating of the death of the tree. Materials containing carbon may be especially useful for radiocarbon dating, and some sediments may also be dated e.g., by thermoluminescence. Increasingly, the recovery of DNA yields more data on the subfossil material. All these remains depend upon the presence of suitable conditions for their preservation through time and changing physical and chemical regimes may winnow out some remains while leaving others. One element of successful geoarchaeology therefore is the understanding of how any remains have come to be in the stratigraphic and spatial locations where they are found and what vicissitudes have affected them through time: their taphonomy.

See also: ARCHAEOLOGY, COLLUVIUM, SOIL PROFILES

Further reading

Goldberg, P. and McPhail, R., 2006. *Practical and theoretical geoarchaeology*. Blackwell, Oxford.

Pollard, A. M. (ed), 1999. *Geoarchaeology: environments, exploration, resources*. Geological Society of London Special Publication no. 165.

GEOCHEMISTRY

Geochemistry is the study of chemical dynamics, interactions, and processes of the earth surface. An important sub-discipline is biogeochemistry, chemical dynamics between living organisms and the environment. Many important chemical changes take place as rocks and soils are formed and changed through volcanic, weathering, and sedimentation processes. The weathering of rocks and the formation of soils and sediments is called diagenesis. Many rocks can be classified as alumino-silicate or carbonate. The weathering and breakdown of these rocks contributes to global silicon and carbon cycles. Clay mineralogy is an important geochemical study. Clays form as alumino-silicate rocks weather. Soil formation, soil erosion and transport, and sediment deposition are important processes of clay mineralogy. The use of particle-reactive radioactive tracers such as Be, Th, Cs, and Pu, which occur in very small concentrations, can be used to determine short-term vs long-term rates of deposition and accumulation. For example, since ^{137}Cs is a bomb product and bomb testing peaked around 1963, the highest concentration of ^{137}Cs in an estuarine soil core occurs at 1963, giving an estimate of the rates of soil accumulation. The geochemistry of organic matter provides insights into the production, dynamics, and decomposition of organic carbon. Trace metal geochemistry concerns itself with the dynamics of metals that occur at very low concentrations in the environment. These trace metals, depending on concentration, can be either toxic agents or required nutrients. Trace metals include such elements as silver, copper, lead, zinc, cadmium, nickel, mercury, and arsenic. Two exciting areas of geochemical study are the deep ocean and on lunar rocks. The sea floor forms at mid-ocean ridges where volcanic activity produces material that then spreads out from the ridges. The geochemical dynamics of hydrothermal vents that occur around the ridges has provided an understanding of the chemical dynamics associated with vent communities (see Oceanography). The geochemical study of lunar rocks collected on the manned moon missions have provided insights into the history of our solar system.

See also: CARBON CYCLE, GEOCHEMICAL CYCLES, GEOLOGY, NITROGEN CYCLE, OCEANOGRAPHY

Further reading

Botkin, D.B. and Keller, E.A., 2007. *Environmental science – Earth as a living planet*. Wiley, New York.

Libes, S., 1992. *An Introduction to marine biogeochemistry*. Wiley, New York.

Nriagu, J. (ed), 1976. *Environmental biogeochemisty*. Ann Arbor Science. Ann Arbor, Michigan.

Rudnik, R.L. (ed), 2005. *The crust*. Elsevier, Amsterdam.

Schlesinger, W.H. (ed), 2005. *Biogeochemistry*. Elsevier, Amsterdam.

GEOLOGY

Geology is defined in the AGI Glossary in 1987 as:

> The study of the planet Earth – the materials of which it is made, the processes that act on these materials, the products formed, and the history of the planet and its life forms since its origin. Geology considers the physical forces that act on the Earth, the chemistry of its constituent materials, and the biology of its past inhabitants as revealed by fossils. Clues on the origin of the planets are sought in a study of the Moon and other extraterrestrial bodies.

Geology is a relatively young science, conceived in 1785 when the Scottish 'father of geology', James Hutton, realized the evolution of the Earth required millions of years. Hutton's works were not widely known until nearly fifty years later, when they were published by the Scottish geologist, Sir Charles Lyell, in 1830.

Geology is commonly regarded as a derivative science, one that is based on the so-called 'primary' sciences: math, physics, and chemistry. Indeed, geology makes broad use of these and other sciences, including biology, but 'geology' is a broad field with many areas of specialization, as follows, all taken from the AGI Glossary, 1987, with modifications, unless otherwise noted:

ECONOMIC GEOLOGY – study and analysis of geologic bodies and materials that can be utilized profitably by man, including fuels, metals, nonmetallic minerals, and water.

ENGINEERING GEOLOGY – application of geologic data, techniques, and principles to the study of naturally occurring rock and soil materials or ground water for the purpose of assuring that geologic factors affecting the location, planning, design, construction, operation, and maintenance of engineering structures, and the development of groundwater resources, are properly recognized and adequately interpreted, utilized, and presented for use in engineering practice (Assoc. of Engineering Geologists, 1969).

FORENSIC GEOLOGY – application of the principles and techniques of geology to the law.

GEOCHEMISTRY – study of the distribution and amounts of the chemical elements in minerals, ores, rocks, soils, water, and the atmosphere, and the study of the circulation of the elements in nature on the basis of the properties of their atoms and ions; also, the study of the distribution and abundance of isotopes, including problems of nuclear frequency and stability in the universe (Goldschmidt, 1954).

GEOPHYSICS – study of the earth by quantitative physical methods.

GLACIOLOGY – the science that treats quantitatively all aspects of snow and ice.

HYDROGEOLOGY – study of subsurface waters and related geologic aspects of surface waters.

IGNEOUS PETROLOGY – study of rocks that solidified from molten or partly molten rock, including the processes leading to, related to, or resulting from the formation of such rocks.

MARINE GEOLOGY – studies of the ocean that deal specifically with the ocean floor and the ocean–continent border, including submarine relief features, the geochemistry and petrology of the sediments and rocks of the ocean floor, and the influence of seawater and waves on the ocean bottom and its materials.

MATHEMATICAL GEOLOGY – all applications of mathematics to the study of the Earth.

METAMORPHIC PETROLOGY – study of the mineralogical, chemical, and structural adjustment of solid rocks to physical and chemical conditions which have generally been imposed at depths before surface zones of weathering and cementation, and which differ from the conditions under which the rocks in question originated.

MINERALOGY – the study of the formation, occurrence, properties, composition, and classification of minerals.

PALAEOBIOLOGY – a branch of palaeontology dealing with the study of fossils as organisms rather than as features of historical geology.

PALAEONTOLOGY – study of life in past geologic time, based on fossil plants and animals and including phylogeny, their relationships to existing plants, animals, and environments, and to the chronology of the Earth's history.

PETROLEUM GEOLOGY – the branch of economic geology that pertains to the origin, migration, and accumulation of oil and gas, and to discovery of commercial deposits of same.

PETROLOGY – study of origin, occurrence, structure and history of rocks, generally subdivided into three broad areas:

PLANETARY GEOLOGY – application of geologic principles and techniques to the study of planets and their natural satellites.

SEDIMENTARY PETROLOGY – study of the composition, characteristics, and origin of sediments and sedimentary rocks.

SEDIMENTOLOGY – study of the processes by which sedimentary rocks form.

SEISMOLOGY – study of earthquakes and of the structure of the Earth, using both natural and artificially generated seismic waves.

STRATIGRAPHY – the science of rock strata; it concerns all characteristics and attributes of *layered rocks*, including original succession and age of rock strata, form, distribution, lithologic composition, fossil content, geophysical properties, and geochemical properties, and their interpretation those layered rocks in terms of their environment or deposition, mode of origin, and geologic history.

STRUCTURAL GEOLOGY – the general architecture, attitude, arrangement, or relative positions of rock masses in a region or area, including fractures, faults, and folds that might be in those rocks.

STRUCTURAL PETROLOGY – study, analysis, and interpretation of the penetrative fabric elements of a rock, such as foliation, lineation, crystal orientations, generally at the mesoscopic and microscopic scales.

TECTONICS – study of assemblage of structural and deformational features of the outer part of the Earth, their mutual relations, origin, and historical evolution.

VOLCANOLOGY – study of the phenomena and causes of volcanic eruptions.

See also: BIOLOGY, EARTH SYSTEM SCIENCE, FOSSILS AND PALALONTOLOGY, SEISMOLOGY

Further reading

Association of Engineering Geologists, 1969. Definition of engineering geology. *AEG Newsletter*, v. 12, no. 4, p. 3.

Bates, R.L. and Jackson, J.A. (eds), 1987. *Glossary of geology*, 3rd Edition, American Geological Institute, Alexandria, Virginia, 788 pp.

Goldschmidt, V.M., 1954. *Geochemistry*. Clarendon Press, Oxford.

Keller, E.A., 1996. *Environmental geology*. Prentice Hall, New Jersey.

Skinner, B.J., 2004. *Dynamic earth: an introduction to physical geology*. Wiley, New York.

Tarbuck, E.J. and Lutgens, F.K., 2007. *Earth: an introduction to physical geology*. Prentice Hall, Upper Saddle River NJ, 9th edn.

GEOMORPHOLOGY

Geomorphology literally means study (Greek *logos*) the shape or form (*morphe*) of the earth (*ge*). The name first appeared in the German literature in 1858 but came into general use, including with the US Geological Survey, after about 1890. For some time the term physiography persisted in North America although that term eventually became used for regional geomorphology. Although originating in geology it became more geographically based with the contributions of W.M. Davis (1850–1934) who developed a normal cycle of erosion, suggested that it developed through stages of youth, maturity and old age; conceived other cycles including the arid, coastal and glacial cycles; and proposed that landscape was a function of structure, process and stage or time. His attractive ideas dominated geomorphology for the first half of the 20th century and arguably provided a foundation for later work and also prompted contrary views. Although alternative approaches, such as that by G.K. Gilbert, were firmly based upon the study of processes, for the first half of the 20th century the influence of Davisian ideas ensured that geomorphology emphasized the historical development of landforms.

In the second half of the 20th century alternative paradigms developed especially that centred on processes so that geomorphology is now generally accepted to be the study of earth surface forms and processes although it is not confined to any one university discipline – in the US it still features in both Geology and Geography departments whereas in the UK it is dominantly found within Geography departments. Other developments since the mid-20th century have emphasized other processes such as periglacial and mass movement, applications of geomorphology, all being aided by greater use of modelling, quantitative and theoretical methods and by new techniques. Establishment of the discipline was aided by creation of societies such as the British Geomorphological Research Group (1960–) which became the British Society for Geomorphology in 2006 and the International Association of Geomorphologists (1985–) and by the establishment of international journals including *Earth Surface Processes and Landforms* (1976–), *Zeitschrift fur Geomorphologie* (1956–), and *Geomorphology* (1989–).

Analogous to the development of other environmental science disciplines, geomorphology has developed a large number of subfields as research became more detailed and reductionist. Geomorphology now embraces three broad approaches which are dynamic which involves understanding of the functional processes of weathering, erosion and transport and deposition, together with their rates of operation; historical, concerned with the last chapter of Earth's history usually contributed during the Cainozoic; and applied geomorphology. The range of these sub-fields (Table 2) can be illustrated according to the domains in which processes are dominant, to analysis or to purpose. Analysis can be achieved not only by focusing upon processes but also upon the way in which combinations of processes occur under different climatic zones (climatic geomorphology), in areas dominated by rock type (karst geomorphology) or by endogenetic processes (structural, tectonic) or according to the role of human activity. Quantitative approaches developed after the mid-20th century with the advent of statistical

Table 2   *Some branches of geomorphology*

Branch of geomorphology	Objective (links to other disciplines and sub-disciplines)
According to process domains	
Aeolian	Wind-dominated processes in hot and cold deserts and other areas such as some coastal zones.
Coastal	Assemblage of processes and landforms that occur on coastal margins.
Fluvial	Investigates the fluvial system at a range of spatial scales from the basin to specific within-channel locations; at timescales ranging from processes during a single flow event to long-term Quaternary change; undertaking studies which involve explanation of the relations among physical flow properties, sediment transport, and channel forms; of the changes that occur both within and between rivers; and that it can provide results which contribute to the sustainable solution of river channel management problems.
Glacial	Concerned with landscapes occupied by glaciers, and with landscapes which have been glaciated because they were covered by glaciers in the past.
Periglacial	Non-glacial processes and features of cold climates, including freeze-thaw processes and frost action typical of the processes in the periglacial zone and in some cases the processes associated with permafrost, but also found in high altitude, alpine, areas of temperate regions.
Hillslope	The characteristic slope forms and the governing processes including processes of mass wasting.
Tropical	Processes, morphology and landscape development in tropical systems associated with chemical weathering, mass movement and surface water flow.
Urban	Processes and morphology in urban environments (urban hydrology, urban ecology).
According to analysis	
Process	Exogenetic and endogenetic processes and the landforms produced.
Climatic	The way in which assemblages of process domains are associated with particular climatic zones. Sometimes extended with crude parameters to define morphoclimatic zones. Three levels of investigation were recognized as: Dynamic – the investigation of processes (as above) Climatic – the way in which contemporary processes are associated with contemporary climatic zones Climatogenetic – allowing for the fact that many landforms are the product of past climates and are not consistent with the climatic conditions under which they now occur
Historical	Analysis of processes and landform evolution in past conditions. Sometimes referred to as palaeogeomorphology and interacting with fields such as palaeohydrology.

Branch of geomorphology	Objective (links to other disciplines and sub-disciplines)
Structural/Tectonic	Study of landforms resulting from the structures of the lithosphere and the associated processes of faulting, folding and warping.
Karst	The processes and landforms of limestone areas which have solution as a dominant process and give rise to distinctive suites of landforms.
Anthropogeomorphology	Study of human activity as a geomorphological agent.
According to purpose	
Quantitative	Use of quantitative, mathematical and statistical, methods for the investigation of landforms, geomorphological processes and form process relationships requiring modelling.
Applied	Application of geomorphology to the solution of problems especially relating to resource development and mitigation of environmental hazards.
Multidisciplinary hybrids	
Hydrogeomorphology	The geomorphological study of water and its effects (fluvial geomorphology; geographical hydrology; hydrology)
Biogeomorphology	The influence of animals and plants on earth surface processes and landform development (ecology)

methods and more advanced computing technology, allowing inclusion of increasingly sophisticated mathematical modelling, have now been assimilated into other branches. Applications of geomorphology have become more extensive including applicable ones that emerge from research, and applied research is undertaken related to specific problems. Increasingly specialized research has created new multidisciplinary fields from overlaps with other environmental science including hydrogeomorphology and biogeomorphology (Table 2).

Also similar to other environment science disciplines a number of ongoing debates are healthy characteristics of the development of any discipline currently raising questions which include:

» Should the discipline be more holistic to counter the fissiparist trend which characterizes many of the sub-branches?

» Should there be a return to evolutionary geomorphology and global structural geomorphology to counteract the emphasis placed upon the investigation of processes? However an alternative view is that plate tectonics is primarily within geology not geomorphology.

» Should geomorphology include planetary geomorphology as the study of the geomorphology of planets other than earth?

» As geomorphology uses different timescales such as steady (tens of years), graded (thousands of years) and cyclic (millions of years) better techniques are required to link the timescales: thus linking studies of process at steady timescale with landform development at graded or cyclic time.

» Should geomorphology include a greater cultural component progressing from studies of human impact? Just as a more society-oriented climatology or cultural climatology is envisioned so we can utilize a cultural geomorphology. This does not detract from existing investigations of form, process and change but follows from them by allowing for differences in culture, human impact, legislative control in relation to future change.

» Can geomorphological systems be explained as the result of continuing processes without catastrophism which may have played a greater role than previously thought? Potential effects of global warming may include the more frequent incidence of high magnitude events, possibly catastrophic ones.

» More opportunities for applied research are available including investigation of potential implications of global change for coasts, flooding, glaciers and ground ice.

» The uncertainty that exists in environmental systems may mean that it is easier to predict than to explain.

Such questions are alternatives but rather show how a pluralist approach may be necessary for this and other environmental science disciplines.

See also: GEOLOGY, HUMAN GEOGRAPHY, HYDROLOGY, PHYSICAL GEOGRAPHY, REDUCTIONISM

Further reading

Baker, V.R. and Twidale, C.R., 1991. 'The reenchantment of geomorphology.' *Geomorphology* 4, 73–100.

Chorley, R.J., Schumm, S.A. and Sugden, D.A., 1984. *Geomorphology*. Methuen, London and New York.

Costa, J.E. and Graf, W.L., 1984. The geography of geomorphologists in the United States. *Professional Geographer* 36, 82–89.

Gregory, K.J., 2000. *The changing nature of physical geography*. Arnold, London.

Gregory, K.J., 2006. The human role in changing river channels. *Geomorphology* 79, 172–191.

International Association of Geomorphologists. http://www.geomorph.org

Slaymaker, H.O. and Spencer, T., 1998. *Physical geography and global environmental change*. Longman, Harlow.

Tinkler, K.J., 1985. *A short history of geomorphology*. Croom Helm, London and Sydney.

GLACIAL GEOMORPHOLOGY

A branch of geomorphology concerned with landscapes occupied by glaciers, and with landscapes which have been glaciated when they were covered by glaciers in the past. The purpose of glacial geomorphology is to provide physically based explanations of the past, present and future impacts of glaciers and ice sheets on landform and landscape development. Although knowledge of glacier accumulation, the physics of flow and the mass balance is necessary glaciological background, glacial geomorphology is concerned with the results of present glacier dynamics and with the way in which glaciers produce landforms at present and provided a legacy in areas that no longer have glacial ice. During the Quaternary period glaciers covered up to one third of the earth's surface whereas they occupy approximately 10% today.

Major subjects of investigation include:

» erosion and entrainment of sediment – ability of glaciers to collect, transport and deposit rock debris determines the geomorphological impact of glaciers and produces landforms including rock drumlins, roche moutonnees, rock basins, cirques and U-shaped valleys;

» depositional processes of glaciers – occur at the glacier margin, beneath the glacier, or beyond the glacier margin. Sedimentary material or till deposited directly by the action of a glacier can arise as a result of several processes (Table 3). Landforms produced cover a range of distinctive features including moraines (subglacial, marginal, stagnant ice features) and drumlins;

Table 3   *Types of till (developed from Owen and Derbyshire, 2005)*

Type of till	Genesis	Particle size
Glaciotectonite	Subglacially sheared sediment and bedrock	Poorly sorted
Comminution	Subglacially crushed and powdered local bedrock	Poorly sorted skewed towards fine
Lodgement	Subglacially plastered glacial debris on a rigid or semi-rigid bed	Poorly sorted with strong up-valley dip
Deformation	Subglacially deformed glacial sediment	Poorly sorted
Meltout	Glacial sediment deposited directly from melting ice	Poorly sorted but may be stratified
Sublimation	Glacial sediment deposited directly from sublimated ice	Poorly sorted with ice foliation preserved
Flow till	Sediment deposited off the ice by debris flow processes and may be classed as debris flow rather than till	Poorly sorted with downslope dip

» fluvioglacial environments – water draining supraglacially, englacially or subglacially can produce a suite of landforms including glacial drainage channels, eskers, kame terraces, and kettle holes. In addition a range of other environments glaciomarine, glaciolacustrine, and glacioaeolian can produce distinctive deposits and features in association with water;

» interpretation of past sequences of glaciation based upon the till sequences and reconstruction of past landform assemblages.

Two approaches receiving attention in recent years have been first glacial processes, based on glaciology, mechanics, chemistry and sedimentology and secondly emphasis upon larger-scale problems both in terms of recognizing and describing landforms and landscapes and in using knowledge of processes at very small scales to relate to much larger-scale problems of landform and landscape development. This latter group has included detailed analyses of processes to derive models of landform and landscape development. Recent developments have also included use of three-dimensional models that include ice flow, basal sliding, erosion and deposition characteristics as well as character of the underlying material.

See also: GLACIOLOGY, GLACIERS, HOLOCENE, QUATERNARY

Further reading

Electronic journal of Glacial Geology and Geomorphology published since 1996.

Evans, D.J., 2004. *Glacial landsystems*. Arnold, London.

Harbor, J.M., 1993. Glacial geomorphology: modelling processes and landforms. *Geomorphology* 7, 129–140.

Owen, L.A. and Derbyshire, E., Glacial environments. In Fookes, P.G., Lee, E.M. and Milligan, G. (eds). *Geomorphology for engineers*. Whittles Publishing, Dunbeath, Caithness. 345–375.

GLACIOLOGY

Glaciology is the interdisciplinary scientific study of the distribution and behaviour of snow and ice on the earth's surface, involving contributions from physics, geology, physical geography, meteorology, hydrology and biology. It is concerned with the structure and properties of glacier ice, its formation and distribution, the dynamics of flow, and the interactions with climate. Three major types of glacier are:

» Ice sheets or ice caps (generally less than 50,000 km² in area) – broad domes which submerge the underlying topography with ice radiating outwards as a sheet. The Antarctic and Greenland are examples of ice sheets with ice caps in Canada, Iceland and Norway.

» Ice shelves – floating ice sheets or ice caps which has no friction with the bed and the ice can spread freely. The largest examples occur in Antarctica.

» Valley or Alpine glaciers – occur in a mountain valley in polar regions and other mountains, have glacier flow strongly influenced by topography, and variations according to the amounts of debris cover.

The World Glacier Inventory contains information for over 67,000 glaciers throughout the world with information on geographic location, area, length, orientation, elevation, and classification of morphological type and moraines (Fig. 2). The International Glaciological Society founded in 1936 provides a focus for individuals interested in practical and scientific aspects of snow and ice with international journals *Journal of Glaciology* and *Annals of Glaciology*. Major subjects of glaciological research include:

» Properties of glacier ice – after the transformation from snow to firn (snow which has survived one summer melt season) to glacier ice involving regrowth of ice crystals and elimination of air passages achieving densities of 0.83 to 0.91 kg.m³ which may occur in a year in valley glacier environments but can take several thousand years on ice sheet environments. Where ice is below the pressure melting point it is known as cold or polar ice, whereas warm or temperate ice contains water and is close to the pressure melting point.

Figure 2   *Corrie glacier below Mont Fort, Valais, Switzerland (Photo K.J. Gregory)*

» Glacier flow – achieved by creep involving plastic deformation within and between ice crystals, basal sliding and bed deformation studied as glacier physics. Movement is characteristically up to 300 m per year but velocities can occasionally attain 1–2 km per year and surging glaciers may periodically have velocities up to 100 times greater than normal with a wave of ice moving downglacier at velocities of 4–7 km per year.

» Mass balance studies – consider the relationship between accumulation processes principally from snow, and ablation loss by surface melting and other output processes including evaporation and calving giving the net balance as the annual difference over a glacier.

» Debris in and on glaciers and its erosional and depsitional role – also investigated by glacial geomorphology.

» Glacier history – include studies of changes in the position of glacier margins including the way in which these may produce large floods as a result of the drainage of glacier-dammed lakes (Jokulhlaup in Iceland). As a result of global warming there is evidence in some parts of the world of a significant recession of ice margins and the separation of large masses of ice from ice sheets and ice caps.

Glaciers can present hazards (Table 4) applicable to a range of timescales which could be exacerbated as a result of global warming.

Table 4   *Glacial and related hazards (developed from Owen and Derbyshire, 2005)*

Type of hazard	Event	Nature	Timescale
Glacier hazards	Avalanche	Slide or fall of large mass of snow, ice and/or rock	Minutes
	Glacier outburst	Catastrophic discharge of water under pressure from a glacier	Hours
	Jokulhlaup	Glacier outburst associated with sub-aerial volcanic activity	Hours–days
	Glacier surge	Rapid increase in rate of glacier flow	Months–years
	Glacier fluctuations	Variations in ice margin due to climate change etc	Years–decades
Glacier lakes	Outburst floods	Catastrophic outburst from a proglacial lake, typically moraine dammed	Hours
Related hazards	Lahars	Catastrophic debris flow associated with volcanic activity and snow fields	Hours
	Water resource problems	Supply shortages, especially during low flow conditions, associated with wasting glaciers climate change etc.	Decades

See also: GLACIAL GEOMORPHOLOGY, GLACIERS

Further reading

Benn, D.I. and Evans, D.J.A., 1998. *Glaciers and glaciation*. Arnold, London.

Boulton, G.S., 1986. A paradigm shift in glaciology. *Nature* 322, 18.

Hambrey, M. and Alean, J., 2004. *Glaciers*. Cambridge University Press, Cambridge.

Owen, L.A. and Derbyshire, E., 2005. Glacial environments. In Fookes, P.G., Lee, E.M. and Milligan, G., (eds) *Geomorphology for engineers*. Dunbeath, Caithness, Whittles Publishing, 345–375.

International Glaciological Society website. www.igsoc.org

Knight, P.G., 1999. *Glaciers*. Cheltenham, Nelson Thornes.

Paterson, W.S.B., 1994. *The Physics of glaciers*. Oxford, Pergamon.

World Glacier Inventory: http://nsidc.org/data/glacier_inventory/

HUMAN ECOLOGY

The first instances of the use of this term were in sociology in the 1920s and 1950s when they were indicative of a school of thought that wanted to explain social facts by a variety of causes, not simply with other social facts. Thus environmental factors might be brought into the sociological domain, though some at least of the environment was man-made (e.g., housing types) rather than 'natural'. Hence areas of cities which had poor communities, run-down housing and were prone to flood, constituted an ecological zone rather akin to an ecosystem in 'pure' ecology. In the natural science community the term has been used by various groups who wished to highlight reciprocal human–environment relations in the way A.G. Tansley (1871–1955) suggested for natural ecosystems. *Homo sapiens* is seen as part of a web that includes the natural world as well as the material human-made world. Detractors have suggested that ecologists using the term have thought of humankind merely as 'interfering' with the natural world or that the idea of humans as behaviourally homogeneous does not do justice to the diversity of human culture. Human ecology has never developed into a major academic discipline: it tends to be a section within departments of anthropology or biological sciences; it does not have a distinctive body of theory nor a set of unequivocally labelled empirical studies. However, work in human ecology might show attention to the flows of energy and materials through humanized ecosystems, as with the work of Vaclav Smil; to the historical evolution of ecosystems with a human component (Smil again); to Malthusian factors such as local and regional population growth (P.R. Ehrlich is a leading scholar); to the relevance of human diseases with environmental connections, or the presence of epidemics in ecosystems which result in rapid and unpredictable change. Not all the above scholars would accept the label 'human ecology' for their work, though they are not likely to be insulted by it.

See also: ECOSYSTEM, CULTURES, POPULATION GROWTH

Further reading

Ehrlich, P.R., Ehrlich, A.H. and Holdren., J.P., 1977. *Ecoscience: population, resources, environment.* Freeman, San Francisco.

Schutkowski, H., 2006. *Human ecology: biocultural adaptations in human communities.* Springer, Berlin.

Smil, V., 1993. *Global ecology: environmental change and social flexibility.* Routledge, London and New York.

HUMAN GEOGRAPHY

It is beyond question that the environment influences human activities. No commercial firm attempts to grow bananas at the South Pole, and most people need oxygen to climb Mount Everest. But in between there are many activities where the physical conditions may influence human actions but not determine them. Increasing amounts of human intervention may spread possibilities: pineapples may be grown at 50'N but only under glass. It is also beyond question that humans can alter their physical environments, in a myriad of ways. They can in fact live at the South Pole provided that enough hi-tech equipment is provided to ameliorate the elements and they can, it seems, affect global climate by changing the composition of the upper atmosphere. So in the course of scholarly development, there have been schools of thought that have been called 'determinist' because they believed that the physical environment was in the end a control on human actions, and the opposite school ('possibilist') who thought that the environment was a series of opportunities for human societies and that they could be realized once the right cultural attitudes and the appropriate technology were applied. Within each of these major divisions there were shades of opinion.

More recent opinion has come to accept certain complexities which make a more nuanced view. One model suggests that the double helix visualization of DNA allows for an intertwining of the physical and the cultural, with the base pairs representing influences of the physical upon the human and vice-versa. The base of the strand might represent a period in the past (such as the start of the Holocene) and the top the recent past. A modification of the original DNA model might then widen the gyre upwards to suggest the growth in human population during these last 10–12 ky. But such images have the disadvantage that they cover over the blips, rapid changes, and operations of chance that have characterized the human tenure of the surface of the planet. One more element of human geography ill served by such a model is the difference between places. There are massive differences in the way of life between industrialized and agricultural societies, between rich and poor and between e.g., North America and the Islamic Middle East. The question of differences can be seen at all spatial scales, in terms of the world as a whole or of most countries or indeed within most major towns and cities. In all of these, the access to material resources of different members of the population is unequal and the impact they have upon local, regional and indeed global environments is also incommensurate. Some human geographers see it as their task to elucidate and explain these differences, whereas others think that their work should be action-oriented.

Looked at historically, and with the environment in mind all the time, human geographies world-wide can be seen to have been shaped by two major influences. The first is access to energy sources. No individual nor society can persist without intakes of energy to enable them to live and reproduce. For hunter-gatherers, this came from hunting animals and gathering plant material, for agricultural societies from planting crops and herding animals and for our industrialized times, from all of these previous eras plus the use of the stored energy in fossil fuels (coal, oil and natural gas) in atomic nuclei and in diffuse energies like those of the wind and tides, along with sunlight that powered pre-industrial agriculture. Every society since the invention of agriculture (perhaps 10–12 ky ago) has striven to have a surplus beyond that needed for basic subsistence so that 'surplus' energy might go into other activities, such as luxury foods, travel, warfare, material possessions, and leisure. Surpluses, too, bring the possibility of population increase and then the necessity for it so as to have the right number of young men aggressive enough to defend the material resource base. So the question of who has access to energy (which may come as food, as technology or even embedded in goods and services) and how this access is controlled is fundamental to the shaping of a human geography. Not for nothing is the word 'power' ambiguous in this context: who is it who travels in their own jets and in heavily armoured Mercedes?

The second major influence is hinted at above, namely the growth of the human population from perhaps 3–4 million in 10,000 BC to something over 6 billion now. Perhaps as important as absolute numbers have been local and regional rates of growth since these are likely to have had notable environmental consequences. The massive increases in European populations in the 19th century, for example, impelled all kinds of forest and grassland manipulations in Asia, Africa and South America in order to provide food and drink for the Europeans; today's rapid increases in numbers in, for example, Egypt put a strain on food supplies that can only be held by intensifying crop production, which usually means the application of more fossil fuel energy in the shape of fertilizers, biocides and new crop varieties. By contrast, the Black Death in 14th century Europe allowed many thousands of hectares of cropland to revert to grassland and in some cases to woodland.

This account places a strong emphasis on the role of technology since this provides the channel down which energy is delivered to make an impact upon the environment. Back one stage is the question of why some societies were so active in developing technology and in adopting new ways, whereas others either rejected it or were forced to do without it. Even here, one school of 19th century thought reckoned that certain climates (especially seasonal regimes like NW Europe and New England) were conducive to active thinking and 'progress' whereas the humid tropics induced only lassitude. But it is now recognized that much more cultural complexity is involved as well as the sheer operation of chance. This throws an interesting element into predictive models: in 8000 BC or thereabouts, hunter-gatherers would have said they were all following a sustainable development pathway, but they were mostly gone not too long after; in AD 1750, solar-powered agriculturalists the world over would have been clear that they had got the situation controlled and would carry on for ever. Yet in 100 years they had nearly all been subject to the impact of industrial technology in one form or another, perhaps as users of machines, perhaps as soldiers brought by steamship. So pervasive has technology become that most societies look to it not only as a source of material resources but as a solution to any anxieties they may have, whether of their own bodies ('take a pill') or of society more widely ('put in CCTV'). But because so much of it has been based on carbon-rich energy sources, that hegemony is being challenged by those who warn of global climatic change. There is thus the irony that in spite of all the escaping from environmental constraints brought by technology (there is no lassitude in Singapore so long as the air conditioning works), the nature of the environment is determining, at the global scale, the limits to human actions.

See also: CULTURES, ENERGY, HOLOCENE

Further reading

McNeill, J.R., 2000. *Something new under the sun. An environmental history of the twentieth century.* Allen Lane, London.

Simmons, I.G., 2007. Living on the Earth. In Douglas, I., Huggett, R. and Perkins,C. (eds) *Companion encyclopedia of geography*. London and New York, Routledge, 2nd edn, 2007, Vol I, 3–15.

Smil, V., 1995. *Energy in world history.* Boulder CO, Westview Press.

HYDROLOGY

Is the study of the different forms of water, their distribution and circulation in the natural environment. Although the science is as old as the ancient civilisations of Egypt, Sumeria and Babylon it was not until the 17th century that key contributions established modern hydrology. Bernard Palissy in 1580 had the first modern understanding of the process of run-off, Pierre Perrault in 1694 showed for the first time that rainfall can sustain river flow, and this established the catchment or drainage basin as the unit of analysis. From these foundations the hydrological cycle covering the continuous movement of all forms of water (liquid, solid and vapour) on, in and above the earth, driven by the sun's radiant energy, was established as a central concept for hydrology. Documentation of the hydrological cycle enabled important progress in hydrology, requiring measurement of hydrological elements such as precipitation, first measured 400 years ago, and river flow measured as early as 3000 BC by the nilometer on the River Nile but gauged by continuous records at a number of stations since the late 19th century. For any area or basin the water balance equation is important because it relates river flow (Q), precipitation (P), evapotranspiration (Et), and changes in storage (ΔS) including changes in storage of soil moisture, ground water and snowpack, in the form:

$$Q = P - Et \pm \Delta S$$

Substantial differences exist between continents (Table 5) and between basins and the water balance equation can be a first stage in estimating water resources. In order to understand how runoff is generated in a drainage basin the Horton overland flow model interpreted water flow either as overland flow, or after infiltration becoming ground water and later emerging as base flow. This was succeeded in the 1960s by an interpretation involving different flow routes through the soil in pipes and above the water table, with additional contributions from dynamic

Table 5 *Water balance of the continents (cm.year^{-1}) (Compiled from several sources; such estimates necessarily vary according to the method of calculation)*

Continent	Precipitation	Evaporation	Runoff
Africa	70	51	18
Asia	62	38	24
Australia	46	41	5
Europe	66	42	24
North America	71	40	31
South America	153	82	71
Antarctica	11	1	10

contributing areas. Greater understanding of how precipitation is translated into river flow, influenced by catchment characteristics of the drainage basin, is important for the analysis of floods, droughts and water supply.

As hydrological investigations have become ever more relevant to policy-making, several subdivisions have become necessary including physical hydrology, which is the detailed measurement, analysis and modelling of hydrological processes; and applied hydrology which is the application of understanding of hydrological processes to their use and management. Links with other disciplines have prompted hydrometeorology, concentrating upon the study of precipitation and evapotranspiration; hydrogeology concerned with ground water, its movement and relations with surface water; as well as glacial hydrology and ecohydrology. Applied hydrology includes analysis and prediction of floods, droughts and water resource estimation. Because rivers and streams play a vital role in environment they can diagnose health not only of the rivers themselves but also of their landscapes. The way in which the hydrological cycle is perceived is important because it conditions decision-making – D.I. Smith suggested in 1998 that it is possible to envisage a hydro-illogical cycle where apathy holds sway and nothing is done during time of sufficient rain but concern arises as a result of drought.

See also: HYDROLOGICAL CYCLE, HYDROSPHERE, INTEGRATED BASIN MANAGEMENT, WATER RESOURCES

Further reading

Barry, R.G., 1969. The world hydrological cycle. In Chorley, R.J. (ed) *Water, earth and man*. Methuen, London, 11–29.

Karr, J.S., 1998. Rivers as sentinels: using the biology of rivers to guide landscape management. In Naiman, R.J. and Bilby, R. E., (eds) *River ecology and management*. Springer Verlag, New York, 502–528.

Kirkby, M.J. (ed), 1978. *Hillslope hydrology*. Chichester, Wiley.

Smith, D.I., 1998. *Water in Australia: resources and management*. Oxford University Press, Melbourne.

LANDSCAPE ARCHITECTURE

This cultural practice is concerned with the design of landscapes and of landscape structures. It was originally associated with English and North American landscape gardeners after the middle of the 18th century including Humphrey Repton who used the term 'landscape gardener' in 1794. The term landscape architecture was invented in 1828 so that in the 19th century the term landscape gardener came to be associated with those who build and may design landscapes, whereas landscape architect came to be associated with those who design and may build landscapes. In the second half of the 19th century F.L. Olmsted's designs for a series of parks, including Central Park in New York, were very influential in the development of landscape architecture as describing a special type of scenery set amongst buildings. The discipline became established after the foundation of the American Society of Landscape Architects in 1899 and the International Federation of Landscape Architects (IFLA) in 1948. Although the roots of the movements leading to a 'design with nature' approach could be traced in Europe back to the mid-19th century, the significance for the environmental sciences became apparent when Ian McHarg, a landscape architect at the University of Pennsylvania, produced a book *Design with Nature* which proposed ideas of wider importance than landscape architecture itself. The ideas were first developed in respect of the city where it was suggested that immeasurable improvements could be ensured in the aspect of nature in the city. *Design with Nature* provided a method whereby environmental sciences, especially ecology, were used to inform the planning process and the intent and language used subsequently appeared in the 1969 National Environmental Policy Act and other legislative instruments in the USA. Landscape architects such as McHarg were prominent in envisioning an ecologically based planning approach, to parallel ecological analogies already offered in innovative building architecture. However although a general ecological foundation was influential, there was insufficient input from hydrology, geomorphology and other environmental sciences, so that, in 1996, McHarg noted that inputs from geography and the environmental sciences were conspicuously absent, because geomorphology could be the integrative device for physical processes whereas ecology was the culminating integrator for the biophysical. Just as it has been suggested that geomorphology can contribute to developing and refining a design science so other environmental sciences should be concerned with environmental design and thus have an awareness of landscape architecture.

See also: ENVIRONMENTAL MANAGEMENT, LANDSCAPE

Further reading

McHarg, I.L., 1969. *Design with nature*. Doubleday/Natural History Press, New York.

McHarg, I.L., 1992. *Design with nature*. Wiley, Chichester.

McHarg, I.L., 1996. *A quest for life. An autobiography*. Wiley, New York.

McHarg, I.L. and Steiner, F.R. (eds), 1998. *To heal the earth: selected writings of Ian L. McHarg*. Island Press, Washington, DC.

Rhoads, B.L. and Thorn, C.E., 1996. Towards a philosophy of geomorphology. In Rhoads, B.L. and Thorn, C.E. (eds) *The scientific nature of geomorphology*. Wiley, Chichester, 115–143.

LIMNOLOGY

Limnology is the study of physical, chemical, geological, and biological aspects of freshwater ecosystems. It includes both basic and applied aspects of running waters (lotic environments) and lakes (lentic environments). Limnology can be considered an ecological science since it involves the interaction of various disciplines to explain processes and structure of freshwater ecosystems. Limnologists (the scientists who study limnology) study a variety of aspects of freshwater ecosystems. The origins and life cycles of lakes has been an important area of study. Lakes form as the result of glacial activity, tectonics, landslides, volcanic activity, dissolution of soils, aeolian processes, river processes (oxbow lakes), and behind shorelines. The structure and function of aquatic ecosystems is studied in limnology. The biota of lakes is highly diverse with representatives of many phyla and can be classified as plankton (small free floating organisms), nekton (actively swimming organisms), benthic (bottom dwelling organisms), and periphyton (organisms growing on substrates). Primary production is the measure of carbon fixation by primary producers and secondary production is the consumption of this production. Analysis of food webs or trophic dynamics describes the trophic or feeding relationships among different organisms in aquatic ecosystems. The study of biogeochemical cycles provides insights into material cycles and their role in the ecology of aquatic ecosystems. For example, nitrogen, phosphorus, silicon, and carbon dioxide are taken up by plants, consumed by animals, and broken down and released by microorganisms during decomposition. Limnologists have described how nutrient enrichment of freshwaters has caused eutrophication which is water quality characterized by algal blooms, low oxygen levels, fish kills, and the development of undesirable species. Other topics studied in limnology include the behaviour of dissolved gases (oxygen, nitrogen, and carbon dioxide), the dynamics of water movements, and interactions between the water and underlying sediments.

See also: ECOLOGY, FLUVIAL ENVIRONMENTS, OCEANOGRAPHY

Further reading

Bronmark, C. and Hansson, L-A., 2005. *The biology of lakes and ponds*. Oxford University Press, Oxford.

Cole, G.A., 1979. *Textbook of limnology*. The C. V. Mosby Company, London.

Wetzel, R., 2000. *Limnology*. Harcourt College Publishers.

Wetzel, R., 2001. *Limnology. Lake and river ecosystems*. Academic Press, New York, 3rd edn.

METEOROLOGY

Meteorology is the study of the atmosphere and processes that produce weather. The majority of Earth's observed weather is located in the troposphere. Meteorology, climatology, atmospheric physics, and atmospheric chemistry are sub-disciplines of the atmospheric sciences. Meteorology and hydrology consist of the interdisciplinary field of hydrometeorology. The term 'meteorology' has a long history stemming from Aristotle's Meteorology, of the fourth century BC. As late as the 19th century, the full extent of the large-scale interaction of pressure gradient force and deflecting force became known as the cause of air masses moving along isobars. By early 20th century, this deflecting force was named the Coriolis Effect (after Gaspard-Gustave Coriolis, who published in 1835 on the energy yield of machines with rotating parts – such as waterwheels).

Early in the 20th century, advances in the understanding of atmospheric physics led to the foundation of modern numerical weather predictions. Lewis Fry Richardson published *Weather Prediction by Numerical Process* in 1922 that described in small terms how the fluid dynamics equations governing atmospheric flow could be neglected to allow numerical solutions to be found. At that time, the number of calculations required was too large to be completed before the advent of computers. At the same time, a group of meteorologists in Norway, led by Vilhelm Bjerknes, developed the model that explained the generation, intensification, and ulti-mate decay (the life cycle) of mid-latitude cyclones, introducing the idea of fronts – or sharply defined boundaries between air masses. The group included Carl-Gustaf Rossby (who was the first to explain the large-scale atmospheric flow in terms of fluid dynamics), Tor Bergeron (who first determined the mechanism by which rain forms), and Jacob Bjerknes.

By the 1950s, numerical experiments with computers became possible. The first weather forecasts derived this way used barotropic (single-vertical-level) models and successfully predicted the large-scale movement of mid-latitude Rossby waves (representing the pattern of atmospheric lows and highs). In the 1960s, Edward Lorenz founded the field of chaos theory, leading to the understanding of the chaotic nature of the atmosphere. These various advances have led to the current use of ensemble forecasting in most major forecasting centres, taking into account the uncertainties that arise due to the chaotic nature of the atmosphere.

In 1960, the launch of TIROS-1, the first successful weather satellite marked the beginning of the age where weather information is available worldwide. Weather satellites, along with more general-purpose Earth-observing satellites that circle the Earth at various altitudes, have become indispensable for the study of a variety of phenomena, from forest fires to El Niño. Meteorological research as treated in http://www.ametsoc.org/ suggests that 'Applied research deals primarily with weather and climate observation, analysis, and forecasting'. In other words, it is research that can be applied to everyday activities and operations. Included in such research are the development of forecast techniques and forecast verification methods, and the performance of diagnostic and case studies. Technique development, among other things, addresses the creation of algorithms for remote-sensing applications used with weather satellites and radars, lightning detectors, and atmospheric profilers and sounders. It also includes the development of techniques that can be applied directly to weather and climate forecasting. Basic research addresses more fundamental atmospheric processes such as the formation of clouds and precipitation, air–sea interactions, radiation budgets, aerosol transport, thermodynamics, and global general circulation. A variety of size and time scales are involved, ranging from a few centimetres to hundreds of kilometres and from minutes to centuries. The end results of basic research often support advances in numerical weather prediction models. Currently, one of the hot issues in basic research is global warming. Models that can deal with global warming, popularly known as the 'greenhouse effect', are becoming more and more sophisticated.

See also: ATMOSPHERIC SCIENCES, CLIMATOLOGY, MICROCLIMATE

Further reading

http://www.ametsoc.org/pubs/books_monographs/index.html

http://www.metoffice.gov.uk/publications/index.html

Aguado, E. and Burt, J.E., 2006. *Understanding weather and climate.* Pearson Prentice Hall, Upper Saddle River NJ.

Richardson, L.F., 1922. *Weather prediction by numerical process.* Cambridge University Press, Cambridge.

MICROCLIMATE

A microclimate is the distinctive climate of a small-scale area (see scale dimensions in Table 6), such as a garden, park, hill, valley, coast, forest, or part of a city. The weather variables in a microclimate, such as radiation, temperature, rainfall, wind or humidity, may be subtly different from the conditions prevailing over the area as a whole and from those that might be reasonably expected under certain types of pressure or cloud cover. There are several controlling attributes that drive microclimate conditions and fall under the categories (1) radiative, (2) aerodynamic, (3) thermal, and (4) moisture (Table 7).

Microclimates exist, for example, near bodies of water which may cool the local atmosphere, or in heavily urban areas where brick, concrete, and asphalt absorb the sun's energy, heat up, and reradiate that heat to the ambient air, or on slopes and valleys where radiation is received differently and shade, temperature, and wind are variable. Microclimates can be created in areas affected by deforestation, afforestation, farming practices, dark-coloured buildings and roadways, or dam construction. Another contributory factor to microclimate is the aspect or slope – south-facing slopes in the Northern Hemisphere and north-facing slopes in the Southern Hemisphere are exposed to more direct sunlight than opposite slopes and are therefore warmer for longer. Upland areas have a specific type of climate that is notably different from the surrounding lower levels. Temperature usually decreases with height at a rate of between 5 and 10 °C per 1000 m, depending on the humidity of the air. This means that even quite modest upland regions can be significantly colder on average than in a valley. Occasionally, a temperature inversion can make it warmer above, but such conditions rarely last for long. With higher hills and mountains, the average temperatures can be so much lower that winters are longer and summers much shorter.

Higher ground also tends to be windier, which makes for harsher winter weather. Hills often cause cloud to form over them by forcing air to rise, either when winds have to go over them or they become heated by the sun. When winds blow against a hillside and the air is moist, the base of the cloud that forms may be low enough to cover the summit. As the air descends on the other (lee) side, it dries and warms, sometimes enough to create a Föhn effect, known as the Chinook on the High Plains of North America. Consequently, the leeward side of hills and mountain ranges is much drier than the windward side. The clouds that form due to the sun's

Table 6 *Scales of climate*

Scale	Length	Area	Locale
Microclimate	1 m – 1 km	1 m² – 1 km²	Local
Mesoclimate	1 km – 100 km	1 km² – 100 km²	Regional
Macroclimate	100 km – 10 000 km	100 km² – 10 000 km²	Continental
Megaclimate	>10 000 km	>10 000 km²	Global

Table 7   *Determinants of microclimate conditions*

Radiative	Albedo, emissivity, surface temperature, surface geometry
Aerodynamic	Roughness, upwind obstacles
Thermal	Conductivity, heat capacity, thermal diffusivity, thermal admittance
Moisture	Status in soil, vegetation impacting evaporation and transpiration

Modified after Oke (1997)

heating sometimes grow large enough to produce showers, or even thunderstorms. This rising air can also create an anabatic wind on the sunny side of the hill. Sunshine-facing slopes (south-facing in the Northern Hemisphere, north-facing in the Southern Hemisphere) are warmer than the opposite slopes. Apart from temperature inversions, another occasion when hills can be warmer than valleys is during clear nights with little wind, particularly in winter. As air cools, it begins to flow downhill and gathers on the valley floor or in pockets where there are dips in the ground. This can sometimes lead to fog and/or frost forming lower down. The flow of cold air can also create what is known as a katabatic wind.

The coastal climate is influenced by both the land and sea between which the coast forms a boundary. The thermal properties of water are such that the sea maintains a relatively constant day-to-day temperature compared with the land. The sea also takes a long time to heat up during the summer months and, conversely, a long time to cool down during the winter. In the tropics, sea temperatures change little and the coastal climate depends on the effects caused by the daytime heating and night-time cooling of the land. This involves the development of a breeze from off the sea (sea breeze) from late morning and from off the land (land breeze) during the night. The temperature near a windward shore is similar to that over the sea whereas near a leeward shore, it varies much more. On the other hand, a sea fog can be brought ashore and may persist for some time, while daytime heating causes fog to clear inland. A lee shore is almost always drier, since it is often not affected by showers or sea mist and even frontal rain can be significantly reduced. When there is little wind during the summer, land and sea breezes predominate, keeping showers away from the coast but maintaining any mist or fog from off the sea.

See also: ATMOSPHERIC SCIENCES, METEOROLOGY

Further reading

Oke, T.R., 1997. Surface climate processes. In Bailey, W.G., Oke, T.R. and Rouse, W.R. (eds). *The surface climates of Canada*. Montreal, McGill-Queen's University Press, 21–43.

Geiger, R. Aron, R.H. and Todhunter, P., 1995. *The climate near the ground*. Rowman and Littlefield, Lanham MD, 6th edn.

Monteith, J.L. and Unsworth, M.H., 1990. *Principles of environmental physics*. Edward Arnold, London.

OCEANOGRAPHY

Oceanography or marine science is the scientific study of the oceans. Although there are names for different parts of the ocean (i.e., Atlantic Ocean, Pacific Ocean, Gulf of Mexico, etc.), the world ocean is in fact a single, interconnected body of water. This world ocean covers about 70% of the earth's surface and contains about 97% of the liquid water on the earth's surface. The average depth of the ocean is about 3800 m and if the surface of the earth were absolutely smooth, the ocean would be about 2700 m.

The science of oceanography is often divided into four general disciplines: geological oceanography, physical oceanography, chemical oceanography, and biological. These divisions are somewhat arbitrary since physical, geological, chemical, and biological processes in the ocean are so interconnected.

Geological oceanography is primarily the study of the dynamic structure of the sediments and crust at the bottom of the ocean. It also involves study of processes in the water that affect and contribute to the underlying structure. Some of the exciting issues that geological oceanographers study are plate tectonics, sea-floor vents, undersea volcanism, the age and history of the oceans, and undersea landslides.

Physical oceanography is the study of the physical dynamics of the oceans including waves, ocean currents, air–sea interactions, sound propagation in water, and the heat budget of the oceans. Two processes widely known by the public in general that physical oceanographers study are tsunamis and surge and waves associated with tropical storms such as hurricanes.

Chemical oceanography is the study of the composition of the sea and the processes that affect that composition. As everyone is aware, the sea is salty. The main component that gives the sea its salty taste is sodium chloride, NaCl, common table salt. But there are many other dissolved and particulate components that are found in sea water. Many of these are biologically and geologically important, so there is a strong interaction with geological and biological oceanography. Some chemical components are used as tracers in the movement of sea water.

Biological oceanography is the study of the biota, or living organisms, in the sea. Biological oceanographers study such things as the composition and distribution of the biota, productivity, and migration of organisms. An important component of biological oceanography is the study of economically important species such as fish and shrimp that make up the world fisheries. Organisms in the sea can be classified in a number of ways and a general way is to refer to those organisms that live in the water as opposed to those that live in or on the bottom. This is somewhat arbitrary, of course, since many organisms spend part of their lives in both environments and because many animals that live in the water feed on the bottom and vice versa. Organisms living in the water can be generally classified as nektonic, those that are free-swimming, and plantonic, those that float with the currents of water. Benthic organisms live on the bottom.

Oceanography can also be defined more specifically by the region of the ocean studied or by a more specific discipline. For example, there are tropical oceanographers, deep sea oceanographers, and blue water oceanographers, wave specialists, and those who date deep sea cores. However, all of these disciplines are highly interactive and interdependent.

Some of the major concepts and findings of oceanography are as follows.

Plate tectonics. One of the most important facts that emerged about the ocean during the 20th century was that the continents slowly moved. At the beginning of the century, most considered that the position of the continents was fixed. By the end of the century, plate tectonics had become the central unifying theme describing large-scale dynamics of the earth's surface, especially ocean basins. Early in the 20th century, the German scientist Alfred Wegener building on earlier suggestions, proposed the theory of continental drift. His conclusions were based on the close fit of the continents into a super-continent 'Pangea' and similar fossils and species on adjacent continents of Pangea. Few accepted the idea of continental drift during Wegener's lifetime (he died in 1930), but evidence accumulated during the century that proved that continental drift was a fact. It was noted that volcanoes and earthquakes were not randomly distributed over the earth's surface but generally occurred in lines associated with mid-ocean ridges and deep trenches. The finding that the age of sediments increased with distance from mid-ocean ridges proved that the sea floor was moving away from these ridges and in some cases flowing back down into the mantle in so-called subduction zones. It is now known that heating due to radioactive decay in the interior of the earth drives giant convection cells that move the tectonic plates around on the surface of the earth.

Productivity of the oceans. Except in coastal areas, the productivity of the oceans is based on photosynthesis by single cell alage called phytoplankton. This productivity is called aquatic primary productivity (APP). APP is not uniform in the surface waters in the ocean but is concentrated in certain zones because of the interaction of physical, chemical, and biological factors. Where light and nutrients (nitrogen, phosphorus, silicon, iron) are abundant, productivity can be high. Productivity is generally higher in coastal areas, upwelling zones, and in high latitudes. In the coastal zone, land drainage brings nutrients that interact with light in shallow coastal waters to support high productivity. In upwelling areas, ocean circulation patterns bring deep, nutrient-rich waters to the surface where high levels of phytoplankton productivity occur. At high latitudes, there is often a strong spring bloom of phytoplankton supported by nutrients brought to the surface by mixing of deeper waters.

Hydrothermal vents. One of the most exciting oceanographic discoveries of the second half of the 20th century was the existence of hydrothermal vents. These vents are formed as ocean water becomes superheated (>300 °C) when it comes in contact with very hot rocks associated with sea floor spreading at mid-oceanic ridges. The hot water contains high levels of dissolved minerals, sulphides. The vents support a rich community of chemosynthetic bacteria that forms the base of a food chain in the hydrothermal vent communities. These ecosystems are composed of a diverse group of animals including crabs, sea anemones, shrimp, and long tube worms. These vents are fairly widespread in the ocean, mostly near mid-ocean ridges.

See also: HYDROSPHERE, ESTUARINE ENVIRONMENTS, OCEAN CIRCULATION

Further reading

There are a number of excellent general oceanography texts that provide an introduction to the subject including:

Barnes, R.S.K. (ed.), 1978. *The coastline*. John Wiley & Sons Inc., Chichester, 356pp.

Coker, R., 1947. *This great and wide sea*. University of North Carolina Press, Chapel Hill (A classic early text).

Fairbridge, R.W., 1980. The estuary: its definition and geodynamic cycle. In: E. Olausson and I. Cato (eds.), *Chemistry and biogeochemistry of estuaries*, John Wiley & Sons Inc., Chichester, 1–36.

Garrison, T., 2007. *Oceanography. An invitation to marine science*. Thomson Brooks, Belmond CA, 6th edn.

Gross, M., 1995. *Principles of oceanography*. Prentice Hall, Englewood Cliffs, New Jersey, 7th edn.

Pinet, P., 2006. *Invitation to oceanography*. Jones and Bartlette Publishers, Boston.

Stowe, K., 1996. *Exploring ocean science*. Wiley, New York, 2nd edn.

Thurman, H., 1994. *Introduction to oceanography*. Macmillan, New York. 2nd edn.

http://www.oceanexplorer.noaa.gov/

http://oceanworld.tamu.edu/home/oceanography_book.htm

PALAEOECOLOGY

Once the study of past ecologies for its own sake, the reconstruction of ecological change during the Pleistocene and Holocene has become an important part of understanding the background to climate change and environmental degradation, especially where it is important to disentangle natural change from that brought about by human action. Palaeoecology may thus provide direct evidence of past biotic communities (e.g., peats which indicate now vanished freshwater mires or saline muddy sediments) that relate to former salt marshes. Equally important, the collections of data about sediments and biota may act as a proxy for other important ecological factors. The presence or absence of warmth-demanding tree species, for instance, may yield an understanding of climatic change and if plotted over a large region, the rate of warming or cooling. The presence and high frequency of the pollen of a weedy species in a dated deposit will be an indicator of the intensity of agriculture at one particular time.

The success of palaeoecology clearly depends first of all on the preservation of materials from the time under investigation: if there are no deposits then nothing can be done except to speculate by extrapolation from somewhere else. Secondly, there must be an understanding of the biophysical processes that have affected the deposit and its contents since it was laid down: its taphonomy. Then comes the key effort to narrow down the nature of the subfossil material. In the case of inorganic sediments, questions must be asked such as whether it has been transported and if so by wind, water, gravity or human activity; if it is biological then can the species be identified and does the assemblage of species give additional information about the natural or human-affected environment? Lastly but critically, can the material be accurately dated by one of the various means now available, such as radiocarbon decay, thermoluminesence, uranium-series or dendrochronology (tree rings).

See also: CLIMATE CHANGE, ECOLOGY, SEDIMENTS

Further reading

Björn E. and Berglund, B.E., (eds) 2003. *Handbook of Holocene palaeoecology and palaeohydrology.* Blackburn Press, Caldwell, NJ.

Brenchley, P.J. and Harper, D.A.T.,1998. *Palaeoecology: ecosystems, environments and evolution.* Chapman and Hall, London.

PALAEOGEOGRAPHY

Geography is the study of the earth and its features, and of living systems of the earth including humans and interactions among living and earth systems. Palaeogeography is the study of the past history of the earth. It has strong links with geological history, evolution, palaeoecology, and the origins of humans. Many topics are included under the heading palaeogeography. Human evolution has strong links with the natural world which affected the pace of evolution and the kinds of interactions that human ancestors had with the environment. As proto-humans as well as *Homo sapiens* spread out over the earth, they encountered a wide variety of environments that strongly affected both culture and biological characteristics. Evolution itself was strongly affected by natural world and the way that it changed. Two examples of this are continental drift and the effects of large meteorites striking the earth. Continental drift resulted in similar species on continents that were formerly linked (such as Africa and South America) as well as allowing speciation after the continents separated. Meteorite impacts decimated life on earth a number of times and resulted in a burst of speciation afterwards. Over the long history of the earth, there have been dramatic changes in the surface of the earth. Drifting continents have changed the size of the oceans, moved continents from tropical to polar regions, and led to the formation of mountains when continental plates collided. Volcanism has also formed mountains and islands. Weathering has worn down mountains and formed broad flood plains and deltas. Deposition in sedimentary basins has recorded such things as temperature and salinity histories. Through all of this living systems have adjusted to these changes by adaptation and evolution. On a shorter timescale, palaeohistory of droughts and fires is recorded in tree rings, some older than 5000 years. More recently, palaeo-records have been used to document human impact on a global scale.

See also: PHYSICAL GEOGRAPHY, GEOLOGIC TIME, PALAEOECOLOGY, TIMESCALES – GEOLOGIC

Further reading

Allmon, W.D. and Bottjer, D.J. (eds), 2001. *Evolutionary palaeoecology*. Columbia University Press, New York.

Huc, Y. (ed),1995. *Palaeogeography, paleoclimate and source rocks*. American Association of Petroleum Geologists. Studies in Geology volume 40, Tulsa, Oklahoma.

http://www.bbm.me.uk/portsdown/PH_065_Palaeo.htm

http://www.geologyshop.co.uk/palaeo~1.htm

Palaeogeography, palaeoclimatology, palaeoecology. An International Journal for the Geo-Sciences. Elsevier, 1965–

Roberts, N., 1998. *The holocene. An environmental history*. Blackwell, MA.

Scotese, C.R., 1997. *Palaeogeographic atlas*. PALEOMAP Progress Report 90-0497, Department of Geology, University of Texas at Arlington, Arlington, Texas.

PALAEONTOLOGY see Fossils and Palaeontology

PHYSICAL GEOGRAPHY

In 2000 a definition of Physical Geography was suggested as:

Physical geography focuses upon the character of, and processes shaping, the landsurface of the earth and its envelope, it emphasizes the spatial variations that occur and the temporal changes necessary to understand the contemporary environments of the earth. Its purpose is to understand how the earth's physical environment is the basis for, and is affected by, human activity. Physical geography was conventionally subdivided into geomorphology, climatology, hydrology and biogeography, but is now more holistic in systems analysis of recent environmental and Quaternary change. It uses expertise in mathematical and statistical modelling and in remote sensing, develops research to inform environmental management and environmental design, and benefits from collaborative links with many other disciplines such as biology especially ecology, geology and engineering. In many countries physical geography is studied and researched in association with human geography.

Although geography is younger than the physical sciences of chemistry and physics, geography is long established because of interest in other lands, associated with exploration and discovery. However the real antecedents of modern physical geography were established in the 19th century, some 200 years ago, and derived largely from the science of geology. Since its foundation in the mid-19th century physical geography developed in association with human geography with awareness of earth spheres, especially the atmosphere first recognized as early as the late 17th century, subsequently the lithosphere, hydrosphere and biosphere, later complemented by others including the pedosphere, cryosphere and noosphere. Knowledge of these spheres provided the foundations for the growth of physical geography in Britain and elsewhere.

A unitary holistic approach prevailed in the 19th century, with influential books produced by Mary Somerville in 1848 and T.H. Huxley in 1877 who advocated physiography; succeeded in the first half of the 20th century by reductionism with the emergence of a series of branches of physical geography including geomorphology, meteorology and climatology, and biogeography. These three areas constituted the three main sub-divisions of physical geography although there were other terms used such as plant geography coined by Schimper in 1903. These remained as the major branches of the subject until 1960 but then as physical geography grew and expanded it became increasingly subdivided into more branches and sub-divisions, each having closer links to other disciplines – a process described as reductionism or fissiparism. In addition, a revolution in physical geography was fostered by the advent of quantification and led to concentration upon five new approaches, focused upon processes, human activity, systems, applied physical geography, and environmental change, although each approach could be traced to much earlier origins.

The continued separation of physical geography, with reductionism, into smaller sub-branches continued until the last two decades of the 20th century but then, encouraged by increasing awareness of hazards and of resources, was complemented by two other branches which could be termed 'global' and 'cultural'. Global approaches are more prominent, as an antidote to investigation of increasingly small-scale spaces and a response to the need for focus upon global change, especially of climate, benefiting from advances in remote sensing. In addition, although cultural themes had been evident since the 1960s especially as a result of research in Mediterranean areas, a theme that may be described as cultural physical geography was increasingly evident because of the intrinsic way in which it was increasingly impossible to separate natural processes from human impact, the two together being essential for the investigation of landscape management.

The structure for physical geography can be envisaged in terms of a physical geography equation embracing morphological elements or results of the physical environment (F), processes operating in the physical environment (P) and the materials (M) upon which processes operate over periods of time t. This indicates how physical geography focuses upon the way in which *processes* operating on *materials* over time t to produce *results* which may be expressed as a landform. The equation can be expressed as:

$$F = f(P, M) \, \mathrm{dt}$$

Investigations take place at five levels:

» **studies of the elements or components of the equation** – study of the components in their own right, focused on the description of landforms, of climate types, of soil character, or of plant communities, and often preparatory to other levels.
» **balancing the equation** – studies of the way in which components are related or balanced at different scales and in different branches of physical geography, often focusing upon contemporary environments and upon interaction between processes, materials and the resulting forms or environmental conditions.
» **differentiating the equation** – includes analyses of the way in which relationships change over time including how one equilibrium situation is disrupted and eventually replaced by another, including impact of climatic change and also of human activity.
» **applying the equation** – when the results of research are directed to applied questions and problems.
» **appreciating the equation** – a cultural physical geography approach embracing ways in which human reaction to physical environment and physical landscape influence how environment is managed and designed.

Despite ever greater specialization, because we have no choice at the beginning of the 21st century, there are many indications of a return to a more integrated physical geography once again with the five overlapping core sub-disciplines of climatology, hydrology, biogeography and ecological processes, Quaternary environmental change and geomorphology (Fig. 3). New structures for physical geography have appeared as illustrated by the titles of papers in the

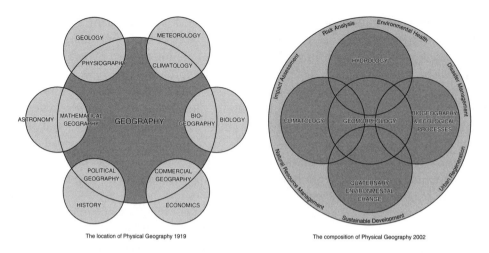

The location of Physical Geography 1919

The composition of Physical Geography 2002

Figure 3 *The structure of Physical Geography in the 20th and 21st centuries. The 1919 view is from Fenneman (1919) and the 2002 structure is after Gregory, Gurnell and Petts (2002)*

collection of branches presented in the USA by Bauer and others in 1999. A more integrated stance has come about because, as the branches of physical geography have advanced, so has each of them had to interrelate with other branches and with aspects of other disciplines – so that multidisciplinary approaches have become increasingly necessary. Just as physical geography is adopting a more holistic approach to physical environment so other disciplines, such as earth system science have developed to fill the void in the range of disciplines arising from earlier research which has occurred and which has been described as a sparsely inhabited niche. Research in physical geography now demonstrates the benefits of collaboration with other disciplines and such multidisciplinary investigations have been exemplified by the need for, and emergence of, multidisciplinary disciplines such as biogeomorphology, geoarchaeology, and ecohydrology and such collaboration is necessitated by investigations of global change. We have to use some form of system closure to limit the fields that we study although we remain aware that no single discipline can do justice to all the implications of major challenges such as those posed by global change. This is one of the justifications for the environmental sciences.

There are a number of challenges for physical geography, many of which may have parallels in the recent development of other environmental science disciplines, and these are indicated in Table 8. If these challenges are met there are many potential opportunities (Table 9) which may also have analogues in other disciplines.

Table 8   *Challenges for physical geography*

Challenge	Implication
Whether a core exists for the discipline	More reductionist approach via sub-disciplines alleviated by a more holistic approach. 'The future of physical geography, like that of most other fields, will be characterized by irreversible divergence and intellectual niche specialization. This deterioration of common frames of reference is the biggest challenge that fragmentation and specialization pose to the practice – as opposed to the institutions – of physical geography' (Phillips, 2004).
Physical geography is unbalanced	More study of land than atmosphere, with oceans receiving very little attention (Gregory, 2001) – the atmospheric and ocean sciences have developed separately.
Acknowledge that now so little of the physical environment is in a 'natural' condition	Cultural underpinning is pertinent throughout physical geography emphasizing the link (e.g., Tickell, 1992) that was once strong, weakened in the second half of the 20th century, but perhaps now needs to be reinforced again.
Utilize the range of new techniques which have revolutionized environmental sciences	Including remote sensing, dating, mass flux tracing, analytical techniques, spatial analysis, LIDAR-based topographic mapping.
Foster good relations with other disciplines	Closer links to other disciplines have occurred through use of common techniques (Gregory, Gurnell and Petts, 2002), the development of Quaternary science (e.g., Richards and Wrigley, 1996), through the creation of new hybrid fields which are often the location for frontier research, through multidisciplinary collaboration in fields such as palaeohydrology (Gregory, 1983), through global investigations, and through a more ethical stance in relation to environmental problems.
Encourage pluralist approaches	As physical geography continues to evolve, and perhaps to reinvent itself to some extent, there should be pluralist approaches with diversity providing a basis for increasing strength. It is no longer possible for all Higher Education Institutions to cover all aspects of physical geography in the same way and it is desirable for there to be a diversity of approach: specialization could intensify further, meaning that different degrees of linkage with human geography and with other disciplines are created including environmental sciences and environmental studies to respond to the infrastructure of institutions and to the expertise of the staff available.
How physical geography relates to human geography and the social sciences	The choice between an environmental earth science approach and a more social human-oriented physical geography (Chorley, 1971) still remains, where the former required the study of physical environment in a scientific manner, divorced from its use and management, whereas the alternative was for a physical geography which included a focus upon human activity (e.g., Goudie and Viles, 1997), upon decision-making and environmental management.
Need to consider ethical implications	Ethics has begun to feature in physical geography: debates engaging philosophers about what is nature have increasing resonance for physical geographers involved in restoring nature, for example in river channel restoration.
Need to investigate holistic and global problems	Effects and potential impacts of global climate change offer a range of valuable investigations.

Table 9 *Opportunities for physical geography. (developed from a table in Gregory, 2004)*

Role and value of Physical Geography	Opportunity
To Physical Geographers	Emphasize holistic approach to the physical environment and relation to other environmental sciences.
To Human Geographers	Develop ways in which physical and human geography interact including risks, hazards, landscape evaluation.
In Education	Emphasize relevance of physical geography contributions made to environmental understanding.
In and by Other Disciplines	Demonstrate singularity of contributions by physical geographers in multidisciplinary teams.
To Managers and Decision-makers	Communicate ways in which physical geography research provides valuable strategic or applied results.
To the General Public	Use every opportunity to explain the value of a contemporary physical geography approach.
Overall	Physical geography is an integral part of geography but because of its role in multidisciplinary teams can be a significant element in other environmental science structures.

See also: ATMOSPHERIC SCIENCES, EARTH SYSTEM SCIENCE, GEOMORPHOLOGY, HUMAN GEOGRAPHY

Further reading

Bauer, B.O., 1999. On methodology in physical geography. Current status, implications and future prospects. *Annals of the Association of American Geographers* 89, 677–679.

Bauer, B.O., Winkler, J.A. and Weblen, T.I., 1999. Afterword: A shoe for all occasions or shoes for every occasion: Methodological diversity, normative fashions, and metaphysical unity in physical geography. *Annals of the Association of American Geographers* 89, 771–778.

Butzer, K.W., 1964. *Environment and archaeology*. Methuen, London.

Chorley, R.J., 1971. The role and relations of physical geography. *Progress in Geography* 3, 87–109.

Goudie, A.S. and Viles, H., 1997. *The Earth transformed*. Blackwell, Oxford.

Gregory, K.J., 1978. A physical geography equation. *National Geographer* 12, 137–141.

Gregory, K.J., (ed) 1983. Background to palaeohydrology. Wiley, Chichester.

Gregory, K.J., 1985. The nature of physical geography. Arnold, London.

Gregory, K.J., 2000. The changing nature of physical geography. Arnold, London.

Gregory, K.J., 2001. Changing the nature of physical geography. *Fennia*, 179, 9–19.

Gregory, K.J., 2004. Valuing physical geography. *Geography* 89, 18–25.

Gregory, K.J., 2005. (ed). Fundamentals of geography. *Physical Geography*. (4 vols) Vol 1, XIX–LVIII.

Gregory, K.J., Gurnell, A.M. and Petts, G.E., 2002. Restructuring physical geography. *Transactions of the Institute of British Geographers* 27, 136–154.

Lane, S.N., 2001. Constructive comments on D.Massey 'Space-time, "science" and the relationship between physical and human geography'. *Transactions Institute of British Geographers* 26, 243–256.

Phillips, J.D., 2004. Laws, contingencies, irreversible divergence, and physical geography. *Professional Geographer* 56, 37–43.

Richards, K.S. and Wrigley, N., 1992. Geography in the United Kingdom 1992–1996. *Geographical Journal* 162, 41–62.

Slaymaker, H.O. and Spencer, T., 1998. Physical geography and global environmental change. Longman, Harlow.

Tickell, C., 1992. The Presidential Address. *Geographical Journal* 158, 322–325.

PHYSIOLOGY

P hysiology concerns the functioning of the different organ systems in an organism, both among themselves and with the environment, that is how different parts of an organism work and carry out life sustaining activities. Physiology reveals and elucidates the dynamic nature of the workings of a living organism. The field is subdivided into many different specialized areas. There is the study of a specific group of organisms, such as human physiology or plant physiology, or of specific organ systems, such as renal physiology or neurophysiology, or of specific processes, such as the physiology of photosynthesis. Physiology often focuses at the cellular or molecular level, because ultimately what an organism does depends on the functioning of its cells (or of the cell for an unicellular organism). Also cellular functioning is dependent on the chemical reactions that take place in them. Ultimately, physiology depends on physics which explains electrical currents among nerves, blood pressure, and the uptake and release of chemicals. The study of physiology and of the structure of an organism, anatomy, must be done together because function reflects structure and vice versa.

Two examples will illustrate the range of topics that physiology encompasses. Human physiologists study the way that toxins, such as cigarette smoke, affect the body and cause cancer, and the response of the body to ways that cancer is treated. Plant physiologists who study coastal wetland plants have explained how plants exist in a stressful environment where there is no oxygen in the soil and salinity gives rise to salt stress. Such plants have special ways of moving oxygen to the root system and dealing with toxins, such as hydrogen sulphide, that are present in anaerobic soils. Many plants also have special mechanisms for excreting excess salt back into the surrounding water.

See also: BIOLOGY, NATURAL PRODUCTION MECHANISM

Further reading

Costanzo, L., 2007. *Physiology*. W.B. Saunders, Philadelphia, 4th Edn.

Marieb, E.N., 2003. *Essentials of human anatomy and physiology*. The Benjamin/Cummings Publishing Co., 7th Edn., Redwood City, CA.

Mendelssohn, I.A. and Morris, J.T., 2000. Eco-physiological control on the productivity of *Spartina alterniflora* Loisel. In Weinstein, M.P. and Kreeger D.A. (eds), *Concepts and controversies in tidal marsh ecology*. Kluwer Academic Publishers, Dordrecht, 59–80.

SEISMOLOGY

Seismology is defined by the AGI Glossary in 1987 as 'The study of earthquakes and of the structure of the Earth, by both natural and artificially generated seismic waves'; a fault as 'a fracture in the Earth's crust along which displacement of the sides has occurred relative to each other, parallel to the walls of the fracture'; and an earthquake as 'the shaking that occurs by the abrupt release of stored elastic strain energy along a fault'.

If a rock is tossed into a pond, then waves ripple out from the point where the rock breaks the surface of the water (Fig. 4). Seismic waves likewise ripple through the Earth when it is broken abruptly along a fault during an earthquake. The earthquake waves emanate outward from the focus (also called the hypocentre) which is the point of breakage on the fault in the Earth's crust (Fig. 5).

Seismic waves are of two principal types (Table 10): *body waves* (including P and S waves), which travel through the Earth, and *surface waves* (called L waves) which travel over the surface of the Earth. P and S waves travel faster than L waves, because the interior of the Earth conducts P and S waves more efficiently than L waves are conducted in the crust.

Seismic waves are recorded on a seismograph as a seismogram (Fig. 6).

Because all the seismic waves originate simultaneously at the focus, the seismologist subtracts the arrival time of the P-wave at the seismograph from the arrival time of the S-wave

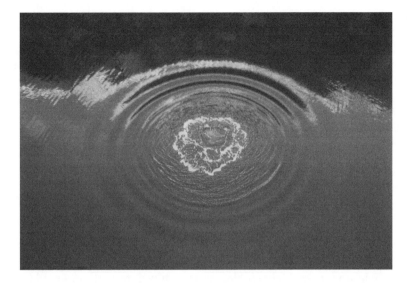

Figure 4 *Waves radiate outward on the surface of a pond from a rock thrown into the pond. Photo by Arthur G. Sylvester, Nov. 1985*

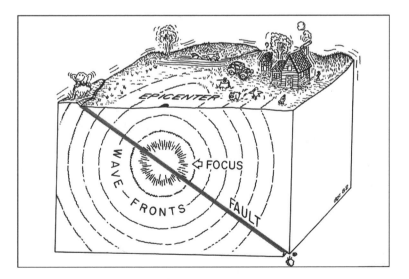

Figure 5 *Seismic waves radiate outward from the earthquake focus. Drawing by Dave Crouch, 1982, with permission. Based on a similar diagram by Gilluly, Waters, and Woodford, 1968. Principles of Geology, 3rd ed., W.H. Freeman, San Francisco, p. 468*

Table 10   *Types of seismic waves*

Body waves	Surface waves
P waves, also called 'pressure' or 'push' waves, vibrate in the direction of energy propagation	*L waves*, also called 'long' or 'longitudental' waves, travel across the surface of the Earth
S waves, also called 'shear' or 'shake' waves, vibrate perpendicular to the direction of energy propagation	

to determine the distance the waves travelled. The S–P calculation is analogous to determining the distance to a lightning strike. Light (analogous to P) travels faster than sound (analogous to S), so because light and sound are produced at the same time and travel the same route, the observer sees the lightning first and then hears the thunder. The seconds between the lightning strike and the arrival of the thunder give an approximate distance in miles to the point of origin of the lightning.

Earthquakes are located by calculating the S–P times at three separate seismographs, determining the focal distance from each, and then finding by triangulation the epicentre, the point on the surface of the Earth above the focus (Fig. 7).

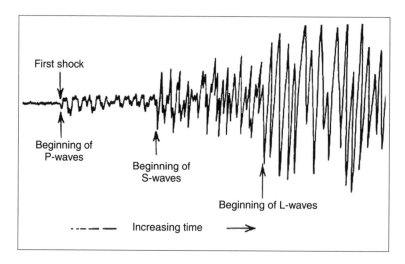

Figure 6   *Idealized trace of an earthquake in a seismogram*

Definition: *Magnitude*: The maximum amplitude of the trace of the earthquake written by a standard seismograph at a standard distance of 100 km from the hypocenter. It is a 'body wave' magnitude if the measurement is of one of the body waves. It is a 'surface wave' magnitude if measurement is of the L-wave trace. Commonly referred to as the 'Richter scale', which is useful for the news media, but now replaced by the Moment Magnitude scale by seismologists.

Definition: *Intensity*: the relative measurement of an earthquake's severity, based on degree of damage and perception by humans to the shaking. The Modified Mercalli scale is subdivided into increments ranging from 0 (imperceptible to humans), through V (felt by all, but little damage), to XII (total destruction).

Definition: *Moment Magnitude*: M_0, is a dimensionless measure of the total amount of stored elastic strain energy that is transformed during an earthquake. Only a small fraction of the seismic moment M_0 is converted into radiated seismic energy Es, which is what seismographs register.

Definition: *Stress*: (AGI Glossary, 1987, with modifications) In a solid, the force per unit area, acting on any surface within the solid. The stress at any point in a solid is a tensor of the second rank defined mathematically by nine values: three to specify the normal components and six to specify the shear components, relative to three mutually perpendicular surfaces through the point.

Definition: *Strain:* the change in shape or volume of a body in response to stress.

Two types of strain are *elastic strain* where a body returns to its original shape after release of stress (e.g., stretch and unstretch a rubber band), and *inelastic strain* where permanent deformation occurs (e.g., stretch a rubber band too far so that it snaps and breaks). Three types of inelastic strain are *translation* (change in position), *distortion* (change of shape), *volumetric strain* (change of size).

The magnitude of an earthquake is determined by measuring the amplitude of the seismic waves on the seismogram and then standardizing the measurement, taking into account the distance from the focus, the density of rocks along the wave path through the Earth, and the type of seismograph. An earthquake has a range of intensities, depending upon the how strongly the shaking is perceived by people and the kind and severity of the damage from place to place.

The source of the seismic energy is the elastic strain energy stored within the rock as a result of stress imposed on it. That stored elastic strain energy is released in several forms of energy: heat, sound, light, and vibrations. It is the vibrational energy that causes the Earth to shake in earthquakes.

Seismologists use reflected and refracted earthquake waves to map the interior of the Earth (see Earth Structure). Petroleum geologists use artificially induced vibrations to map subsurface strata in their search for oil and gas.

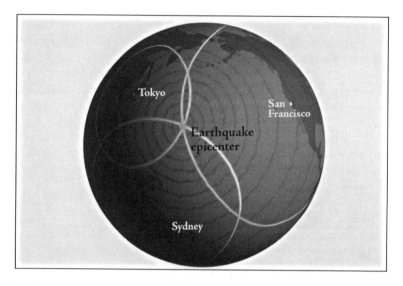

Figure 7   *An earthquake epicentre is triangulated in the middle of the Pacific Ocean from S–P differences determined at seismographs at San Francisco, Tokyo, and Sydney. Diagram from Collier, M., 1999. A Land in Motion. University of California Press*

***See also:* GEOLOGY, TECTONICS**

Further reading

Ben-Menahem, A., and Singh, S.J., 2000. *Seismic waves and sources*. Dover Publications, Minneola NY, 2nd edn.

Bullen, K.E., and Bolt, B.A., 1985. *An introduction to the theory of seismology*. CUP, Cambridge, 4th edn.

Scholz, C., 2002. *The mechanics of earthquakes and faulting*. CUP, Cambridge.

http://earthquake.usgs.gov/regional/neic/ for North America

http://www.ngdc.noaa.gov/wdc/usa/seismol.html for the world

SOIL SCIENCE

From the Greek word, *pedon* (soil, earth) is derived the science of pedology which covers the whole field of soils and their uses. However, for convenience, the material can be split into pedology, which is the basic science of soils and their influence on plants and water; soil classification and distribution which depends on spatial scale; soil distribution, soil degradation, which deals with processes such as erosion and acidification; and soil management, which is the application of pedology to the human use of soils. The basic object of study is the soil, which is a hybrid component of the lithosphere. It is formed from the dynamic interaction of its substrate (e.g., weathered rocks or an unconsolidated sediment, angle of slope, drainage) with the atmosphere (in the form of temperature and precipitation for instance) and with living and recently dead organic matter. Recently dead living tissues are the main component of humus. The biota involved are usually plants and microorganisms, with bacteria having especially important roles in soil formation. Most animals are of less relevance but their faeces may be a key input into the levels of organic matter and nitrogen, for example.

As with most planetary entities involving life, the soil is complex and the approach taken by many soil scientists has been dominated by the identification and measurement of various subsystems within the soil unit. Thus the standardization of various soil metrics has required international agreement: a vocabulary that transcends national descriptions of colour, or morphology of the soil profile is needed, just as the taxonomic names of biota have to be worldwide to be useful. Once a working vocabulary of measurements has been established then many features of soils can be described and explanations sought in both the laboratory and field. Most of the normal divisions of the natural sciences can be applied to soils when relevant: biology, chemistry and physics all find their place. Soil biology, for example, may focus on the role of plant root systems in the physiology of soils since they aerate the soil mass and also supply it with organic matter when they die. The roots may also be a substrate for bacteria or fungi. Some species of bacteria live in nodules on plant roots and fix atmospheric nitrogen for the metabolism of the plant; other species (the genus *Nitrosomas* is one such) live freely in the soil but give off their nitrogen into the soil body. Nitrogen is an essential feature of soil chemistry but not the sole matter of interest. The balance between acidity and alkalinity is an important descriptive feature of soils, as well as confirming its suitability or otherwise for a variety of uses. It is usually measured in terms of the activity of hydrogen ions in solution (pH) where 7.0 is neutral; acid soils fall below that number and alkaline soils above it. Soil physics spans the physical description of soil particles (size and shape), soil aggregates, the storage and transport of water, gases, heat, and dissolved substances.

A major focus of immediate importance for soil science is in the study of the dynamics of carbon within soils since this is a biospheric pool relevant to the problems of the impact of increased concentrations of carbon dioxide (and other greenhouse gases) in the atmosphere. Obviously, soil uses which result in the oxidation of soil carbon and its release as a gas to the air will enhance carbon levels in the atmosphere, just as soil management which stores carbon in soils will fix some of it away from potential effects on climate. The pool of soil carbon totals about 1.6×10^{18} g, compared with the atmosphere at 0.66×10^{18} g and land plants at 0.95×10^{18} g, so it is not a negligible quantity should more of it be released, as when deforestation takes place.

See also: DEFORESTATION, GREENHOUSE GASES, SOIL CLASSIFICATION AND DISTRIBUTION, SOIL DEGRADATION, SOIL MANAGEMENT LITHOSPHERE

Further reading

Hillel, D., Hatfield, J.L., Powlson, D.S., Rosenweig, C., Scow, K.M., Singer, M.J. and Sparks, D.L. (eds) 2005. *Encyclopedia of soils in the environment*, 4 vols. Elsevier, Amsterdam.

Wild, A., 1993. *Soils and the environment: an introduction.* CUP, Cambridge.

***STRATIGRAPHY* see** Geology
***ZOOGEOGRAPHY* see** Biogeography

PART II

ENVIRONMENTS

INTRODUCTION

We all have our ideas about environment which Einstein is reputed to have described as 'everything that isn't me'; we all probably agree that 'environments' in the plural relates to surroundings in terms of objects, conditions or circumstances. Three hundred years ago the word environment(s) was probably not used very much, if at all, but nowadays we hear more and more about environments, all qualified by an increasing range of adjectives: including those spaces that we function in (office, desktop, social, outdoor, creative, narrative, virtual, intelligent, multisensory, art, bluespace), those areas described according to location and how we do things (working, business, learning, virtual learning, personal learning), as well as those that have an element of evaluation (healthy, healthcare, controlled, managed, themed, magnetic). Even more familiar are the ways in which we now characterize world environments (temperate, tropical, alpine, highland, mountain, dryland, urban), sometimes according to continent (African) or to the past (Quaternary), which have led increasingly towards evaluations of environments (protected, polluted, responsive, extreme, harsh, aquatic, hazardous, wilderness). Some discussion has also focused on ideal environments, providing clues about objectives for ways in which environments might be planned, designed, managed, or restored.

These examples of 40 environments are just some that have now been identified; how have we reached this multi-environment situation over 300 years? In order to recognize and characterize environments we all use senses of sight, touch, hearing, smell separately then collectively, but each individual may perceive environments differently. Rudyard Kipling's words 'what should they know of England who only England know' underline the fact that only by experiencing other environments can we recognize and appreciate the characteristics of the one most familiar to us. The earth as the Blue Planet (Fig. 8) was appreciated only with the advent of space travel giving us a distant view of the earth from space. (Earth has been referred to as the Blue Planet due to the abundant water on its surface and/or the colour of its atmosphere.) Recent space exploration was the culmination of phases of travel to satiate curiosity about other places, lands and environments dating from the 16th century. Our experience affects our perception of environments in different ways and our language reflects what exists or is perceived to exist in our environment; if we have never seen mountains can we imagine mountain environments? Environments necessarily overlap with the culture of societies: the countryside became fashionable in contrast to the unattractive environments of towns produced by the industrial revolution and coastal environments were 'discovered' in the early 19th century when seaside holidays became fashionable.

Figure 8   *Cyclone Catarina, 26th March 2004. Photographed by International Space Station Crew. Credit NASA/International Space Station Crew*

However it was not only travel of various scales and exploration that began to raise awareness of environments but also the growth of scientific enquiry. By the end of the 17th century botany (the study of plants) as well as the investigation of animals became popular pastimes for the middle classes of England, France and Germany, subsequently catalyzed by the study of natural history, for example by Gilbert White (1720–1793), and by the beginnings of modern classification and taxonomy by Linnaeus (1770–1778). Although such developments accompanied growth of physics and chemistry, the environmental sciences did not really begin to grow until the 19th century, with geology as one of the first followed by ecology and physical geography. In the latter part of the 20th century knowledge of environments increased enormously with technological advances, especially in measurement, computing and remote sensing; they have unlocked information about environments that could not have been envisaged even a few decades ago.

This section cannot include references to all types of environments but introduces some of those that are most important to the environmental sciences. To the adjectives used for environments in, and of, the Blue Planet we should perhaps add 'sensitive' and 'fragile'. Although not separate entries in this section, many contemporary environments may be increasingly sensitive to human impact and more fragile as a consequence of global climate change. The conservation movement in the west is generally thought to have gained enormous impetus with the book written in 1864 by George Perkins Marsh entitled *Man and nature* in which he set out to 'illustrate the doctrine that man is, in both kind and degree, a power of a higher order than any of the other forms of animated life …'. We now know so much more about environments of the Blue Planet but do we need to reinforce an environmentalist movement to ensure that environments are not changed irreparably – now and in the future?

ANTARCTIC ENVIRONMENT

Antarctica is the largely ice-covered continental land mass surrounding the South Pole, for the most part occurring within the Antarctic circle (66½ deg S), having a total area of 13,900,000 km², equivalent to an area twice the size of Europe. Some 98% of Antarctica is covered by ice with an average thickness of 1.6 km. Islands joined by ice make up the western part which has high mountains with the highest in the Ellsworth Mountains, Vinson Massif, at 4892 m. The Antarctic peninsula was first seen in 1820, the first landing and the first overwintering were made in 1895, and the South Pole was first reached by Amundsen in November 1911 and Scott in January 1912. One Antarctic research station on the South Orkney Islands has maintained a continuous meteorological record since its establishment in 1904. The environment is the coldest on the Earth with an average temperature of −45 °C at the South Pole and with less than 5 cm annual precipitation, thus classified as polar desert where hurricane force winds and blizzards occur. Each 24 hours remains dark in June (during the Antarctic winter) and light in December (during the Antarctic summer).

The Antarctic ice sheet, over an area of 12 million km² has a volume of nearly 30 million km³ of ice containing 65% of the world's fresh water, which would produce a world sea-level rise of 70 m if it were all to melt. The Antarctic ice sheet is about 10 times larger than that which covers Greenland and has existed for at least 34 million years. The East Antarctic ice sheet forms a broad dome over 4000 m in altitude, drained around its margins by a series of radial ice streams with some leading into rock-bound glacial troughs. The West Antarctic ice sheet comprises three contiguous domes rising to 2000 m, partly founded on bedrock below sea-level; it is also drained by ice streams with many flowing to ice shelves, which are floating sheets of ice occupying coastal embayments and flowing seawards at speeds of several hundred metres per year. The Ross ice shelf, an area as large as France, is fed by outlet glaciers from both East and West Antarctica. The Antarctic ice sheet and surrounding ice shelves and sea-ice are believed to play a significant role in global climate by influencing the amount of energy absorbed from the sun, as well as affecting sea-ice formation, and atmospheric and ocean circulation. The low temperatures and high elevations of the ice-covered continental interior produce a steep equator-to-pole temperature gradient, which is the driving force for winds and ocean currents. In addition to geologic, glaciologic, atmospheric and oceanographic research collecting data on the contemporary ice sheet, ice shelves and sea-ice, measurements can be made to provide insight into how the ice sheet may have reacted to climate change in the past. Ice cores have provided records from ice sheets up to 1 million years; the Vostok ice core 3623 m long has yielded a record covering more than 500,000 years giving variations in mean temperature based on analysis of isotope ratios of O_2 and H_2. These results have been supplemented offshore by the Deep Sea Drilling Project and onshore by the Dry Valley Drilling Project both in the 1970s. The new ANDRILL (ANtarctic geological DRILLing) Program, funded by the National Science Foundation (NSF) of the USA, aims to extend Antarctic palaeoclimate records through time intervals that are thus far poorly sampled in Antarctica, while also providing information on the tectonic history of the continental margin.

Variability in Antarctic temperatures and the extent of glacial and sea-ice affects climate systems throughout the globe. Consequently, many researchers believe it is critical to understand how the ice sheet will react to current and future global warming. Some of the scenarios from the International Panel on Climate Change predict that within a few centuries, atmospheric greenhouse gas concentrations, most notably carbon dioxide, could be higher than when the ice sheets formed on Antarctica more than 30 million years ago. The Poles are the Earth's super-sensitive early-warning systems – what has been referred to as 'the miner's canary' by Chris Rapley when he was Director of British Antarctic Survey. As in the Arctic regions, there are documented effects on Antarctica from global warming; in 2005, a mass of ice comparable in size to the area of California briefly melted and refroze. This may have resulted from temperatures rising to 5°C. The US space agency NASA reported it as the most significant Antarctic melting in the past 30 years.

Since 1908 when claims were first made to national sovereignty over parts of Antarctica it has been necessary to regulate international relations. Following the International Geophysical Year of 1957–1958 the 12 nations active in the Antarctic at that time initiated the Antarctic treaty (implemented 1961) which is a necessary international legal agreement designed to ensure that member countries work together in Antarctica for peaceful and scientific purposes. It prohibits military activity, except in support of science; prohibits nuclear explosions and the disposal of nuclear waste; promotes scientific research and the exchange of data; and holds all territorial claims in abeyance. The Treaty applies to the area south of 60° South Latitude, including all ice shelves and islands, which amounts to about 10% of the world's land surface. The Treaty has subsequently been augmented, for example by the Protocol on Environmental Protection to the Antarctic Treaty (Madrid, 1991) effected on 14 January 1998.

See also: ARCTIC ENVIRONMENTS, GLACIERS, GLACIOLOGY, ICE CORES

Further reading

Arctic, Antarctic and Alpine Research (AAAR) Published by INSTAR (Institute of Arctic and Alpine Research) since 1999 and was published as *Arctic and Alpine Research* from 1969.

Council of Managers of National Antarctic Programs website, www.COMNAP.aq

Hansom, J.D. and Gordon, J., 1998. *Antarctic environments and resources: a geographical perspective*. Longman, London.

Scientific Committee on Antarctic Research website, www.SCAR.org

ANTHROPOGENIC IMPACT, NOOSPHERE

A nthropogene, a Russian term for the period during which man has been an inhabitant of the earth, indicates that anthropogenic impact is the impact of human activities upon physical environment. We now realize how substantial such impacts have been (e.g., Fig. 9) and there are many which are obvious or direct including cloud seeding, soil conservation, deforestation, or the creation of new lakes and reservoirs by the damming of rivers. Less obvious at first are the many less well known, or indirect, effects which include changes to precipitation character and amounts as a result of the existence of urban areas; the incidence of soil erosion as a result of grazing pressure; the introduction of new plant species; or the ways in which geomorphological processes have been changed for example by modifying the active layer in permafrost areas or modifying river processes downstream of dams and reservoirs. Indications of the extent of human impact are illustrated in Table 11.

The implications of such changes were not fully appreciated until the mid-20th century although George Perkins Marsh had written a book in 1864 that later came to be regarded as a foundation for the conservation movement, a seminal work on *Man's role in changing the face of the Earth* published in 1956 was succeeded in 1990 by *The Earth as transformed by human action* affectionately known by the six editors as 'ET', a volume of 720 pages with four main sets of chapters. Anthropogenic landscape was first used by Tansley in 1923 to describe stable plant communities created and maintained by human action termed anthropogenic formations. As a result of increased awareness of human impacts in the late 20th century a number of new

Figure 9 *Substantial open cast mining of magnetite iron ore at Zhelenogorsk, Russia. (Photo K.J. Gregory)*

Table 11 *Examples of magnitude of human impact (from Gregory, 2005)*

Impact on	Overall impact
Atmosphere	Rise of ocean temperatures by an average of 0.5 °C in 40 years
	Temperatures expected to rise by between 1.4 °C and 5.8 °C by 2100
Hydrosphere	In view of 39,000 dams higher than 15 m on rivers few of the world's rivers remain unregulated 11% of freshwater runoff of USA and Canada withdrawn for human use.
	The Aral sea, which was the world's fourth largest lake has lost more than 40% of its area, 60% of its water volume and its level has fallen by more than 14 m as a result of increased irrigation which has decreased inflow of water to the lake. The Kok-Aral dam completed in 2005 has helped restoration and the Northern Aral Sea has helped restore 13% of the surface area of the sea, reduced salinity and returned the coastline to some stranded villages.
Pedosphere	250 million ha of the world affected by strong or extreme soil erosion
Geosphere	About 75% of earth's sub-aerial surface bears the imprint of human agency
Biosphere	In early to mid 1980s 0.3% of earth's sub-aerial surface deforested each year

terms have been adopted including Noosphere (defined as a new geological epoch initiated as a consequence of man's activity by Trusov in 1969) and homosphere (defined as the biosphere modified by *Homo sapiens* by Svoboda in 1999). Specific areas of investigation have developed including urban physical geography and earth hazards, and in many cases knowledge of the magnitude of human impact can be a useful input when devising management strategies, particularly of a sustainable character.

There has been a tendency to think of human impacts as deleterious and negative but there are many that are positive including ways of mitigating earth hazards. However the greenhouse effect and global warming is now recognized to be one of the most serious examples of human impact but continues to show that it is not easy to convince all policy-makers of the verity of human impact and that even when it is established the indirect effects may still take time to identify and assess.

See also: CULTURES, GLOBAL WARMING, HUMAN IMPACTS ON ENVIRONMENT, NATURE, URBAN ENVIRONMENTS

Further reading

Goudie, A.S., 1993. *The human impact on the natural environment.* Blackwell, Oxford.

Gregory, K.J., 2005. Editor's Introduction. In Gregory, K.J., *Fundamentals of physical geography.* Sage Publications, London, 4 vols, volume 1, xix–lviii.

Tansley, A.G., 1923. *Introduction to plant ecology*. Allen and Unwin, London.

Thomas, W.L. (ed), 1956. Man's role in changing the face of the earth. University of Chicago Press, Chicago.

Turner, B.L., Clark, W.C., Kates, R.W., Richards, J.F., Mathews, J.T. and Meyer, W.B. (eds), 1990. *The Earth as transformed by human action*. Cambridge University Press, Cambridge.

ARCTIC ENVIRONMENTS

The region surrounding the Earth's North Pole and usually thought of as comprising all areas north of the Arctic circle (latitude 66½° N). The word 'arctic' derives from the Greek *arktos* meaning bear, and the area includes the ice-covered Arctic ocean, parts of Canada, Greenland, Russia, Alaska, Iceland, Norway, Sweden and Finland, a total area of 14,056,000 km², equivalent to twice the size of Australia. This is an environment with 24 hours of darkness in December (the Arctic winter) and 24 hours of daylight in June (during the Arctic summer), the warmest monthly mean temperature does not exceed 10°C and the mean monthly temperature of the coldest month is below 0°C. The tree line, the northern limit of trees, approximates to the southern boundary of continuous permafrost and to the northern limit of the Subarctic. Much of the sea has sea-ice several metres thick which forms when water temperatures fall to −1.9°C, called fast ice when attached to land and pack ice when floating free under the influence of currents and wind. Land areas have treeless uplands and marshy plains called tundra, sometimes referred to as the Low Arctic, and polar semi-deserts (<25% plant cover) and polar deserts (<5% plant cover) called the High Arctic, found on islands within the Arctic basin including Svalbard, Canadian Arctic islands and northern Novaya Zemblya. Chemical contaminants from industrial areas of Europe and Asia can accumulate in the Arctic atmosphere giving a reddish-brown Arctic haze, especially in winter and spring; the contaminants can then be deposited on snow, ice and tundra affecting Arctic fauna.

Arctic environments are important for their natural resources including hydrocarbons, and also for the sensitive role which the cryosphere plays in global climate. The Arctic has experienced large and rapid changes of vegetation in the recent geological past, and in relation to climate change is thought to give early warning of change for the planet; it is already exhibiting rapid changes in vegetation and sea-ice, and shows clear positive feedbacks between sea-surface conditions, ecosystems and the atmosphere.

See also: CRYOSPHERE, GLOBAL WARMING, PERIGLACIAL ENVIRONMENTS, PERMAFROST

Further reading

French, H.M., 1996. *The periglacial environment*. Addison Wesley Longman, London

1ACE (Arctic Climates and Environments) see http://www.wun.ac.uk/ace/index.html

ARID ZONES/DESERTS

Regions of Earth that are arid and semi-arid have relatively low to very low amounts of annual precipitation. In some of the most arid regions, a number of years may pass before measurable precipitation occurs. Earth's arid and semi-arid regions are sometimes called dry lands and they occupy about one third of Earth's land surface. More importantly, about one-fifth of the human population on earth lives in arid and semi-arid regions and as a result, these environments are particularly important to people. Related to this is the fact that global warming is occurring and producing changes in the world's arid and semi-arid environments. In many areas with warming, it is expected that arid lands will expand and with them desert processes, or what is called desertification. Before discussing the important process of desertification, it is necessary to consider what deserts actually are and where they occur.

Irrespective of where a desert is located, it is characterized by low to very low amounts of annual precipitation. The largest deserts on Earth are located in two belts between 15° and 30° latitude north and south of the equator. The global pattern of atmospheric circulation on earth produces these belts. The processes that operate are related to the fact that at the equator more warm air exists, and this air rises and moves north and south toward the cooler poles. At approximate latitudes of 15° and 30° north and south of the equator, there are areas of descending air producing semi-permanent cells of high pressure. Low pressure is associated with rising air and rainfall and high pressure with descending air, or heavy air, and low precipitation. The global circulation pattern is shown on Fig. 10.

The global circulation of masses of air moving from the equator toward the poles explains the major deserts of the world such as those in Africa, the Middle East and parts of India, South America, North America, and Australia. However, the global circulation patterns do not explain all of the deserts on Earth. Semi-arid and arid lands that form deserts in North America, as well as in Central Asia, result because these lands are far inland in continents where rainfall does not often reach. Also as storms move inland up a flank of a mountain range toward the crest they may be blocked by the high mountains producing a 'rain shadow' on the other side of the mountain with the result that little precipitation is produced. Some of the highest mountains on Earth may also not receive much precipitation in the form of rain or snow because as water-laden air moves inland, it loses that water to precipitation when it first reaches significant topography, often well before the actual crest of a large mountain range such as the Himalaya.

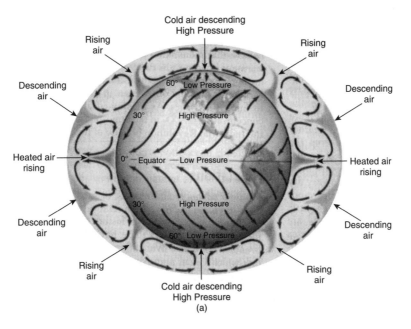

Figure 10 *Global pattern of atmospheric circulation*

Thus in India, the highest areas of rainfall are not in the highest mountains, but well in front of the highest topography (Fig. 11). At any rate, the mountain range blocks the precipitation from prevailing storm tracks and so semi-arid and arid regions develop. There are also polar deserts that result because of permanent areas of high pressure in polar regions. Some of the major concerns of global warming will be changes in climate patterns. In some regions, deserts will expand into what were formally wetter areas, but in other areas, precipitation will increase. For example, snowfall is expected to increase in some polar regions as a result of global warming with an increase in water vapour. Increased snowfall will lead to expansion of glaciers in some

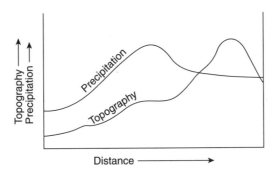

Figure 11 *Relation between precpitation and relief in India*

polar regions while alpine glaciers in the mountains of North America, South America, and Europe retreat or melt.

This short discussion of the origin of semi-arid and arid regions and deserts leads to consideration of the important topic of environmental science related to desertification. The simplest definition of desertification is the transformation of land from one state to another that more closely resembles a desert. This transformation is related to human-induced processes that produce desert-like conditions with a more desert-like landscape. Desertification as such does not refer to the major long-term climatic changes that occurred during the Pleistocene (last 1.65 million years) that had many changes that resulted in expansion as well as contraction of arid and semi-arid lands around the globe. Today desertification is becoming a global problem in the semi-arid and arid regions of Earth. When we think of desertification, we think of degradation of the land related to loss of soil (soil erosion as well as loss or change of vegetation). Of particular concern is the fact that when we are concerned about soil erosion and changes of vegetation, we are talking about potential damages to the human food production that may be related to malnutrition and famine. Land degradation from desertification may be so extensive that the productivity of the land is lost and may be very difficult if not impossible to recover on the time span important to humans. This results because with soil erosion, the rate of soil erosion may be a much more rapid process than the rate of soil formation. When soils are lost and vegetation degraded, the problem may become chronic.

The causes of desertification have been known for decades and are related to several varieties of poor land use practice.

» Overgrazing by stock animals on grasslands that causes increase in soil erosion and loss of productivity of the land.
» Excessive agriculture that continuously disturbs the land resulting in accelerated soil erosion. Such erosion may be a slow insidious process as nutrients are removed from soil and the topsoil itself slowly is lost.
» Deforestation, which involves the removal of trees by a variety of processes including clear cutting, which removes almost all trees. When the trees are removed, the hydrology changes and soils are exposed to erosion. In some cases, erosion is so severe that re-establishment of forest is a long and difficult process. Associated with loss of trees on steep slopes is the increase of landslide activity as roots of cut trees decay, reducing soil strength, and increasing rates of hillslope erosion.
» Irrigation practices that over time result in soils becoming saltier and less productive. As irrigation water is added, the water may leave by evaporation or subsurface flow, leaving behind higher concentration of salts. Unless irrigation is properly managed and used, large regions of land may become less productive and in some cases unproductive with salt deposits at the surface.

The role of environmental science, and particularly the more applied aspects of that science, is of particular importance in preventing and minimizing, and in some cases reversing, desertification. Some specific aspects that may be applied are:

» Ensuring that high-quality land, whether it is for crops or grazing animals, is adequately protected by sound conservation practice. This is more important than attempting to improve low quality land that is not cost effective.

» Apply proven techniques of range management so that land is protected from overgrazing by livestock or game animals including deer, elk in North America, and elephants and other animals in Africa.

» Apply known measures of land conservation to agricultural areas so that soil resources are preserved and sustained.

» Use environmental technology to both increase crop production as well as allowing poor lands to be returned to less intensive activities such as wildlife preservation, forestry, or grazing by livestock.

» Use appropriate technology with ecological restoration to restore a variety of landscapes from rivers, sand dunes, and wetlands, as well as severely disturbed lands from activities such as mining.

See also: EARTH SURFACE PROCESSES

Further reading

Dregne, H.E., 1983. *Desertification of arid lands*. Vol. 3: advances in desert and arid land technology and development. Harwood Academic Publishers, Chur, Switzerland.

Goudie, A., 2001. *The nature of the environment*, Blackwell Scientific, Oxford, 4th edn.

Grainger, A., 1990. *The threatening desert*. Earthscan Publications, London.

Mainguet, M., 1994. *Desertification*, Springer-Verlag, Berlin, 2nd edn.

Oberlander, T.M., 1994. Global deserts: a geomorphic comparison. In *Geomorphology of desert environments*, ed. Abrahams, A. D. and Parsons, A. J. (eds) Chapman and Hall, London, 13–35.

ATMOSPHERE

The atmosphere is a layer of gases (and liquids/particles) surrounding the Earth's surface held in place by gravity. The atmosphere is composed of 78% nitrogen, 21% oxygen, 0.97% argon, 0.04% carbon dioxide, and small amounts of other gases and water vapour. The atmosphere shields the planet by reducing temperature extremes between day and night and by absorbing dangerous Sun's rays. There exists no definite boundary between outer space and the atmosphere. Three-quarters of the atmosphere's mass is within 11 km of the surface. The average temperature of the atmosphere at the surface of Earth is 15 °C (*ca.* 60 °F). Generally, atmospheric pressure and mass drop by 50% at an altitude of about 5 km. At sea level, the average atmospheric pressure is about 101.3 kilopascals (1013.3 mb or about 14.7 lb/in²). Even at heights of 1000 km, the atmosphere is still measureable. Most of the atmosphere by mass is below 100 km, although there are auroras and other atmospheric effects in the rarefied region above this level. Below 100 km, the Earth's atmosphere has a uniform composition and is called the homosphere. However, above 100 km, composition varies with altitude. Without mixing, the density of a gas decreases exponentially with increasing altitude, at a rate that depends on the molecular mass. Higher mass constituents, such as oxygen and nitrogen, reduce faster than lighter constituents such as helium, molecular hydrogen, and atomic hydrogen. The layer with varying composition is called the heterosphere. The atmosphere is dominated successively by helium, molecular hydrogen, and atomic hydrogen as the altitude increase. The altitude of the heterosphere and the layers it contains varies significantly with temperature. The density of air at sea level is about 1.2 kg/m³. Natural variations of the barometric pressure occur at any one altitude as a consequence of weather. This variation is relatively small for near-surface altitudes, but much more pronounced in the outer atmosphere and space due to variable solar radiation. In addition, atmospheric density decreases as the altitude increases. According to the National Center for Atmospheric Research, 'The total mean mass of the atmosphere is 5.1480×10^{18} kg with an annual range due to water vapour of 1.2 or 1.5×10^{15} kg depending on whether surface pressure or water vapour data are used; somewhat smaller than the previous estimate. The mean mass of water vapour is estimated as 1.27×10^{16} kg and the dry air mass as $5.1352 \pm 0.0003 \times 10^{18}$ kg'. The history of the Earth's atmosphere prior to one billion years ago is uncertain. The modern atmosphere is sometimes referred to as Earth's 'third atmosphere', in order to distinguish the current chemical composition from two notably different previous compositions. The original atmosphere was primarily helium and hydrogen. Heat from the molten crust and the sun, in addition to solar wind, diffused this atmosphere. About 4.4 billion years ago, the surface of the Earth cooled enough to form a crust, still with volcanoes which released steam, carbon dioxide, and ammonia. This signifies the early 'second atmosphere', which was primarily carbon dioxide and water vapour, with some nitrogen but virtually no oxygen. This second atmosphere had approximately 100 times as much gas as the current atmosphere, but as it cooled much of the carbon dioxide was dissolved in the seas and precipitated out as carbonates. The later 'second atmosphere' contained nitrogen, carbon dioxide, and probably 40% hydrogen. It is generally believed that the greenhouse effect, caused by high levels of carbon dioxide and methane, kept the Earth from freezing. Temperatures were probably in excess of 70 °C (158 °F), until some 2.7 billion years ago. Bacteria were able to convert carbon dioxide into oxygen, playing a major role in oxygenating the atmosphere. Photosynthesizing plants would later evolve and convert more carbon dioxide into oxygen. Over time, excess carbon became locked in fossil fuels, sedimentary rocks, such as limestone, and animal shells. Oxygen reacted with ammonia to release nitrogen; in addition, bacteria would also convert ammonia into nitrogen. Most of the modern day nitrogen is due primarily to sunlight-powered photolysis of ammonia released over the eons from volcanoes. As more plants appeared, the levels of oxygen dramatically increased, while carbon dioxide decreased. With the appearance of an ozone layer, lifeforms were protected from ultraviolet radiation. This oxygen–nitrogen atmosphere is the 'third atmosphere'. Some 200–250 million years ago, up to 35% of the atmosphere was oxygen. This modern atmosphere has a composition which is maintained by oceanic blue-green algae as well as geological processes. Oxygen does not remain naturally free in the atmosphere, but tends to be consumed by inorganic chemical reactions, as well as by animals, bacteria, and land plants at night, while CO_2 tends to be produced by respiration and decomposition and oxidation of organic matter. Oxygen would vanish within a few million years due to chemical reactions and CO_2, which dissolves easily in water, and would be gone in millennia if not replaced. Both are maintained by biological productivity and geological forces over millions of years.

See also: ATMOSPHERIC SCIENCES, EARTH SPHERES, TROPOSPHERE

Further reading

Barry, R.G. and Chorley, R.J., 1987. *Atmosphere, weather, and climate. Methuen*, London. 5th edn.

Lutgens, F.K. and Tarbuck, E.J., 2001. *The Atmosphere*. Prentice Hall, Upper Saddle River, N.J. 8th edn.

Material partially surveyed from 'Earth's Atmosphere' Wikipedia.

BIOMES

The biome is the largest land ecosystem which is usually designated and as such is often depicted on world-scale maps. The diagnostic characteristic is the life form of mature stands of a vegetation type: a deciduous forest biome takes its name from the dominance of deciduous trees, though the exact species in the forest will vary from continent to continent. This life form exercises a degree of control over the rest of the ecosystem: it forms the structure for the non-dominant plants and for the animals, and is the essential photosynthetic layer which dominates the supply of chemical energy to the rest of the ecosystem. It may well organize the flow of radiant energy as well as in terms of how much light penetrates e.g., a forest canopy and reaches the forest floor. The vegetation and the climate are major influences on the soil profile and so some biomes have a characteristic soil type that accompanies the identifiable vegetation and animal communities. The overriding control of the distribution of biomes is world climate, notably the incidence of solar radiation and the effectiveness of precipitation. Some biome have wide limits: the tropical forests cover a wide range of precipitation but a narrow temperature span; grasslands are delimited by moisture but can tolerate a wide variety of temperature regimes. A biome is thus a major world region which integrates a number of factors into a complex but recognizable whole. Attempts to subdivide biomes into smaller units using the same principles result in e.g., the World Wildlife Fund's 867 'terrestrial ecoregions' which are areas of land or water that contain a geographically distinct assemblage of natural communities that share a large majority of their species and ecological dynamics and similar environmental conditions and interact ecologically in ways that are critical for their long-term persistence.

The world maps which depict biome distribution often mislead. Of necessity, they show boundaries between biome types as sharp lines, when the reality is more often of a gradual transition. Also, many of the vegetation types are simply not there since they have been transformed entirely by human activity or have been modified in ways not immediately apparent. The tropical rainforests of South America for example are shown as mature communities when it is likely that they are in a state of succession after widespread depopulation of the lowland tropics that followed European colonization and conquest (Fig. 12). Thus a world biome map is more properly a map of potential natural vegetation: what would be there if human influence was removed and the subsequent succession was telescoped. Today's reality is better conveyed by a map of land cover, which recognizes the coexistence of both the more natural biome and the modified surface of the earth at human hands.

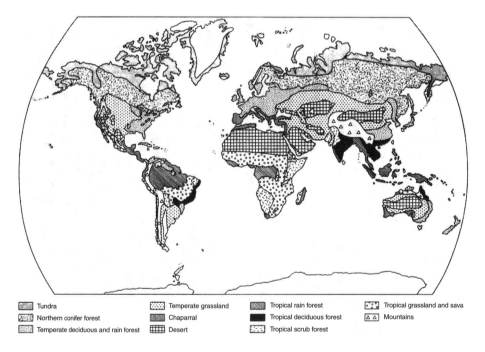

Figure 12   *World map of major biome distribution*

See also: ECOSYSTEM, SOIL PROFILE, VEGETATION AND ITS CLASSIFICATION

Further reading

http://www.nationalgeographic.com/wildworld/terrestrial.html

Smith, P., Addison, K. and Atkinson, K., 2002. *Fundamentals of the physical enviroment* 3rd edn. Routledge, London and New York.

Simmons, I. G., 1997. *Humanity and Environment. A Cultural Ecology*. Addison Wesley Longman, Harlow.

BIOSPHERE

In 1875 the Austrian geologist Eduard Suess (1831–1914) proposed the name 'biosphere' for the zone of the planet Earth which contained life in all its forms together with the conditions promoting life, such as the presence of minerals, water, light and other factors. This interaction would now be described as ecology. The concept was elaborated in 1926 by V.I. Vernadsky who thought it was a stage on the road to the development of the noosphere. The biosphere which houses life runs from a sub-oceanic stratum of rocks some 6 km below sea-level and 300 m below the sea floor to the highest peaks of mountain ranges at about 8800 m ASL. Compared with the atmosphere and the lithosphere this is quite a narrow zone and the quantity of living matter per unit area or volume is small at the extremes, where it is largely confined to relatively simple organisms like bacteria. The quantity of living matter is highest in places such as the land–sea interface (especially estuaries and coral reefs), and forests of tropical environments and the whole is the product of about 3.5 million years of evolution.

The diversity of life is remarkable. The simplest creatures, such as bacteria, have a low level of differential function within the single cell of which they are composed. At the other extreme, mammals are the most complex organisms, with high levels of differentiation of tissues into organs, and a high degree of adaptation to environment, the species *Homo sapiens* is probably the most developed example.

The biosphere is much influenced by the other spheres with which it interacts. The lithosphere, for example, is crucial in determining the distribution of land and sea, the height of the land masses and the depth of the oceans. At regional and local scales the type of rock substrate will affect many plants and animals via the type of soil which is developed: there are plants for example which are confined to areas with a limestone substrate (calcicoles), just as there are acidophiles which will tolerate very low pH values. But it is not a one-way process. Life affects the lithosphere, for example, in the weathering of rocks colonized by lichens or bacteria and in the development of soils. The atmosphere too may be modified by life processes as when bacteria at sea give off particulates which then form cloud nuclei and give rise eventually to onshore rainfall.

The biosphere is not at equilibrium. It is always in a state of flux in response to external forcing functions whether these are at cosmic scale, such as fluctuations in solar radiation or in the 'wobble' of the Earth's axis, or at planetary scale such as the climatic and land cover changes of the ice ages of the Quaternary. Adaptations can be remarkably fast as in the spread of warmth-seeking vegetation in the early Holocene but many have been speeded up and changed in direction by the actions of *Homo sapiens*.

See also: ATMOSPHERE, EARTH SPHERES, ECOLOGY, HOLOCENE, LITHOSPHERE, QUATERNARY

Further reading

Smil, V. 2003. *The Earth's biosphere. Evolution, dynamics and change.* MIT Press, London and Cambridge, MA.

CLIMATE CLASSIFICATION

Climatologists have developed methods to organize atmospheric observations into classifications of phenomena. There are basically three types of classifications used in climatology. First, there are empirical systems such as the Köppen classification which is an empirical system based on observations of temperature and precipitation (shown below is a recent version labelled the Köppen-Geiger Climate Classification) (Fig. 13). These are two of the easiest measurable climate characteristics and long-term historical data exists for the Earth. Climates are grouped based on annual averages and seasonal extremes. Second, there are genetic classification systems, e.g., synoptic types, based on the recurrence of various weather types that are more directly controlling the character of climates. A genetic system relies on information about the climatic elements of air masses, pressure systems, solar radiation, etc. Third, there are more applied classification systems that are developed for a particular climate-related problem. The Thornthwaite climate classification system is based on potential evapotranspiration, which groups climates based on water requirements. Research conducted by C.W. Thornthwaite and associates attempted to formulate a water budget technique for agricultural purposes that assesses water demand under varying environmental conditions.

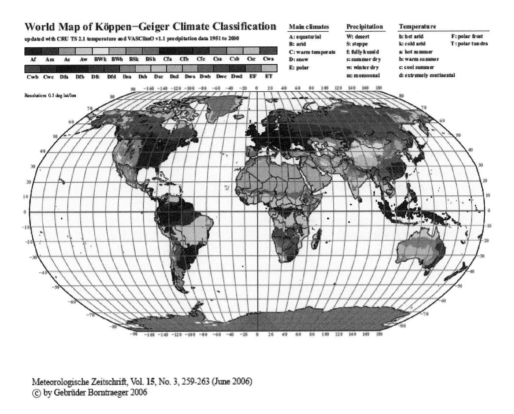

Meteorologische Zeitschrift, Vol. 15, No. 3, 259-263 (June 2006)
© by Gebrüder Borntraeger 2006

Figure 13 *The Köppen – Geiger System (http://koeppen-geiger.vu-wien.ac.at)*

The Köppen system recognizes five major climate types based on the annual and monthly averages of temperature and precipitation. Each type is designated by a capital letter.

A – Moist Tropical Climates are known for their high temperatures and for abundant amounts of rainfall year-round.

B – Dry Climates are characterized by limited rainfall with a huge daily temperature range. Two sub-groups, S – semi-arid or steppe, and W – arid or desert, are used with the B climates.

C – In Humid Middle Latitude Climates, land/water differences play a major role. These climates have warm/dry summers and cool/wet winters.

D – Continental Climates are generally found in the interior regions of massive land. Total precipitation is not high, with the other characteristic being widely varying seasonal temperatures.

E – Cold Climates describe this climate type perfectly. These climates are part of areas where permanent ice and tundra dominate, with only about four months of the year having above-freezing temperatures.

Sub-groups presented below are given a lower case letter that distinguishes specific seasonal characteristics of temperature and precipitation.

f – Moist with adequate precipitation during all months, with no dry season. This letter usually accompanies the A, C, and D climates.

m – Rainforest climate in spite of short, dry season in monsoon-type cycle. This letter only applies to A climates.

s – Dry season in the summer of the respective hemisphere (high-sun season).

w – Dry season in the winter of the respective hemisphere (low-sun season).

To further denote variations in climate, a third letter was added to the code.

a – Hot summers where the warmest month is over 22 °C (72 °F). These can be found in C and D climates.

b – Warm summer with the warmest month below 22 °C (72 °F). These can also be found in C and D climates.

c – Cool, short summers with less than four months over 10 °C (50 °F) in the C and D climates.

d – Very cold winters with the coldest month below -38 °C (-36 °F) in D climate.

h – Dry-hot with a mean annual temperature over 18 °C (64 °F) in B climate.

k – Dry-cold with a mean annual temperature under 18 °C (64 °F) in B climate.

See also: ATMOSPHERIC SCIENCES, CLIMATOLOGY, CLIMATE ZONES

Further reading

Aguado, E. and Burt, J., 2006. *Understanding weather and climate*. Prentice Hall, Upper Saddle River, NJ.

Hoffman, J., Tin, T. and Ochoa, G., 2005. *Climate: the force that shapes our world and the future of life on Earth*. Rodale Books, London.

CLIMATE ZONES

Climate zones are areas of similar climatic characteristics such as temperature and moisture, length of the solar day, and latitudinal distance from the equator. Climate zones are general and do not account for local scales features such as hills or mountain ranges, large lakes, and seas or oceans. The most commonly used classification is the Köppen climate classification system to define basic zones (Fig. 14). This classification depicts climates into primary classifications and numerous sub-classifications (see CLIMATE CLASSIFICATION). At the equator is the tropical or equatorial zone, a belt of relatively low atmospheric pressure and heavy rainfall associated with converging winds and thunderstorms. At about 30° north and south of the equator is a subtropical or arid climate of dry, descending air, associated with

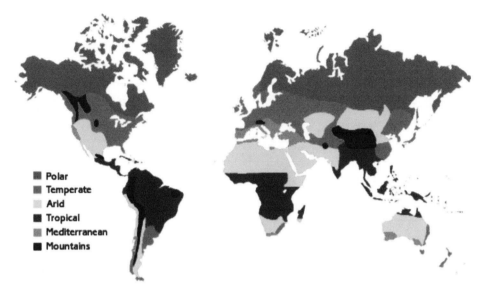

Figure 14 *Climate Zones (UK Met Office)*

high atmospheric pressure and clear skies. In the Northern Hemisphere, this zone is centred over the Sahara in Africa and is sometimes called the Azores High. Daytime surface temperatures can often exceed 40 °C, but extensive heat loss due to lack of cloud cover at night brings down temperatures close to freezing.

High temperatures and little rainfall typically represent the desert climate, which is commonly found in the subtropical zone. During the Northern Hemisphere summer, the subtropical zone moves northward to influence the Mediterranean region. Mediterranean climates are characterized by hot, dry summers, but much cooler and wetter winters than subtropical climates. Trade winds exist between the subtropical and equatorial zones – north-easterly in the Northern Hemisphere and south-easterly in the Southern Hemisphere. These regions are much drier than the equatorial zone, but receive more rainfall than the desert climates. In the north and south mid-latitudes, there is a belt of cyclonic low pressure that arises from the convergence of cold polar easterly winds and warm subtropical wind that produce a temperate zone. In the Northern Hemisphere, cyclonic depressions are known to develop in the North Atlantic and North Pacific. These regions are known respectively as the Icelandic and Aleutian Lows. They are characterized by relatively mild, moist winds that bring frequent cyclonic precipitation (rain and snow), in particular along the west-facing side of continents. This precipitation develops along warm and cold fronts, where cold air from the polar easterlies forces the warm, moist air of the westerlies to rise and cause precipitation. At the highest latitudes in the polar regions, cold air sinks producing high atmospheric pressure. More local variations in climate occur in the zone labelled 'Mountains', (see Figure 14), where elevation and mountain orientation cause substantial changes of temperature and moisture.

See also: ATMOSPHERIC SCIENCES, CLIMATE CLASSIFICATION

Further reading

Oliver, J.E., 1991. The history, status, and future of climate classification. *Physical Geography* 12, 235–251.

COASTS

S
ome of the most dynamic environments on Earth are where the ocean meets the land. The coastal landscape, because it is so dynamic, is capable of very rapid change. It is not unusual, in terms of one person's lifetime, to see many coastal changes in a particular area. Sea cliffs retreat, coastal sand dunes erode, new inlets to barrier islands form, and beaches significantly widen or narrow.

Coastlines of the world may be divided into those that are emerging and those that are experiencing submergence (Fig. 15). Emergent coastlines may be related to active tectonic processes that uplift the land, volcanic eruption that builds new land into what was the sea, and rapid deposition of sediment that allows the land to move seaward, as for example active deposition on a delta. Submergent coastlines may also be related to tectonic processes that cause the land to sink as well as other processes of subsidence, some of which may be human induced. For example, if coastal wetlands are drained, the soils may dry out and compact over time causing subsidence. With the loss of coastal wetlands, the land loses an important buffer to coastal erosion. Fresh and saltwater marshes buffer the impact of water and waves further inland.

From an environmental perspective, the biggest concern in coastal areas today is coastal erosion. Coastal erosion is a world-wide problem today for several reasons, including that sea-level is rising in response to global warming. Most of the rise is a result of thermal expansion of warming oceans, but melting of glacial ice also contributes water to the ocean basins. It seems apparent that some low-lying islands in coming decades may completely disappear as the sea-level rises and inundates them from below and all sides.

A major environmental question is how are we going to respond to coastal erosion that is bound to increase in both magnitude and intensity in the future. Several options are available to try to minimize coastal erosion.

» The so-called hard solution, which involves building structures such as seawalls.
» The soft approach, which involves a plan to retreat from erosion or nourishing beaches with sand.
» Some combination of the hard and soft approach depending on specific conditions.

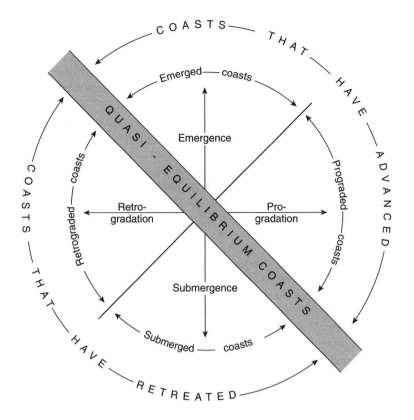

Figure 15 *Genetic classification of Coastlines. Coasts are categorized along two axes from emerged to submerged, and from prograded to retrograded. (After Valentin, 1970. Paper read at the Symposium of the IGO Commission on Coastal Geomorphology: Moscow)*

What solutions we apply to a particular erosion problem will in part reflect our values. For example, when we build a sea wall, the beach generally narrows over a period of decades and there is also a loss of biodiversity of coastal ecosystems. This must be balanced at a particular site with what structures and infrastructure must be protected from erosional processes. In order to make decisions concerning future coastal erosion, it is necessary to recognize some general principles.

» Coastal erosion is a natural process. Problems occur when we build too close to the active coastal area where erosion is more likely to occur. It is recognized that erosion is naturally occurring on most coastlines and so we need to encourage human activities that are more compatible with expected changes. Some of these activities may include recreation in coastal parks rather than building homes right on frontal sand dunes or coastal bluffs.

» We need to recognize that when we choose a hard control structure to minimize erosion, we will cause changes to the coastal environment. When we interfere with coastal processes, then we can expect other changes to also occur. For example, if we interfere with the flow of sand along a coast, then erosion is likely to occur in the direction the sand was initially moving.

» When we choose to protect the coastline from erosion through building engineering structures, we admit that what we are doing is protecting the development, not the beach. These structures are often built to protect the property of a few people, often at the expense of the public at large.

» Engineering structures such as sea walls that are designed to control erosion may in fact eventually damage a beach. A series of hard structures to protect coastline will eventually produce a coast that scarcely resembles a natural environment.

» Coastal engineering and building of walls and other structures to retard erosion sets in force a path of development that is difficult if not impossible to reverse. With time, structures may fail and they are generally replaced by larger and more expensive structures. Given enough time, the coast and environment will be irreversibly changed. This may result in loss of natural coastal areas for future generations. There are many opportunities for applying principles of environmental science to coastal problems. Of particular importance is the linkage of physical and biological systems leading to a better understanding of coastal ecosystems and what is necessary for them to be sustainable. For example, we are painfully learning the importance of coastal wetlands to protecting inland areas from wind and waves. Choosing to work with coastal processes rather than against them will result in lower cost and losses in the coastal zone in years ahead.

See also: COASTAL MANAGEMENT

Further Reading

Flanagan, R., 1993. Beaches on the brink. *Earth* 2(6), 24–33.

McDonald, K.A., 1993. A geology professor's fervent battle with coastal developers and residents. *Chronicle of Higher Education*, 40(7), A8–89, A12.

Lipkin, R., 1994. Weather's fury. In *Nature on Rampage*. Smithsonian Institution, Washington, DC, 20–79.

Neal, W.J., Blakeney, W.C., Jr., Pilkey, O.H., Jr. and Pilkey, O.H., 1984. *Living with the South Carolina shore*. Duke University Press, Durham, NC.

Pilkey, O.H. and Dixon, K.L., 1996. *The Corps and the shore*. Island Press, Washington, DC.

CRYOSPHERE

The cryosphere is that part of the Earth's surface where water is in a solid form, usually as snow or ice, and includes glaciers, ice shelves, snow, icebergs, and snow fields. Cryogenic processes involve freeze–thaw processes which include frost wedging (by shattering, riving, scaling and splitting), frost heaving, frost creeping, frost sorting, nivation, and solifluction or gelifluction. Periglacial environments are affected by cryogenic processes, which include the croystatic pressures developed on freezing of the active layer and cryturbation which is the process whereby soils, rock and sediments in the active layer can be deformed as a result of the cryogenic processes.

The snow, ice, frozen ground and sea-ice of the cryosphere can play a significant role in the global climate system and therefore in climate change. Permafrost science or geocryology is the study of frozen, freezing and thawing terrain. Monitoring of data for Northern Hemisphere snow cover, mountain glacier fluctuations, sea-ice extent and concentration, changes in ice shelves, and global sea-level, current permafrost conditions is provided by the National Snow and Ice data Center (NSIDC) State of the Cryosphere giving the status of snow and ice as indicators of climate change. Permafrost, which preserves a data archive of temperature changes, is important in relation to climate change because it translates effects to natural ecosystems converting birch forests to low-lying wetlands for example, and it can facilitate further climate change if organic carbon stored in the upper layers of permafrost is released to the atmosphere as carbon dioxide and methane, leading to intensified climate warming.

See also: CLIMATE CHANGE, EARTH SPHERES, PERIGLACIAL ENVIRONMENTS, PERMAFROST

Further reading

French, H.M., 1996. *The periglacial environment.* Addison Wesley Longman, London.

NSIDC (National Snow and Ice data Center) http://nsidc.org/sotc

Slaymaker, O., 2007. *The cryosphere and global environmental change.* Blackwell, Oxford.

Washburn, A.L., 1979. *Geocryology: a survey of periglacial processes and environments.* Arnold, London.

DELTAS

Deltas are coastal landforms formed by the deposition of sediments generally in shallow waters of the inner continental shelf. After a deltaic landmass becomes subaerial, vegetative growth becomes important in the formation of deltaic soils. The input of materials to deltas is not constant over time, but occurs in pulses that vary spatially and temporally. These pulsing events are arranged in a hierarchical manner and produce benefits over different time and spatial scales. They range from daily tides, weekly winter frontal passages, annual river floods, strong storms such as hurricanes, great river floods, and formation of new delta lobes each 500 to 1000 years. Major deltas are listed in Table 12.

Two important processes affect delta wetland survival in deltas: subsidence and accretion. Most deltas have high rates of subsidence (up to 1 cm/yr) causing relative sea-level rise (RSLR) to be > eustatic sea-level rise. If the rate of accretion of the deltaic plain is not ≥ RSLR, then wetlands will not survive. Human activity has increased subsidence in some deltas. Despite the high rates of natural RSLR, deltas greatly increased in area over the past several thousand years due to river input. The input of river water not only adds sediments, but brings fresh water which lowers salinity stress and nutrients which increases the productivity of deltas. Sea-level rise is predicted to increase by 3–4 times over the next century. Sea-level rise is leading to wetland loss in several coastal areas. This means that the rate of accretion in deltas must increase in order for wetlands to survive.

Deltas are very important ecologically and economically and the majority of the world's coastal wetlands and marine fisheries are in deltas. But many deltas are in crisis because

Table 12 *Major world deltas*

Delta	Receiving Sea	Area (Km² x 10³)
Amazon	Atlantic Ocean	467
Ganges	Bay of Bengal	106
Mekong	South China Sea	94
Yangtze	East China Sea	67
Lena	Laptev Sea	44
Hwang Ho	Yellow Sea	36
Indus	Arabian Sea	30
Mississippi	Gulf of Mexico	29
Volga	Caspian Sea	27
Usumacinta Grijalva	Gulf of Mexico	22
Orinoco	Atlantic Ocean	21
Irrawaddy	Bay of Bengal	21`

various human impacts have led to deterioration. Human activities have severely diminished the ability of deltas to survive, especially with increasing sea-level rise, by systematically reducing pulsing events at all important spatial and temporal scales. Dams reduce sediment and water to deltas (i.e., more than 95% for some rivers), river channels in deltas have been closed, and dikes reduce input of river water to delta wetlands. Many delta wetlands have been impounded or reclaimed. These changes have resulted in wetland loss, eutrophication, salinity intrusion, and reduced fish catches.

Sustainable management of deltas should be based on a return to the natural functioning of deltas, but in a controlled manner. This means that the natural pulsing energies must be utilized rather than diminished. River input to the delta can be accomplished through controlled inputs to different parts of the deltaic plain. The benefits include marsh creation, reduction of salinity stress, and enhanced estuarine productivity. In some river systems, it may be necessary to remobilize sediments trapped in reservoirs so that they can be transported to the delta. The development of sustainable management plans for a delta should be done in an integrated way as part of a holistic management strategy for the entire delta using ecological engineering and restoration ecology.

See also: ECOLOGICAL ENGINEERING, RESTORATION OF ECOSYSTEMS, SEA-LEVEL CHANGE, WETLANDS

Further Reading

Day, J., Martin, J., Cardoch, L. and Templet, P., 1997. System functioning as a basis for sustainable management of deltaic ecosystems. *Coastal Management*, 25, 115–154.

Day, J., Psuty, N. and Perez, B., 2000. The role of pulsing events in the functioning of coastal barriers and wetlands: Implications for human impact, management and the response to sea-level rise. In Weinstein, M. and Dreeger, D. (eds), *Concepts and controversies in salt tidal marsh ecology*. Kluwer Academic Publishers, Dordrecht, The Netherlands, 633–660.

Day, J.W., Pont, D., Hensel, P. and Ibañez, C., 1995. Impacts of sea-level rise on deltas in the Gulf of Mexico and the Mediterranean: The importance of pulsing events to sustainability. *Estuaries* 18, 636–647.

Ibañez, C., Curco, A., Day, J.W. and Prat, N., 2000. Structure and productivity of microtidal Mediterranean deltaic coastal marshes. In Weinstein, M.P. and Kreeger, D.A. (eds), *Concepts and controversies in tidal marsh ecology*. Kluwer Academic Publishers, Dordrecht, The Netherlands, 107–136.

Rybczyk, J.M., 2005. Deltaic ecology, In Schwartz, M.L. (ed.), *The encyclopedia of coastal science*. Springer, Dordrecht, The Netherlands, 359–361.

Penland, S., Kulp, M.A., 2005. Deltas. In Schwartz, M.L. (ed), *The encyclopedia of coastal science*. Springer, Dordrecht, The Netherlands, 362–367.

Yáñez-Arancibia, A., 2005. Middle America, Coastal Ecology and Geomorphology. In Schwartz, M.L. (ed), *The encyclopedia of coastal science*. Springer, Dordrecht, The Netherlands, 639–645.

DRAINAGE BASINS see Processes

EARTH SPHERES

Earth and its environment can be thought of as composed of a series of spheres, rather like concentric shells or layers, although some are not discrete and overlap with others. Such division of Earth's environments into a series of spheres is a useful way of subdividing material and information as the basis for study by different disciplines. Because air, land, water, and living things are the main ingredients of environmental systems it was the spheres of atmosphere, lithosphere, hydrosphere and biosphere which were first identified by Suess in 1875. At that time this provided a way of identifying those spheres of the Earth's environment concerned with air, rocks, water and life, thus providing the subject matter for the environmental sciences. The atmosphere had been identified as early as the late 17th century, was complemented by the other three in the 19th century, later supplemented by others concerned with the soil (pedosphere), ice (cryosphere) and human activity (noosphere). So many spheres were eventually suggested that in 1997 Richard Huggett proposed we may have become 'sphere crazy'.

These spheres provided a convenient framework as the environmental sciences grew and expanded in the first half of the 20th century, a time when the major sub-divisions of environmental disciplines were becoming identified. Particular disciplines were primarily concerned with one or more spheres in the way that meteorology was concerned with the atmosphere, geology with the lithosphere, and biology with the biosphere. Some disciplines necessarily transcend several of the spheres in the way that physical geography is concerned with the surface of the earth and so with the interactions between several spheres, particularly between the atmosphere, lithosphere and biosphere. Later disciplines were often first developed as subdivisions of existing ones in the way that pedology or soil science was concerned with the pedosphere, subsequently becoming a separate discipline in view of its importance for soil survey and agricultural land planning. Similarly, hydrology focused on the hydrosphere, developed as an integral part of several disciplines including civil engineering, geology, physical geography, geomorphology, and ecology but also emerged as a separate discipline in view of its importance in relation to flow measurement, water resources and flood management.

For just the same reasons that led to the formulation of environmental science it has been necessary to focus upon interactions between the original spheres with specific spheres, such

Table 13   *Some Earth spheres that have been suggested*

Sphere: those in bold have separate entries	Approximate date when suggested if known	Definition: see more details under separate entries
Magnetosphere		A magnetosphere is the region around an *astronomical object* in which phenomena are dominated or organized by its *magnetic field*. The Earth is a huge magnet, its magnetic influence extends far into space and it is surrounded by a magnetosphere
Celestial sphere		The whole universe beyond the atmosphere – the Sun, Moon, and stars, as well as the asteroids and the little bits of dust that make meteors when they hit the atmosphere
Cosmosphere		Makes up space which encompasses and surrounds the Earth. It is everything beyond Earth – the stars and the vast outer space
Exosphere		The region above 700 km, at which height atoms may begin to escape into space
Thermosphere		The layer of the atmosphere above the mesopause 80 km above the earth's surface in which temperature increases with height.
Ionosphere		The layer of the atmosphere where free ions and electrons occur, sometimes reserved for the belt of high electron density between 100 and 120 km above the Earth's surface
Mesosphere		A layer between the stratosphere and the ionosphere
Stratosphere		The stable layer about 10-50 km above the troposphere in which temperature is largely independent of altitude and averages −60 °C.
Troposphere		The part of the atmosphere between the Earth's surface and the tropopause (at 8–16 km above Earth's surface) which marks a sharp change in the lapse rate of temperature change with altitude above the surface of the Earth.
Atmosphere	Late 17th century	The gaseous envelope of air surrounding the Earth and maintained by gravitational attraction, without a clear upper limit but about 200 km maximum
Heterosphere		The upper portion of a two-part division of the atmosphere (the lower portion is the homosphere) where *molecular diffusion* dominates and the chemical composition of the atmosphere varies according to chemical species. There is no convective heating at this height, the material found in the heterosphere is layered according to its mass.
Homosphere		The portion of the Earth's atmosphere, up to an altitude of about 80 km above sea-level, in which there is continuous turbulent mixing, and hence the composition of the atmosphere is relatively constant. Homosphere is also sometimes used for the biosphere as modified by human activity.

(Cont'd)

Sphere: those in bold have separate entries	Approximate date when suggested if known	Definition: see more details under separate entries
Hydrosphere	1875	Water body of the Earth in liquid, solid or gaseous state and occurring in fresh and saline forms
Geosphere	1980	Sometimes used for the lithosphere but more usually for several spheres (e.g., lithosphere + hydrosphere + atmosphere, or core, mantle and all layers of crust) or for the zone of interaction on or near the earth's surface involving the atmosphere, hydrosphere, biosphere, lithosphere, pedosphere and noosphere.
Relief Sphere	1982	Used for the totality of the Earth's topography
Toposphere	1995	Representing the interface of the pedosphere, atmosphere and hydrosphere
Landscape sphere	1983	Zone of interaction at or near the Earth's surface involving atmosphere, hydrosphere, biosphere, lithosphere, pedosphere and noosphere.
Cryosphere	1980	That part of the Earth's surface where water is in a solid form, usually as snow or ice, and includes glaciers, ice shelves, snow, icebergs, and snowfields
Noosphere	1930s	The realm of human consciousness in nature or the 'thinking' layer arising from the transformation of the biosphere under the influence of human activity. In the original theory of *Vernadsky* (1863–1945) the noosphere is the third in a succession of phases of development of the Earth, after the *geosphere* (inanimate matter) and the *biosphere* (biological life). Vernadsky began to use the term in the mid-1930s. Some regard the noosphere as synonymous with the anthroposphere.
Anthroposphere		Including human activity and constructions which is the human population, including our cities, bridges, dams and roads – everything we build
Geoecosphere	1995	The sphere in which other spheres (biosphere, troposphere, atmosphere, pedosphere and hydrosphere) interact.
Biosphere	1875	The zone in which living organisms occur on Earth thus overlapping with the hydrosphere, lithosphere and atmosphere
Ecosphere		The living and non-living components of the biosphere
Pedosphere	1938	The outermost layer of the *Earth* that is composed of *soil* and subject to *soil-forming processes*. It includes the dynamic interaction that occurs at the interface of the *lithosphere, atmosphere, hydrosphere* and *biosphere*.

Sphere: those in bold have separate entries	Approximate date when suggested if known	Definition: see more details under separate entries
Lithosphere	1875	The rocks of the Earth's crust and a portion of the upper mantle, a zone which is up to 300 km thick and is more rigid than the asthenosphere below.
Asthenosphere		A deformable zone within the upper mantle of the earth extending from 50 to 300 km below the surface of the Earth sometimes to a depth of 700 km. Characterized by low-density, partially molten rock material chemically similar to the overlying lithosphere. The upper part of the asthenosphere believed to be the zone involved in *plate movements* and *isostatic* adjustments.
Barysphere		The interior of the Earth beneath the lithosphere, including both the mantle and the core. Sometimes used to refer only to the core or only to the mantle.
Centrosphere		The central core of the Earth. Also sometimes called barysphere.

as the geosphere, sometimes used to express such an interaction. Recent suggestions have often been for spheres which represent dynamic interactions, such as those between the land surface and the biosphere, expressed as the geoecosphere, and the landscape sphere was suggested to connote that sphere which involves interaction between processes above, on and immediately below the surface of the earth.

The spheres that have been suggested, many included in Table 13, should be helpful and not viewed too rigidly. They should assist in defining disciplines and research investigations but indicate the need to focus on the subject of each sphere, the interaction between spheres, and the way in which they can all combine together. The spheres have not always been defined or used in exactly the same way. Increasingly human activity is responsible for modifications of the processes operating within, and the character of, particular spheres so that the noosphere was introduced to represent the thinking layer arising from the transformation of the biosphere under the influence of human activity occurring outside and above the biosphere. The term anthroposphere was introduced for similar reasons.

See also: ANTHROPOGENIC IMPACT, ATMOSPHERE, BIOLOGY, BIOSPHERE, CRYOSPHERE, EARTH STRUCTURE, GEOLOGY, HYDROSPHERE, LITHOSPHERE, TROPOSPHERE

Further reading

Huggett, R.J., 1997. *The evolving ecosphere*. Routledge, London.

EARTH STRUCTURE

The Earth consists of several shells around a central core; from the Earth's surface inward, they are: lithosphere, asthenosphere, mantle, outer core, and inner core (Fig. 16).

The lithosphere is made up of the crust and the uppermost mantle. The outer crust comprises less than 1% of the Earth's mass. The continents are imbedded in the outer crust and are up to 33 km thick, whereas the ocean floors are about 7 km thick. The lower crust is about 160 km thick consisting of solid rock mainly of basaltic composition. Eight elements make up 99% of the Earth's crust: oxygen, magnesium, aluminium, silicon, calcium, sodium, potassium, and iron.

The mantle makes up about 70% of the Earth's mass. It is a rock layer about 2800 km thick that reaches from the base of the lower crust to about half the distance to the centre of the Earth. Parts of this layer become hot enough to liquefy and become slow-moving molten rock or magma akin to basalt in composition. Convection in this layer drives the movements of crustal plates in plate tectonics. Silicon, oxygen, aluminium, and iron are the principal elements in the mantle.

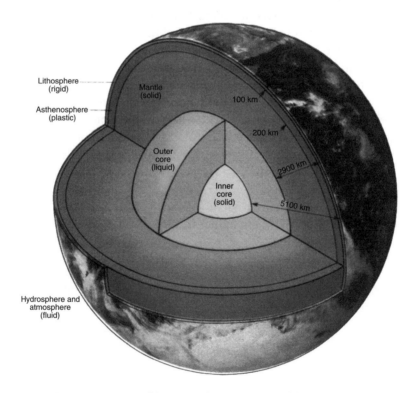

Figure 16 *Structure of the interior of the Earth. Source of diagram unknown*

Table 14 *Dimensions and properties of internal layers of the Earth's interior (Walker and Cohen, 2006)*

Layer	Depth to boundaries (km)	Fraction of volume	Mass (in 1027 g)	Density (g/cm³)
Continental Crust	0. to 33	0.0155	0.05	2.67–3.0? (2.84 avg)
Mantle	33 to 2898	0.8225	4.05	3.32–5.66 (4.93 avg)
Core	2898 to 6371	0.1620	1.88	9.7–12.3? (10.93 avg)
Total Earth	6371	1.0	5.90	5.5

The Earth's outer core is a mass of molten iron and nickel about 2270 km thick that surrounds the solid inner core. The outer core makes up about 30% of the Earth's mass. It generates electrical currents that produce the Earth's magnetic field.

The inner core is a mass of solid iron about 1270 km thick with a temperature of about 7000 °F. Although such temperatures would normally melt iron, immense pressure keeps it in a solid form.

Seismic waves from large earthquakes are used to map the interior of the Earth (Fig. 17). Seismometers on the Earth's surface, depending on their distance from the earthquake, receive

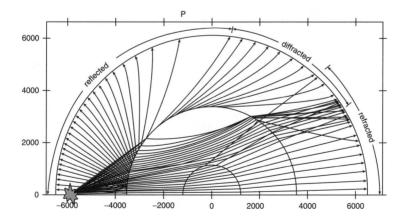

Figure 17 *Schematic slice through the centre of the Earth, showing how wave fronts radiate outward from an earthquake focus (star). Seismic waves, represented here as rays by the arc-shaped arrows, emanate from the quake. X and Y axes indicate distance from the centre of the Earth in kilometers; concentric lines indicate boundaries of the mantle and inner and outer cores. Source of diagram unknown.*

waves that have been reflected off, diffracted around or refracted across the core–mantle boundary. A careful examination of these waves reveals the structure of this boundary.

The properties of each of the interior shells is inferred from the behaviour of earthquake waves as they travel through the Earth. For example, the rays of P-waves travel in arc-shaped paths through the mantle, but those that intersect the core–mantle boundary are reflected and refracted in such a way that seismographs do not record P-waves in a 'shadow zone' between 103° and 143° latitude from the earthquake's epicentre (Fig. 18). From this information, seismologists are able to determine the depth from the earthquake hypocentre to the core–mantle boundary. Seismologists conclude that the outer core is molten, because S-waves, which are incapable of passing through liquids, are not received by any seismographs more than 103° of latitude from the earthquake epicentre.

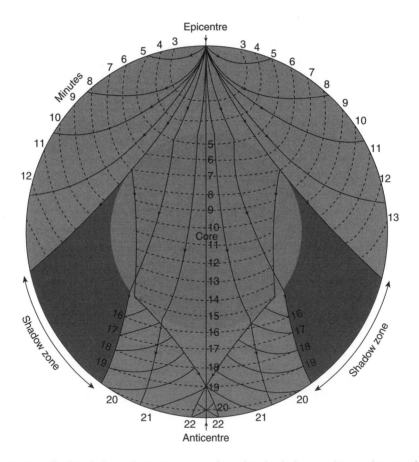

Figure 18 *P-wave ray paths through the Earth. P-waves are not detected in the shadow zone because they are refracted at the core-mantle boundary. Source of diagram unknown*

Table 15 *Density of common substances and elements*

Substance	Density (in g/cm³)
Pinewood	0.35–0.50
Water	1.0
Brick	1.84
Sand	2.2
Aluminium	2.70
Iron	7.8
Silver	10.5
Mercury	13.6
Gold	19.3

The density of the Earth's interior layers is determined by arrival times to seismographs at different locations on the Earth's surface. P-waves travel faster in dense rock than in less dense rocks, so if a P-wave arrives at a seismograph station earlier than predicted, then it has passed through denser rocks than one that arrived at the expected time at another station. After analysis of many earthquakes over many years, seismologists have concluded that the average density of the lithosphere is 2.9 g/cm³, and that of the mantle is 3.5 g/cm³. Because the average density of surface rock is only around 3.0 g/cm³, then denser rocks must exist within Earth's core. If the density of the entire Earth is 5.5 g/cm³, then the core must have a density of 10.9 g/cm³, about the same as silver (Table 15).

See also: EARTH SPHERES, GEOLOGY, ROCK TYPES

Further reading

Jacobs, J.A., 1992. *Deep interior of the Earth*. Springer, Karato, S., 2003., *The dynamic structure of the Deep Earth: An interdisciplinary approach*. Princeton University Press, Princeton.

Walker, J.D., and Cohen, H.A., (compilers), 2006. *The Geoscience Handbook*. American Geological Institute, Alexandria, Virginia.

ESTUARINE ENVIRONMENTS

Estuaries can be very broadly defined as that portion of the Earth's coastal zone where there is interaction of ocean water, fresh water, land, and atmosphere. Large estuarine zones are most common in low-relief coastal regions, especially the broad coastal plains of Europe and the east coast of North America. They are much less common in uplifted coastlines like the Pacific edge of North and South America. Pritchard's early definition, that 'an estuary is a semi-enclosed coastal body of water having a free connection with the open sea and containing a measurable quantity of sea salt', was very perceptive but is not now generally used. Instead, the later modified definition by Cameron and Pritchard, where 'an estuary is a semi-enclosed body of water having a free connection with the open sea and within which the sea water is measurable diluted with fresh water deriving from the land drainage', is more commonly adopted but is in reality a much too restrictive definition, since many types of inland coastal marine-connected systems are excluded. Fairbridge (1980, p.1) made a significant change in the ecological approach by defining an estuary as:

> an inlet of the sea reaching into a river valley as far as the upper limit of tidal rise, usually being divisible into three sectors: (a) a marine or lower estuary, in free connection with the open ocean; (b) a middle estuary, subject to strong salt and freshwater mixing; and (c) an upper or fluvial estuary, characterized by fresh water but subject to daily tidal action.

Probably coastal lagoons are associated geomorphologic features in the coastal zone, close to the classical view of an estuary, and this is a clear concern of the definition of coastal lagoon given by Lankford, namely: 'a coastal zone depression below mean high high wave MHHW, having permanent or ephemeral communication with the sea, but protected from the sea by some type of barrier'.

From an ecological point of view, however, coastal lagoons and estuaries constitute a similar type of ecosystem, and Day and Yáñez-Arancibia (1982, p. 11) emphasized the concept of lagoon-estuarine ecosystems, as:

> a coastal ecotone, connected to the sea in a permanent or ephemeral manner. They are shallow bodies of water, semi-enclosed of variable volumes depending on local climatic and hydrologic conditions. They have variable temperatures and salinities, predominantly muddy bottoms, high turbidity, and irregular topographic and surface characteristics. The flora and fauna have a high level of evolutionary adaptation to stress and environmental pulsing conditions, and have originated from marine, freshwater and terrestrial sources. This biota is directly important to man for what is yield, and to many marine and freshwater organisms which use the system. In this natural condition, the lagoon-estuarine ecosystem incorporates a balanced network of physical gradients, environmental pulsing, and biotic interrelationships.

The advantage of this concept is that it points out that salinities, mixing waters, environmental pulsing, and biota adaptation, are key concerns for understanding estuarine environments.

Although many other criteria, such as geomorphologic structure, water balance, ecological characteristics coupling physical and biological processes, circulation-mixing, and marine-fluvial processes, have been used to define and classify estuarine systems, the geomorphologic approach remains easy and very attractive. Therefore Kerfve proposed that inland coastal ocean-connected water can advantageously be organized into six categories, namely: (1) *Estuary*: an inland river valley or section of the coastal plain, drowned as the sea invaded the lower course of a river during the Holocene sea-level rise, containing sea water measurably diluted by land drainage, affected by tides, and usually shallower than 20 m; (2) *Coastal Lagoon*: an inland water body, usually oriented parallel to the coast, separated from ocean by a barrier, connected to the ocean by one or more restricted inlets, and having depths which seldom exceed a couple of meters; a lagoon may or not be subject to tidal mixing, and salinity can vary from that of a coastal freshwater lake to hyper-saline system, depending on the hydrologic balance; and formed as a result of rising sea level during the Holocene or Pleistocene and the building of coastal barriers by marine processes. (3) *Fjords*: a glacially scoured inland marine area with sea water measurably diluted by land drainage and surface layer, consisting of high salinity waters in deep basins, affected by tides, and usually measuring several hundred metres in depth. (4) *Bay*: a coastal indentation, usually the result of faulting or other tectonic or regional geologic processes, strongly affected by tides, and exhibiting salinities ranging from oceanic to brackish, depending on the amount of land drainage relative to oceanic exchange. (5) *Tidal River*: an inland river valley, drowned as the sea invaded the lower river course during the Holocene sea-level rise, containing only fresh water, but subject to tidal sea-level variations and sometimes reversing tidal currents in the downstream section. (6) *Strait*: an inland marine waterway, connecting two oceans or seas; characteristics of sea straits with respect to circulation, salinity distribution, tidal processes, and water depth vary widely between straits.

See also: COASTS, ESTUARINE DEPENDENT OR RELATED SPECIES, PULSING

Further reading

Cameron, W.M. and Pritchard, D.W., 1963. Estuaries, In Hill, M.N., (ed), *The Sea, Volume 2*, Wiley, New York, 306–324.

Day, J.W. and Yáñez-Arancibia, A., 1982. Coastal lagoons and estuaries: ecosystem approach. *Ciencia Interamericana* OAE. Washington, DC. 22 (1–2), 11–25.

Day, J.W., Hall, C.A.S., Kemp, W.M. and Yáñez-Arancibia, A., 1989. *Estuarine ecology*. Wiley, New York.

Fairbridge, R.W., 1980. The estuary: its definition and geodynamic cycle. In: Olausson, E. and Cato, I. (eds), *Chemistry and biogeochemistry of estuaries*. Wiley, New York, 1–36.

Hobbie, J.E. (ed), 2000. *Estuarine science: A synthetic approach to research and practice.* Island Press, Washington, DC.

Kjerfve, B., 1989. Estuarine geomorphology and physical oceanography, In Day, J.W., Hall, C.A.S., Kemp, W.M. and Yáñez-Arancibia, A., 1989. *Estuarine Ecology.* Wiley, New York, 47–78.

Kjerfve, B., 1990. *Manual for investigation of hydrological processes in mangrove ecosystems.* UNESCO/UNDP Regional project (RAS/86/120), COMAR-UNESCO.

Kerfve, B., 1994. Coastal lagoon processes. In Kjerfve, B. (ed) *Coastal lagoon processes.* Elsevier, Amsterdam, Oceanography Series, 60, 1–8.

Lankford, R.R., 1977. Coastal lagoons of Mexico their origin and classification. In Wiley, M. (ed), *Estuarine Processes, Volume 2: Circulation, Sediments, and Transfer of Material in the Estuary.* Academic Press, New York, 182–215.

Pritchard, D.W., 1952. Estuarine hydrography. In Landsberg H.E. (ed), *Advances in Geophysics, Volume 1.* Academic Press, New York, 243–280.

Pritchard, D.W., 1967. What is an estuary: physical standpoint. In Lauff, G.H. (ed), *Estuaries.* American Association for the Advancement of Science, Publication 83, Washington DC, 3–5.

FLUVIAL ENVIRONMENTS

Fluvial refers to things that are of, or in, a river, whereas fluviatile refers to the products of a river. As rivers are the major land arteries of the hydrological cycle and have provided an important locus for human settlement the associated environments are of considerable significance in at least four ways.

The *river channel environment* refers to the processes and characteristics within the channel cross-section. The river flow, measured by discharge (volume of water per unit time, for example in cubic metres per second) reflects the size of the drainage basin upstream, the climate over the basin and the drainage basin characteristics (rock and soil type, topography, vegetation and land use) which collectively determine the rate of transformation of water received as precipitation into discharge. In addition to water flow, sediment is transported through the river channel environment as bedload or as suspended sediment and solutes are conveyed in solution. Characteristics of the river channel include bedforms and sedimentary structures, alluvial channels and bars, and also the habitats provided for plants. Lotic ecosystems in rivers and streams are studied by limnology. Aquatic ecology varies from one river environment to another and there are trends in the downstream direction as described by the river continuum concept which expresses the way that plants, biota (shredders, grazers, collectors and predators) occur along the river profile. From a point in the river environment there is a gradation of scales of river environment progressing from within channel, to channel unit, river reaches, valley segments, zones, and finally the complete drainage basin.

The *valley environment* associated with the river is associated with fluvial environment. The floodplain is the valley floor area adjacent to the river channel and may include the hydraulic floodplain, inundated at least once during a given return period, and the genetic floodplain largely composed of alluvial sediments adjacent to a channel (Fig. 19).

Populated fluvial environments have featured since the first agricultural communities gravitated to the flat fertile lands adjacent to rivers and subsequently rivers were vital for industrial location. Hydraulic civilizations were established more than 4000 years ago, many localized developments occurred prior to the industrial revolution which utilized river water for mills, power generation, irrigation and water supply, succeeded in the 20th century by flow regulation, flood control, and conservation projects.

Modified fluvial environments have been the result of extensive modification by human activity which has directly affected some environments, for example by extracting sand and gravel from the channels, and has indirectly affected others by changing fluvial processes. River discharge and sediment transport has been altered, downstream of dams or below urban areas for example, so that river environments have altered with some channels decreasing in size and character and others increasing significantly. Early management measures used hard engineering techniques to restrict erosion or degradation of the channel environment but more recently there have been attempts to adopt a more holistic approach which considers all aspects and fluvial environments upstream and downstream from the problem reach. Some fluvial environments have channels which were engineered in a severe way or where meandering channels were straightened, have subsequently been the subject of restoration schemes involving restoration of sinuous more natural channels.

River environments are extremely varied, ecologically rich habitats possessing a range of geomorphological and biological processes and features and in addition to their economic significance they afford many opportunities for recreation (Penning-Rowsell and Burgess, 1997).

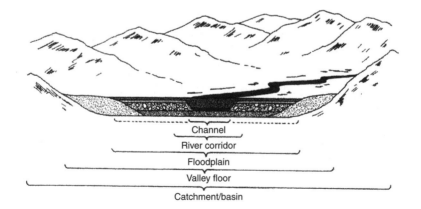

Figure 19 *Spatial scales in the fluvial system and drainage basin. From Downs and Gregory (2004) based on Newson (1997)*

See also: DRAINAGE BASINS, HYDROLOGICAL CYCLE, LIMNOLOGY,

Further reading

Bridge, J.S., 2003. *Rivers and floodplains*. Blackwell Publishing, Malden, MA, USA.

Downs, P.W. and Gregory, K.J., 2004. *River channel management*. Arnold, London.

Gregory, K.J., 2006. The human role in changing river channels. *Geomorphology* 79, 172–191.

Newson, M.D., 1997: *Land, water and development: sustainable management of river basin systems*, Routledge, London, 2nd edn.

Penning–Rowsell, E. and Burgess, J., 1997. River landscapes: Changing the concrete overcoat? *Landscape Research* 22, 5–12.

FOSSIL FUELS

We live in the age of oil, which is one of the fossil fuels (all of which are forms of stored solar energy). Plants are collectors of solar energy and this energy is converted to chemical energy through photosynthesis. Thus we can see the connection between solar energy and fossil fuels. The main fossil fuels that we use today, such as coal and oil, were created as a result of incomplete biological decomposition of dead organic material. Most of the organic material involved in the creation of the fuels was land and marine plants. The organic material that was not completely decomposed slowly was converted through heat and pressure to fossil fuels. The transformation takes anywhere from hundreds of thousands to a few million years for oil to tens or even hundreds of millions of years for coal.

On a world-wide basis, the fossil fuels provide 90% of the energy consumed. The age of fossil fuels, and in particular oil, has fueled economies for many decades, but we now are facing its eventual decline. Of particular importance to environmental science is what is known as 'peak oil', which is the time when one half of the total amount of oil within the earth will have been extracted and consumed. Following the peak, there may be a gap between production and demand, leading to large increases in the price of oil. Of course, we have much more coal and natural gas that may take up some of the demand, but there are problems of potential limitations with natural gas as there are for oil, and coal has its own particular problems. In particular, we are concerned with global warming and burning of vast amounts of coal will release even more carbon dioxide into the environment, which may further warm the planet.

Abundant cheap, plentiful fossil fuels have allowed our civilization to make tremendous advances in agriculture and technology. The human population on earth today of about 6.6 billion people depend heavily on use of chemical fertilizers and technology through machinery to produce our food. Most of these depend upon oil and natural gas to a lesser extent. Thus when a crunch comes in the availability of oil, repercussions will spread throughout society.

If we wish to be proactive rather than reactive to potential shortages of oil, we need to prepare now for the likelihood that production rates of oil will fall sometime in the mid-21st century. In order to avoid major disruption to societies around the world, we need to start with an education programme now to understand the potential impacts of peak oil and prepare for them. Unfortunately today, we are often acting in ignorance of the coming energy crisis and only a few countries are making plans for the transformation to alternative energy. This is not to say that there are not large incentives and programmes in alternative energy. In fact, wind and solar power are the fastest-growing source of energy today in the world.

It appears that our transition from oil to other fuels will likely be a bumpy ride with respect to prices of energy in the future. Natural gas can serve as one transition, as may coal if we develop the technology to burn it cleaner and find ways to sequester carbon dioxide so that it will not further enhance global warming. It also seems likely that we will turn again to nuclear energy in the future, as that source of energy does not enhance global warming. However, we will need to develop newer reactors that are safer than some we have used in the past. Finally, the transition will include major increases in alternative energy. The rapid growth we are experiencing today in wind and solar energy, (doubling every two to three years) is a good sign, but alternative energy sources only provide a few per cent of our energy needs.

The final message for the environment is the age of oil is changing and as a result of peak oil, prices will increase and demand will go up and we will need to plan very carefully for our future energy supplies.

See also: LITHOSPHERE, MINERALS, NATURAL RESOURCES

Further reading

British Petroleum Company, 2006. *BP statistical review of world energy.*

Edwards, J.D., 1997. Crude oil and alternative energy production forecast for the twenty-first century: the end of the hydrocarbon era. *American Association of Petroleum Geologists Bulletin* 81(8), 1292–1305.

Maugeri, L., 2004. Oil: Never cry wolf—When the petroleum age is far from over. *Science* 304, 1114–1115.

Youngquist, W., 1998. Spending our great inheritance. Then what? *Geotimes* 43(7), 24–27.

GLACIAL ENVIRONMENTS

Glacial environments refer to environments at, or near, the margins of glaciers and include a variety of forms and processes associated with cold climate phenomena. Included is permafrost, which is perennially frozen ground, that melts in the upper layers for part of the year. When we speak of glacial environments, we may also include paraglacial conditions, or cold conditions associated with cold climate, as well as proglacial conditions, or land found in front or at the margins of glacial ice.

From a perspective of the glacial environments, the ground that is permanently frozen containing permafrost is perhaps of most significance to environmental science. The area covered by permafrost, whether it is continuous permafrost or discontinuous, is widespread around the globe, particularly at the high latitudes above about 60° (Fig. 20). As mentioned, there are two main types of permafrost that are defined in terms of the aerial extent of the frozen ground. Discontinuous permafrost is found further south then continuous frozen ground and is characterized by scattered areas of thawed ground in what is a more predominantly frozen area. Closer to polar regions, continuous permafrost is characterized by frozen ground being almost everywhere except beneath deep lakes or rivers.

Permafrost has been extensively studied because people have long lived in areas where it occurs. Much of Alaska and northern Russia has extensive permafrost and people interact with the landscape where it occurs.

Permafrost has an upper active layer that thaws during the summer over the perennially frozen ground below. The thickness of the active zone is variable and depends upon factors including the surface topography and slope, the amount of water, and the exposure of the site to solar energy. Vegetation that covers the surface is also important because vegetation can act as a thermal cover for the soil. When vegetation is removed in areas with permafrost, the insulation is no longer present and as a result, more permafrost may melt. Sometimes when vegetation is removed on a road or a track where vehicles have crossed the land, deep ruts develop as the melting of the permafrost goes deeper and deeper.

There are a variety of special problems related with permafrost where people wish to construct buildings and infrastructures such as roads, railroads, airfields, and pipelines. A particular problem is the finer grain soils, which may hold a lot of frozen water, and may fall by sliding even on gentle slopes with partial or complete melting of the permafrost to deeper layers.

From an environmental perspective, building of roads and crossing areas with permafrost can be disruptive to the land and the ecosystems found on it. As a result, development for energy on the North Slope of Alaska in the oil fields has to take particular care so that the environment is not damaged. One method is to construct roads in the winter of ice that upon melting in the spring and summer become invisible. Pipelines carrying oil may be elevated above the ground surface and insulated well so that there is no melting of the permafrost and animals may migrate beneath the pipeline. These methods along with others have been suggested for the north slope of Alaska where the Arctic National Wildlife Refuge is located. Most of the

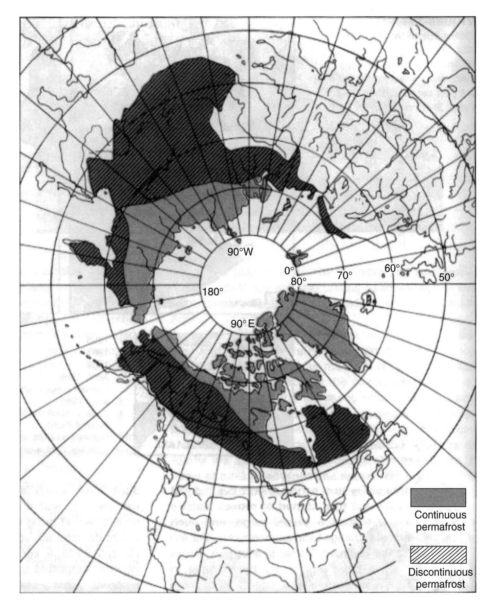

Figure 20 *Extent of permafrost zones in the Northern Hemisphere. (From Ferrians, Kachadoorian, and Greene, Permafrost and Related Engineering Problems in Alaska U.S Geological Survey Professional Paper 678, 1969)*

challenge to developing oil fields in the north slope of Alaska and other sensitive areas with permafrost is to find ways of developing the oil that do not damage the environment and the pristine ecosystems that are often found in these areas.

See also: GLACIAL GEOMORPHOLOGY, GLACIERS, GLACIOLOGY

Further reading

Gillespie, A.R., Porter, S.C. and Atwater, B.F., 2004. *The Quaternary period in the United States.* Developments in Quaternary Science 1. Elsevier, Amsterdam.

Pelto, M.S., 1996. Recent changes in glacier and alpine runoff in the North Cascades, Washington. *Hydrological Processes* 10, 1173–1180.

GLACIERS

A glacier is a land-bound mass of moving ice. The glacier itself is composed mostly of glacial ice, which forms through transformation of snow to glacial ice over a period of several years to several hundred years. However, glaciers also contain variable amount of water, gases (mostly trapped air), as well as sediment and other materials. The transformation of snow to glacial ice begins as snow is packed together by the overlying pressure and recrystallizes to a more granular material that is still porous ice. Over time, the compaction continues with recrystallization of the ice until eventually dense, blue glacial ice is formed. The density of glacial ice is approximately 0.9 g/cm³, which is about 150 times as dense as new snow. In order for a glacier to form, a number of years of snowfall must avoid melting and accumulate to the point that the weight of the overlying snow goes through the transformation described above and glacial ice forms. Today the Earth is still in the Pleistocene ice ages and the Last Glacial Maxima occurred about 20,000 years ago when glacial ice covered about 30% of the land surface of Earth. For comparative purposes, glaciers today cover about 10% of the land areas of Earth. The Last Glacial Maxima is only the most recent glaciation and several others were present during the Pleistocene, which began about 1.8 million years ago. Presently, we are in a time period known as an interglacial, when the abundance of glacial ice is relatively low. During interglacial times, sea-levels tend to be high as less water is stored on land in glaciers. During the Last Glacial Maxima, sea-levels were approximately 120 metres lower than they are today.

Glaciers, when they cover large regions of land, are called continental glaciers, ice caps or ice sheets (Fig. 21). Ice sheets and caps are found at high latitudes and an example today would be the Greenland ice cap (Fig. 22). At lower latitudes or higher elevations, smaller glaciers, known, as alpine glaciers may be present. A number of alpine glaciers confined to valleys are located in North America and Europe. One of the major environmental consequences of global warming

Figure 21   *Margin of the Antarctic ice cap. Stockphoto*

Figure 22   *Midgard Glacier, East Greenland, Stockphoto*

in recent decades is accelerated melting of alpine glaciers. Around the world, the glaciers in mountain areas are generally retreating. Many of the glaciers in Glacier National Park in the United States have already disappeared and the remainder may be gone in only a few decades. Similar stories are told also in the Alps in Europe. Melting of glaciers has additional environmental significance because a number of locations around the world use glacial meltwater as a source for irrigating crops and for urban consumption. For example, La Paz, which is the capital of Bolivia at an elevation of 3.4 km, is dependent upon glacial meltwater for its water supply. The source of glacial meltwater is particularly important in years when rainfall is relatively low and as a result, there is less runoff. During these times, the melting glaciers provide a major source of water. In Europe and other locations, melting glacial water is sometimes used to generate electrical power. A power station in Switzerland is dependent almost entirely upon meltwater from glaciers for the electricity it produces for the Swiss railway system.

With global warming, there will be less snowfall and more rain and as a result, quicker runoff. As a result, it will be a challenge to catch and store water in reservoirs. In order to be proactive with respect to global warming and less snowfall and more rain, we will have to rethink our utilization and storage of water resources.

See also: GLACIAL ENVIRONMENTS, GLACIAL GEOMORPHOLOGY, GLACIERS, GLACIOLOGY

Further reading

Benn, D.I., and Evans, D.J.A., 1998. *Glaciers and glaciation*. Arnold, London.

Boulton, G.S. and Hindmarsh, R.S.A., 1987. Sediment deformation beneath glaciers: Rheology and geological consequences. *Journal of Geophysical Research*. 92, 9059–9082.

Ensminger, S.L., Evenson, E.B., Alley, R.B, Larson, G.J., Lawson, D.E., and Stasser, J.C., 1999. Example of the dependence of ice motion on subglacial drainage system evolution: Matanuska Glacier, Alaska, United States. In *Past and Present*, Geological Society of America Special Paper 337, 11–22.

Kamb, B., 1987. Glacier surge mechanism based on linked cavity configuration of the basal water conduit system. *Journal of Geophysical Research*. 92, 9083–9100.

Martini, P.I., Brookfield, M.E., and Sadura, S., 2001. *Glacial geomorphology and Geology*. Prentice Hall, Upper Saddle River, NJ.

HYDROSPHERE

The Earth sphere which includes water in fresh or saline form in a liquid, solid or gaseous state, consists chiefly of oceans, but technically includes all water in the world, embracing inland seas, lakes, rivers, and underground waters. Approximately 70.8% of the Earth is covered by water and only 29.2% is landmass, so that the abundance of water is a distinguishing feature of Earth as the 'Blue Planet' in the solar system. The water of the hydrosphere occurs in several stores of which the oceans are the largest including more than 97% of the world's water, more than 0.6% in ground water, some 1.65% in ice sheets and glaciers, with very small amounts in the atmosphere, on land in lakes and seas (0.0148%) and rivers (0.00012), and in the soil (0.0055) and in atmospheric vapour (0.00096). Water in the Earth's system is largely constant, although volcanic activity emits water vapour from the Earth's interior and it has been estimated that the minerals in the mantle/lithosphere may contain as much as 10 times the water as in all of the current oceans, though most of this trapped water will never be released. Transfers that occur within, and between, the various stores are represented by the hydrological cycle and estimates of rate of exchange which indicates residence time in particular stores are shown in Table 16.

Within stores movement of water takes place and in the case of the oceans movement is driven by differences in temperature and in salinity. Because warm water is less dense or lighter it will tend to move upwards, whereas dense colder water tends to sink. Salt water is also denser so that it tends to sink, while fresh or less salt water being less dense tends to rise towards the surface.

Table 16 *Water volumes in stores of the hydrosphere. Based upon various sources. Note that estimates of volumes and percentages vary quite significantly from one source to another*

Hydrosphere store	Volume of water (10^6 km^3)	Percentage	Order of rate of exchange (years)
Oceans	1370	97.25	3000
Ground water	9.5	0.68	5000
Ice sheets and glaciers	24	1.65	8000
Lakes and seas	0.125	0.01	
Soil moisture	0.08	0.0055	1
Rivers	0.0012	0.0001	0.031
Atmospheric vapour	0.013	0.001	0.027 c.10days
Water in living organisms	0.0006	0.00004	
Lakes and rivers	0.2	0.04	
Surface water on land	0.28	0.019	7

Circulation patterns of ocean water are also horizontal, involving ocean currents which are surface water movements largely driven by major global wind systems, which affect the upper ocean layers down to about 100m, and are responsible for energy transfer towards the poles thus balancing the differences in receipt of solar radiation and being a major element in the world's heat balance. The broad circulation pattern of ocean currents is clockwise in the Northern Hemisphere and anticlockwise in the Southern Hemisphere. There is coastal upwelling along the western margins of continents where cold, nutrient-rich deeper water rises when surface currents flowing towards the equator are weak.

See also: EARTH SPHERES, ENERGY FLUX, HYDROLOGICAL CYCLE, LAKES,

Further reading

Berner, E.K. and Berner, R.A., 1987. *The global water cycle: geochemistry and environment.* Prentice Hall, London.

Niller, P.P., 1992. The ocean circulation. In Trenberth, K.E. (ed), *Climate system modelling.* Cambridge University Press, Cambridge, 117–148.

KARST

The term 'karst' refers to karst topography, which is peculiar to and dependent upon chemical weathering of rocks (generally limestone and marble) and diversion of surface waters to subterranean routes. The best examples of karst topography are developed upon soluble rocks such as limestone that are dense, thin bedded and well jointed. That is, there are many surfaces along which water may move and dissolve the limestone, producing sinkholes at the surface and subterranean features such as caverns. Sedimentary rocks cover much of the surface of Earth and limestone is present in about a quarter of the areas where sedimentary rocks are exposed. As a result, karst topography is very common and karst areas are found in many areas of Earth including Europe, North America, Africa and Asia.

The surface of the land in karst areas may be pitted by chemical weathering of the rock that produces sinkholes that vary in size from several to several hundreds of metres in diameter (Fig. 23). The sinkholes result from several processes, including chemical weathering at the surface of the limestone where the water from the sinkhole may be diverted to underground routes. Sinkholes may also be produced by collapse of surface material over underground cavern systems. This process can form spectacular collapse sinkholes.

Figure 23 *The Mitchell karst plateau in southern Indiana with numerous sinkholes.(Samuel S. Frushour, Indiana Geological Survey)*

The chemical weathering of limestone involves weak carbonic acids that are sufficient to cause dissolution along joints and fractures within the rock that may enlarge to form cavern systems at or near the ground water level (water table). Through time, with changes in ground water levels, different levels of caverns develop. When the ground water level falls below the level of a cavern, water moving through and downward or laterally may produce a variety of cave forms such as stalactites and stalagmites or flowstone (Fig. 24).

The most common, large landform associated with karst topography is the karst plain, which is an elevated area with sinkholes and subterranean cavern systems. In some cases, spectacular karst topography forms very steep hills known as tower karst that is famous in China and also found in the Caribbean island of Puerto Rico. The towers are residual limestone left after the bulk of the rock has been removed by chemical weathering processes.

Karst areas have a variety of environmental problems related to the particular hydrology associated with karst areas. Because the water infiltrates quickly from the surface to the subsurface, agricultural areas may have problems in holding soil water. Surface streams in karst areas, when they do exist, may become polluted by a variety of sources that quickly pollute the subsurface waters as streams disappear into cavern systems. In the past, some sinkholes have been used as waste disposal sites. When water from these sinkholes enters the subsurface, then the ground water is polluted. Homes using septic systems also have problems because the wastewater may quickly enter the general ground water flow system of the area.

Another hazard associated with karst areas is subsidence or collapse. Large collapse sinkholes may suddenly form if there is a collapse over part of a cavern system. As such, collapse features have swallowed parts of agricultural land and in some cases, urban lands. For example, a large collapse sink formed rapidly in only three days in Winterpark, Florida. The collapse swallowed part of a community swimming pool, several businesses, a house, and several automobiles in a dealership (Fig. 25).

Figure 24 *Cave Formations in Carlsbad Caverns, New Mexico. (Bruce Roberts/Photo Researchers, Inc.)*

Figure 25 *The Winter Park Florida sinkhole that grew rapidly for three days, swallowing part of a community swimming pool as well as several businesses, houses and automobiles. (leif Skooplors/Woodfin Camp and Associates)*

See also: EARTH SURFACE PROCESSES

Further reading

Ford, D., and Williams, P.W., 1989. *Karst geomorphology and hydrology*. Unwin Hyman, Winchester, Mass.

Palmer, A.N., 1991. Origin and morphology of limestone caves. *Geological Society of America Bulletin* 103, 1–21.

Ritter, D.F., Kochel, R.C. and Miller, J.R., 2002. *Process geomorphology*. McGraw Hill, Boston.

White, W.B., 1988. *Geomorphology and hydrology of karst terrains*. Oxford University Press, New York.

LITHOSPHERE

The solid portion of the Earth, as compared to the atmosphere and the hydrosphere. In plate tectonics, a layer of strength relative to the underlying asthenosphere for deformation at geologic rates. It includes the crust and part of the upper mantle and is of the order of 100 km in thickness (Dennis and Atwater, 1974, p. 31).

The lithosphere constitutes the rocky, solid outer layer of the Earth, consisting of both the crust and the uppermost part of the upper mantle. It overlies the asthenosphere, which constitutes the remainder of the mantle down to the outer core, like the shell of an egg. Ranging from 50 to 200 km in thickness, the lithosphere is thin under the oceanic crust and thick under the continental crust (Table 17)

The lithosphere constitutes only 1.55% of the volume of the Earth and the minutest fraction of the Earth's thickness. Earthquakes occur in the lithosphere because it is strong and brittle relative to the more ductile asthenosphere.

Oceanic basalt covers 71% of the lithosphere, whereas the continents are made mostly of granitic rocks together with lesser sedimentary and metamorphic rocks derived from granitic and volcanic rocks. The prevalence of basalt in the lithosphere is reflected by the abundances of elements (Table 18) that make up the main minerals (plagioclase plus olivine, pyroxene) (Table 19) in basalt.

Beneath the basaltic ocean floor, the lithosphere consists of mafic and ultramafic plutonic rocks that represent the refractory mafic residuum left when basaltic magma partially melted from the asthenosphere and rose into the crust where it ponded in small subvolcanic magma chambers and dikes. These mafic and ultramafic rocks crop out sparsely on the Earth's surface

Table 17 *Physical properties of the Lithosphere*

Quantity	Value
Mass of the crust	2.36×10^{22} kg
Crust fraction of Earth's volume	1.55%
Area	5.10×10^{14} m²
Land area	1.48×10^{14} m²
Continental area (incl. margins)	2.0×10^{14} m²
Water area	3.62×10^{14} m²
Oceans area (excluding margins)	3.1×10^{14} m²
Mean land elevation	8.75 m
Mean ocean depth	3794 m
Thickness of the crust	0 to 33 km
Mean thickness of continental crust	40 km
Mean thickness of ocean crust	6 km
Mean surface heat flow	87 mW/m²
Mean continental heat flow	65 mW/m²
Mean oceanic heat flow	101 mW/m²

Table 18 *Most abundant elements in the Earth's crust, 99% of which is made up of the first eight elements in this list*

Element	Atomic number	Symbol	Abundance in Earth's crust	
			Wt %	Vol.%
Oxygen	8	O	46.44	91.97
Silicon	16	Si	28.30	0.80
Aluminium	13	Al	8.41	0.77
Iron	26	Fe	5.21	0.68
Calcium	20	Ca	4.58	1.48
Magnesium	12	Mg	2.81	0.56
Sodium	11	Na	2.28	1.60
Potassium	19	K	1.50	2.14
Titanium	22	Ti	0.43	0.03
Manganese	25	Mn	0.08	<0.01
Phosphorous	15	P	0.06	<0.01
			100.10	

Table 19 *Most abundant minerals in the Earth's crust*

Mineral	Weight %
Plagioclase	39
K-feldspar	12
Quartz	12
Pyroxenes	11
Micas	5
Amphiboles	5
Clay minerals and chlorites	4.6
Olivine	3
Calcite and aragonite	1,5
Dolomite	0.5
Magnetite	1.5
Others (e.g., garnets, kyanite, apatite)	4.9
Total	100.0

today just because they have to be uplifted from such deep depths. Evidence of their existence at depth comes from refractory enclaves brought up by the basaltic magmas.

The lithosphere is in constant flux. Cycles of tectonism, volcanism, erosion, and sedimentation cause formation, deformation, consumption and reformation of the crust. Oceanic lithosphere forms at the mid-ocean ridges; it is then transported by convection in the asthenosphere to convergent plate margins where it is subducted into the mantle. Because the process of oceanic lithosphere formation and consumption is so quick relative to continental lithosphere, most of the oceanic basalt is very young – less than 200 million years – whereas some of continental crust is as old as four billion years.

Geoscientists are anxious to learn much more about the lower crust and upper mantle parts of the lithosphere, and to that end have tried to sample it directly by deep drilling into the oceanic crust where the depth to mantle is little more than 5 km. In fact, the entire deep-sea drilling programme was spawned by an unsuccessful attempt to drill a sampling hole through the oceanic crust to the mantle. To know what composes the lithosphere beneath the continents far deeper than technology is capable, geoscientists must resort to indirect seismologic means, or by analyzing the chemistry of igneous melts that have gone through the continental crust to the surface – or have now been exposed at the surface by erosion. One day technology may build a 'terramobile' that, like a submarine, can burrow around under the ocean floors and continents to bring back samples we can study – in the laboratory under the microscope.

See also: EARTH SPHERES, EARTH STRUCTURE

Further Reading

Simkin, T., Tilling, R.I., Vogt, P.R., Kirby, S.H., Kimberly, P. and Stewart, D.B., 2006. *This dynamic planet: world map of volcanoes, earthquakes, impact craters, and plate tectonics*. US Geological Survey Geologic Investigations Series Map I-2800, 1 two-sided sheet, scale 1:30,000,000. Also on the web at: http://www.minerals.si.edu/tdpmap

PERIGLACIAL ENVIRONMENTS

Periglacial, a term first used by Lozinski (1909) to describe frost weathering conditions in the Carpathians, is used for the type of climate and the climatically controlled surface features and processes adjacent to glaciated areas. Subsequently, periglacial environment was extended to apply to non-glacial processes and features of cold climates, including freeze–thaw processes and frost action typical of the processes in the periglacial zone and in some cases the processes associated with permafrost, but also found in high altitude, alpine, areas of temperate regions. Periglacial environments at present cover approximately 20% of the Earth's land surface, and were extensive in the past occurring extensively in parts of central Siberia and Alaska and the Yukon in western North America which were not glaciated for much of the Quaternary but instead experienced intense periglacial conditions. Approximately a third of the Earth's land surface has been subject to periglacial conditions at some time.

Processes unique to the periglacial environment include the formation of permafrost, the development of thermal contraction cracks, the differential thawing of permafrost to give thermokarst, and the formation of ice wedges and injection ice. Associated processes, not confined to periglacial environments, include ice segregation, seasonal frost action, frost or cryogenic weathering, and rapid mass movement. Where permafrost occurs very distinctive landforms occur including the polygonal patterns, usually 20–30 m across, on the surface together with ice-cored hills or mounds called pingos, and peat mounds which include ice lenses called palsas. When permafrost melts, and especially when it is affected by human activity, the features that can develop are named thermokarst because of their similarity to landforms in limestone areas. These include a suite of features ranging from small depressions to thaw lakes and larger basins called alas. Frost action in periglacial environments is responsible for frost wedging erosion of exposed bedrock, giving rise to angular rock fragments which create talus slopes, patterned ground and may leave angular tors on hilltops and hillslopes together with altiplanation terraces. Particularly above the permafrost table (the level below which permafrost does not thaw out) the result of spring thaw and summer melting of snow and of ground ice gives large amounts of water in the soil and so mass movement called solifluction or congelifluction occurs on most slopes even those with angles as low as 1 or 2°. Periglacial environments have marked seasonal contrasts: in winter low temperatures mean that surface processes are comparatively minor, the spring thaw sees melting of snow and of ground ice and the building up of transfers of material over the surface by solifluction together with large breakup floods along the rivers. In summer slope processes continue with active layers increasing in thickness but lower discharges in rivers, followed by autumn which sees the progressive onset of colder conditions once more.

Periglacial environments have distinctive soils, often described as tundra soils, which are often comparatively thin because there has been insufficient time for their development and because material moves over the slopes by frost creep, solifluction and water action. Ecological communities, including those known as tundra, are characterized by low growing shrubs, grasses, lichens and mosses often interspersed with wetter sites and bogs including string bogs, having productivity values which are typically less than 10% of those of the tropical rain forests and less than 20% of those in the boreal forest.

Because former periglacial environments extended into what are now temperate areas there are still many landforms and deposits, in Europe and parts of North America for example, that are now relict periglacial features including patterned ground of polygons and stone stripes, blockfields or felsenmeer, as well as deposits that often have national or local names such as 'head' or 'coombe rock' in southern England.

As cold-climate, non-glacial phenomena are not unique to cold climates the term (geo)cryogenic environments can be used. In such environments frost weathering processes are instigated by the expansion of water in rocks on freezing, by nival processes which depend upon snow in weathering under snow patches and in avalanches, as well as processes associated with ground ice.

See also: PERMAFROST

Further reading

Clark, M.J. (Ed), 1988. *Advances in periglacial geomorphology.* Wiley, Chichester.

French, H.M., 1996. *The periglacial environment.* Addison Wesley Longman, London.

PERMAFROST

Permafrost is ground in which a temperature lower than 0° has existed continuously for two or more years whether water is present or not. In areas of severe winters the depth of permafrost can be greater than 1400 m in Siberia (Lena and Yara river basins), more than 700 m in the Canadian Arctic islands, and more than 600 m in Alaska (Prudhoe Bay). The lower limit of permafrost is determined by the increase of temperature with depth beneath the Earth's surface (the geothermal gradient). There are spatial variations in permafrost so that in the Northern Hemisphere a zone of *continuous permafrost* occurs everywhere except under deep lakes in the northern part of the zone; a *discontinuous permafrost* zone occurs with mean annual soil surface temperature between −5 and 0 °C covering between 50 and 90% of the landscape and north of about 55 °N in Canada; and a *sporadic permafrost* zone where permafrost occurs under less than 50% of the landscape and tends to be preserved at increasingly scattered sites such as north-facing slopes or peat bogs. Permafrost has existed in Arctic areas for large parts of the Quaternary and it is estimated that it takes 100,000 years for permafrost to develop to depths greater than 500 m.

Landscapes underlain by permafrost have very distinctive features and processes which depend upon the annual cycle of permafrost melting and freezing (Fig. 26). In the winter the ground is frozen from the surface downwards but in the spring as thawing takes place from the surface downwards the water released cannot infiltrate because the permafrost table (the level below which permafrost does not thaw) provides an impermeable layer. An active layer develops during the spring and summer months, with a thickness typically between 0.6 and 4 m according to air temperature, slope, drainage, soil and vegetation characteristics. Subsequently, in autumn and winter, the active layer freezes from the surface downwards. As water expands by 9% volume on freezing, stresses are produced together with high pore water pressures which can lead to extrusion of water to the surface which will immediately freeze as an icing. If the winter is not severe then freezing may not occur down to the permafrost table so that an unfrozen

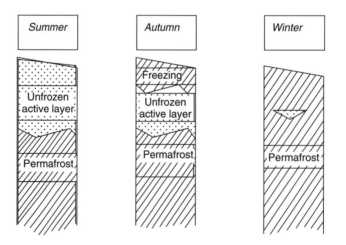

Figure 26 *The annual cycle in permafrost areas. In the summer thawing of the active layer occurs down to the permafrost table; in the autumn the ground freezes from the surface downwards with an unfrozen layer below and above the permafrost; in winter the ground is frozen from the surface downwards although there may be lenses of unfrozen ground (taliks) enclosed in the frozen ground*

layer (talik) persists until the following spring. Repetition of this pattern each year can produce complex vertical patterns of frozen and unfrozen ground.

Such annual changes have a significant influence upon landscape features affected by permafrost including patterned ground, pingos and palsas. They also constrain human activity because to avoid melting the permafrost the ground below buildings has to be insulated by building structures on piles or on a thick gravel pad. Pipelines also have to be insulated so that they do not sink into the permafrost and fracture. Thermokarst phenomena can develop where permafrost is melted.

In addition to annual changes, there can be trends over a number of years and for example in the Yukon the zone of continuous permafrost has moved poleward by 100 km since 1899 and the years since 2000 have produced record thawing of permafrost in Siberia and Alaska. Permafrost thawing as a result of global warming could also release hydrocarbons including methane which are themselves significant greenhouse gases. Recent measurements indicate a rise in permafrost temperatures and although it will take several centuries to millennia for permafrost to disappear completely, in areas which are now actively warming causing thawing or permafrost negative impacts can occur fairly soon because the highest ice content in permafrost usually is found in the upper few tens of metres.

See Map of permafrost distribution under GLACIAL ENVIRONMENTS entry (p. 131).

See also: PERIGLACIAL REGIONS

Further reading

French, H.M., 1996. *The periglacial environment*. Addison Wesley Longman, London.

Pewe, T.L., 1991. Permafrost. In Kiersch, G.A. (ed), *The heritage of engineering geology: The first hundred years*. Geological Society of America, Boulder, Co, 277–298.

Romanovsky, V., Burgess, M. Smith, S. Yoshikawa, K., and. Brown, J, 2002. Permafrost

Temperature Records: Indicators of Climate Change, *EOS, AGU Transactions*, Vol. 83, No. 50, 589–594, December 10, 2002.

The International Polar Year (IPY) is a large scientific programme focused on the Arctic and the Antarctic from March 2007 to March 2009. See website.

SOIL PROFILES

S oil formation begins with the breakdown of rock into regolith. Continued weathering and soil horizon development process leads to the development of the vertical display of soil horizons, which is called the soil profile (Fig. 27).

O Horizon: At the top of the profile is the O horizon, which is mostly organic matter: in a forest this would be mostly leaf litter, on rangeland it would be dead grass and animal faeces. It becomes decomposed and is called humus. The humus transfers nutrients such as nitrogen and potassium, to the soil, aids soil structure and keeps up the levels of soil moisture.

A_1 Horizon: Beneath the O horizon is the A_1 horizon. In this horizon organic material mixes with inorganic products of weathering and is usually dark in colour. If soil water is moving downwards in the profile then there is a transfer of inorganic and organic substances. If water moves up the profile (as in dry climates with irrigated soils) then inorganic materials may move up the profile and cause salinization.

A_2 Horizon: Like the A_1 Horizon, the removal of clay particles, organic matter, and metal oxides (e.g., iron and aluminium) occurs in this horizon. It can also be a level of deposition for upward-moving minerals.

B Horizon: The B horizon accumulates downward moving material. This may be as a dense layer in the soil, which may be a continuous solid horizon or pan. If there is moderate upward movement of minerals in the profile in dry climates then some finer material may crystallize out in this horizon.

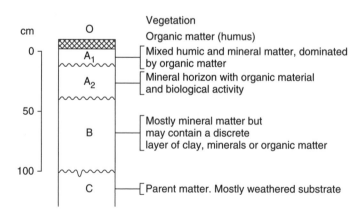

Figure 27 *An idealized soil profile (from I.G. Simmons)*

C Horizon: This is simply weathered substrate such as rock or regolith in 'natural' soils. In soils developing over e.g., mine wastes, it is the basic material which forms the parent material of the soil, just as if it were weathered granite or glacial till.

See also: SOIL SCIENCE, SOIL MANAGEMENT

Further reading

Ashman, M.R. and Puri, G., 2002. *Essential soil science*. Blackwell Publishing, Oxford.

Bridges, E.M., 1997. *World soils*. Cambridge University Press, Cambridge, 3rd edn.

STRATIGRAPHY

Stratigraphy is 'the arrangement of strata, especially as to geographic position and chronologic order of sequence' (AGI Glossary, 1987); stratification is rock layering formed during deposition by changes of some kind in the materials being deposited or in the conditions

Figure 28 *Thick white limestone beds overlain and underlain by thin-bedded shale. These beds have been tectonically tilted after they were originally deposited horizontally. Deep Spring Valley, California. Photo by Arthur G. Sylvester, July 2000*

of deposition, also called 'bedding'; and a stratum is a rock layer so formed (Dunbar and Rodgers, 1954, p. 97), also called a 'bed'.

Stratigraphy is the study and description of layered rocks (Fig. 28). Groups of these kinds of rocks – which are usually sedimentary in nature – are called strata or, less formally, beds. Beds, in turn, are described according to how they differ from adjoining beds (grain size, fabric, composition, primary colour); their shape; their thickness; their lateral extent; their internal structures; and the nature of their contacts with adjacent beds.

Figure 29   *Thick sandstone interbedded with thin siltstone beds. Palm Spring Formation, Mecca Hills, California. Scale given by geologist in lower middle of picture. Photo by Ivar Midkandal, 2004*

Stratification is produced, layer upon layer, by several kinds of changes during deposition that clearly demonstrate the passage of time. The most obvious change is in composition. In Figure 30, the thick beds are limestone, whereas the thin beds are shale. Stratification is also produced by abrupt changes in grain size, as from silt to sand (Fig. 29), caused by changes in sediment supply or change of rock supply at the source. A period of erosion or a pause in deposition also cause differences between two conformable beds. Other textural characteristics, including colour, grain roundness, degree of cementation, and variations in the arrangement or fabric of rock particles may contribute to the stratified aspect of a sedimentary rock sequence.

The lateral extent and regularity of stratification vary greatly. In some varieties of shale, thin beds are remarkably even (Fig. 30), but in others like limestone, beds may pinch and swell or are broken in nodules. The lateral persistence and regularity of stratification reflects the persistence and regularity of the deposition agent. Stream or wind currents that operate in channels or over small areas produce irregular strata that are not persistent, whereas atmospheric winds or deep sea currents distribute sediment evenly and persistently over large areas.

Stratigraphy and the two principles behind it (Table 20) are absolutely fundamental in geologic studies, in that they provide the geologist with a basic reference datum. For example, the geologist is able to assume that rocks s/he sees standing vertically in a road cut were once horizontal, because that is how strata are originally deposited. Then the geologist must determine how those strata came to be vertical. Similarly, if a sequence of beds abruptly terminates within an outcrop, then the geologist may assume that they have been cut by a fault after they were deposited, because the beds had to have been originally continuous or thinned out to a feather edge.

Figure 30 *Even and persistent white sandstone interbedded with gray shale beds, Allegheny Plateau, western Pennsylvania. Photo by Arthur G. Sylvester, 1980*

Table 20 *Stratigraphic principles*

Principle	Definition
Original horizontality	Strata are originally deposited in horizontal layers.
Superposition	In a sequence of layered rocks that has not been overturned, any layer is older than the layer next above (Dunbar and Rogers, 1957, p. 110, with modifications)
Lateral continuity	A stratum maintains its lateral continuity unless it is truncated by a fault or thins out to a feather edge.

See also: GEOLOGY, ROCK TYPES

Further reading

Bally, A.W. (ed), 1987. *Atlas of seismic stratigraphy*: AAPG Studies in Geology No. 27.

Boggs, S. Jr., 2001, *Principles of sedimentology and stratigraphy*, 3rd edn., Prentice-Hall, Inc., NJ.

Dunbar, C.O. and Rodgers, J., 1957. *Principles of stratigraphy*. John Wiley and Sons, Inc., New York.

Winchester, S., 2001. *The map that changed the world: William Smith and the birth of modern geology*. HarperCollins, London.

TEMPERATE ENVIRONMENTS

The Greek intellectual Aristotle (384–322 BC) identified the temperate zone in between the frigid and the torrid zones of the world. As later scientists, especially climatologists and human geographers, have confirmed, this is a zone highly favourable to many human activities and hence has become one of the most transformed parts of the Earth by anthropogenic means. Deforestation, followed by conversion of land to grazing, arable or urban land has affected much of the temperate zone. The zone is defined by the climates present between the tropics of Cancer and Capricorn at 23.5 degrees and the line of latitude at 66.5 degrees, both north and south of the equator, though there is a much greater land mass in the temperate zone of the Northern Hemisphere. In this zone, seasonal climates without great extremes occur (see Climate Classification) and are of two basic types: the maritime and the continental. The latter mostly occur in the centres of the great land masses, but are extended to e.g., the east coast of North America because the western cordillera's north–south configuration cuts off maritime air from the Pacific. In Eurasia, the Alps–Caucasus chains run east–west and allow maritime air further into the continent. The interaction of continental and maritime air masses is not always easily predictable and so the climates of the temperate zone show more variability than their neighbours to north and south. Five sub-zones can be identified (Table 21) each associated with particular combinations of environmental hazards.

Within this regime, during the Holocene, there developed distinguishable suites of soils, vegetation and animal communities. Most temperate zones were either glaciated or affected by periglacial conditions during the Pleistocene, and so recolonization from biotic refugia took place. Typically, a succession of tundra, low scrub, coniferous forest and deciduous forest occurred, with soils to match. Skeletal soils characterized the tundra, highly differentiated podzols underlay the conifers, and a less horizonated brown-earth, with diffused minerals and organic material was found under deciduous forest. This pattern was differentiated by regional variations, as in mountainous regions, around the Mediterranean or in California where adaptation to hot dry summers became the norm. Many of these plant communities were subject to management by fire from hunter-gatherers for some thousands of years, until agriculture supplanted the foragers as the main way of life. Large tracts of land were deforested in Britain before the Romans arrived in the first century BC, and there was a great wave of deforestation in western and central Europe in medieval times. In Mediterranean lands, where the natural vegetation was mixed evergreen and deciduous forest of oaks, pine, beech and cedars, deforestation began as early as 4600 years ago. Temperate North America was wooded from the Atlantic coast as far west as the Mississippi when the first Europeans arrived, but lost more woodland in the following 200 years than Europe had lost in the previous 2000.

Paradoxically although this zone has the imprint of many former climates and environmental conditions, it has often been thought to be the norm against which other landscapes should be considered, and it has also stimulated research which has donated many of the foundations for environmental understanding. Temperate areas inspired the normal cycle of erosion, the importance of the drainage basin, and the basis for hydrological understanding of runoff generation. Contemporary temperate environments, which contain some of the most densely populated parts of the Earth's surface, can be thought of as the domain of rain and rivers operating on landscapes which often contain the legacy of a variety of past environmental processes. In most parts of the temperate zone the energy available to environmental processes is less than in other world areas, although the equivalent of a great increase in energy can be released by human activity. Sediment transfers are fairly slow and storage of sediment on slopes and in flood plains is common, except where human activity, especially by deforestation, has accelerated change.

Many of the zone's soil types proved suitable for rain-fed agriculture, and irrigation proved possible in river valleys, so that the accumulated wealth that led to intercontinental empire-building by Greece, Rome, Spain, France, Portugal and England in pre-industrial times was based here, though a strong cultural impulse not necessarily traceable to physical features was also present. Similarly, the advent of steam-based industrial power took place in this zone and at first spread exclusively within it. Once more, a deterministic interpretation (seasonal climate

Table 21   *Sub-divisions of the temperate environment (Developed from a table in Gregory, 2005). Zones 1–4 were originally forested but have been substantially transformed by human activity. Zones 1 and 5 are transitional to other climatic zones*

Zone (including)	Climate	Surface processes	Major hazards and problems which may occur
1. Zone on Quaternary permafrost (north Canada, Russia)	Severe winters, may be associated with periglacial zone	Permanently frozen ground beneath land surface may be continuous or discontinuous, and is residual from the Quaternary and not forming at present	Wildfires; frost or ice storm; snowstorm; subsidence, windstorms Modification of surface affects thermal regime and can lead to thermokarst features with surface collapse.
2. Maritime zone of middle latitudes (SE USA, NW and Central Europe, China)	Maritime without severe winters. No large seasonal variations in temperature or humidity	Chemical erosion limited by moderate temperatures, some frost action but penetration rarely reaches bedrock. High angle slopes can be stable where still covered by forest	Accelerated erosion, avalanches; soil heave and collapse, floods; landslides on devegetated slopes; coastal erosion Many ancient deposits over landscape. Flooding may increase downstream of vegetation changes
3. Continental zone of middle latitudes (Mid West of USA, Russia)	Severe winters and seasonally distributed precipitation	Heavy showers and snowmelt can produce higher streamflow rates than in zone 2, mechanical processes more important as frost penetration is great and can reach bedrock. Chemical erosion limited by winter frost.	Drought, severe thunderstorms, Hailstorms; snowstorms, landslides when vegetation removed. Downstream flooding increases when vegetation changed and other catchment characteristics altered
4. Mediterranean zone of middle latitudes (southern Europe, California, South Africa, SW Australia)	Seasonal precipitation, mild winters, warm/hot summers. Frost uncommon at low elevations	Alternation of wet and dry conditions can induce landslides. Seasonal streamflow regime can give high seasonal discharges which elevate course debris and rapid dissection and gullying where vegetation removed or degraded	Soil erosion; floods, high spatial and temporal variability; high sediment yields along rivers, earthquakes, volcanic eruptions; landslides and sheet erosion where vegetation removed. Increased flooding downstream, and gully development may occur
5. Subdesert steppes and prairies (Great Plains, S. Russia, Turkey)	Summer rainstorms, dry cold, severe winters	Transitional to temperate deserts with some frost action in winter. Wind action, occasional sheet wash and gullying	Drought; tornadoes; soil erosion; deflation encouraged by removal of vegetation. Gullying where land ploughed

means lively people) is inadequate, especially once buffers between humans and climate become widespread (such as air conditioning) in hotter zones and central heating in the 'frigid' zone. Many of the 'natural' vegetation types, soils, river regimes, slope facets and indeed lower atmosphere gaseous and particulate composition have been altered by human societies in the last 10 ky, but with a marked acceleration after AD 1950. It is also the case that this zone is the largest contributor to the 'greenhouse gases' (notably from China and the USA) which appear to be generating global warming.

See also: CLIMATE CLASSIFICATION, HUMAN GEOGRAPHY, SOIL CLASSIFICATION, BIOMES

Further reading

Gregory, K.J., 2005. Temperate environments. In Fookes, P.G., Lee, E.M. and Milligan, G., (eds), *Geomorphology for engineers*. Whittles Publishing, Dunbeath, Caithness, 400–418.

The millennium ecosystem assessment: http://www.millenniumassessment.org/en/Synthesis.aspx

Simmons, I.G., 2008. *Global environmental history 10,000 BC to AD 2000*. Edinburgh University Press, Edinburgh.

Thomas, W.L. (ed), 1956. *Man's role in changing the face of the earth*. Chicago University Press, Chicago.

TROPICAL ENVIRONMENTS

The tropical zone generally occurs between the Tropic of Cancer (23.3° N) and the Tropic of Capricorn (23.3° S). At the summer and winter solstice, the sun is directly over these latitudes. About 50 million km² of land occurs within the tropics. The tropical zone can be divided between the inner tropics or equatorial zone and the outer tropics or the intertropical zones.

Climate and seasonality are dominated by the interaction of the intertropical low and subtropical high, the east to west dynamics of the tropical atmosphere, and the distribution of land and sea. The outer tropics are characterized by wet and dry seasons with intrusions of temperate weather systems such as cold fronts. Mean annual temperature at sea level is generally >18 °C and seasonal variations in temperature and mean daily solar radiation are small compared to temperate and polar regions.

Tropical climates have been divided into several zones. Wet zones are often nearer the equator and have rainfall in the range of 2000 to 10,000 mm/yr and include montane, superwet, and wet climates. The Monsoon zone is located mainly around the Indian Ocean and has two seasons with pronounced differences in rainfall. The wet-dry zone is located in intertropical and subtropical areas not influenced by Monsoons. The semi-arid zone occurs in intertropical areas where morphological factors (such as mountain ranges) or wind circulation patterns lead to reduced rainfall. Arid tropical zones generally are associated with lower latitude parts of the vast subtropical deserts.

The intertropical convergence zone (ITCZ), the primary factor affecting tropical weather, occurs where the sun is directly overhead and there is maximum heating. This causes high evaporation and rising air masses of the tropical low. As the air rises, it cools, and there is high rainfall. The ITCZ moves between the two tropics during the year. The equatorial zone generally has high rainfall for most of the year, but the intertropical and subtropical zones have wet-dry seasons. The dry air masses move poleward, and descend at around 30 °N and S, leading to the subtropical high and arid conditions in the subtropical desert regions. The north and south intertropical and subtropical zones are affected by strong tropical storms (hurricanes and typhoons). Because of this and rains associated with the ITCZ, there is high freshwater runoff and associated sediment input to the coastal ocean in the tropics.

The distribution of tropical rainforests, mangroves, and coral reefs fall largely between the two tropics. The highest biodiversity in the world occurs in the tropics. The types, abundance and productivity of vegetation are strongly influenced by precipitation. In montane climates, high-diversity montane rainforests occur. In wet climates, there are high-diversity lowland rainforests. Evergreen seasonal forests grow in wet seasonal climates. In wet-dry climates, there are semi-evergreen and deciduous seasonal forests and savannas and thorn forests in semi-arid regions.

Tropical coastal ecosystems include mangrove forests, submerged aquatic vegetation, and coral reefs. Many of the world's largest deltas occur in the tropics. In the outer tropics, coastal systems are generally smaller and coastal lagoons often close during the dry season. In arid tropical areas, the intertidal zone is often composed of salt flats. Coral reefs occur in high salinity coastal areas.

See also: ECOSYSTEM, WETLANDS

Further reading

Day, J.W, Hall, C.A.S., Kemp, W.M. and Yáñez-Arancibia, A., 1989. *Estuarine ecology.* Wiley, New York.

Garrison, T., 2007. *Oceanography an invitation to marine science.* Thomson Brooks/Cole, Belmont, CA., 6 edn.

Longhurst, A.R. and Pauly, D., 1987. *Ecology of tropical oceans*. Academic Press Inc., San Diego, CA, 407 pp.

Mann, K.H., 2000, *Ecology of costal waters with implications for management*, 2nd edn., Blackwell Science Inc., Malden, Massachusetts, 406 pp.

Mitsch, W.J. and Gosselink, J.G., 2000. *Wetlands*. Wiley, New York. 3 edn.

Osborne, P. L., 2000. *Tropical ecosystems and ecological concepts*. Cambridge University Press, Melbourne, Australia, 464 pp.

Tapia Garcia, M. (ed.), 1998. *The gulf of tehuantepec: The ecosystem and the resources*. UAM-Iztapalapa, UAM Press Mexico DF.

Valiela, I., 1995. *Marine ecological processes*. Springer-Verlag, New York, 2 edn.

Yáñez-Arancibia, A. (ed.), 1985. *Fish community ecology in estuaries and coastal lagoons: Towards an ecosystem integration*. UNAM Press Mexico DF.

Yáñez-Arancibia, A. and Day, J.W., 1988. *Ecology of coastal ecosystems in the Southern Gulf of Mexico*. Organization of American States, Washington, DC. UNAM Press Mexico DF.

Yáñez-Arancibia, A. and Lara-Dominguez, A.L. (eds), 1999. *Mangrove ecosystems in tropical America*. Instituto de Ecologia A.C., IUCN World Conservation Union, NOAA National Marine Fisheries Service, INECOL Xalapa Press Mexico.

TROPOSPHERE

The word troposphere stems from the Greek 'tropos' for 'turning' or 'mixing'. Troposphere is the lowermost portion of Earth's atmosphere. It is the densest layer of the atmosphere and contains approximately 75% of the mass of the atmosphere and almost all the water vapour and aerosol. The troposphere extends from the Earth's surface up to the tropopause where the stratosphere begins. The region of the atmosphere where the lapse rate changes from positive (in the troposphere) to negative (in the stratosphere), is defined as the tropopause. The tropopause is the boundary region between the troposphere and the stratosphere. By measuring the temperature change with height through the troposphere and the stratosphere, one identifies the location of the tropopause. In the troposphere, temperature decreases with altitude. In the stratosphere, the temperature increases with altitude. The depth of the troposphere is greatest in the tropics (about 16 km) and smallest at the poles (about 8 km). The troposphere is the most turbulent part of the atmosphere and is the part of the atmosphere in which most weather phenomena are seen. Jet aircraft fly just above the troposphere to avoid turbulence. The pressure of the atmosphere is highest at the surface and decreases with height. This is because air at the surface is compressed by the weight of all the air above it. At higher levels the

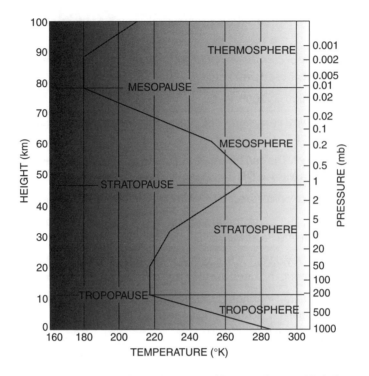

Figure 31   *Earth's atmosphere in cross-section, showing the position of the Troposphere, roughly the lowest 1 km of air. (http://www.metoffice.gov.uk/education/secondary/teachers/atmosphere.html#main)*

weight of the air above is smaller, so the air is compressed less and has a lower pressure. In the troposphere, the temperature decreases with height at an average rate of 6.4 °C for every 1 km increase in height. This decrease in temperature is caused by adiabatic cooling – as air rises, the atmospheric pressure falls and so the air expands. In order to expand, the air temperature decreases (due to conservation of energy). Temperatures decrease at middle latitudes from approximately. +17 °C at sea level to approximately. −52 °C at the beginning of the tropopause. At the poles, the troposphere is thinner and the temperature only decreases to−45 °C; whereas at the equator, the temperature at the top of the troposphere can reach −75 °C (Figure 31).

See also: ATMOSPHERE, ATMOSPHERIC SCIENCES, EARTH SPHERES

Further reading

http://www.metoffice.gov.uk/education/secondary/teachers/atmosphere.html#main

URBAN ENVIRONMENTS

According to the World Health Organization (WHO) in 1990 an urban area can be defined as a 'man-made environment encroaching and replacing a natural setting and having a relatively high concentration of people whose economic activity is largely non-agricultural' whereas the US Census bureau definition is areas with more than 386 people per km². Why are environmental scientists interested in such urban environments?

First because so many people live in them. Just 2% of the world's surface is built up, but since 2000 these areas include >50% world's population, compared with just 14% in 1920 and 2% in 1850; it is estimated that they could include 60% by 2025. Some countries such as Australia, New Zealand, Singapore and the UK, already have >80% urban populations. In 2000 there were 60 cities of 5 million or more inhabitants, all consuming quantities of energy (electricity, transport fuel), of food, materials and land, generating waste, and requiring investment in buildings and infrastructure. The 'ecological footprint' represents the environmental impact of cities in terms of the amount of land required to sustain them and, if the Earth's population consumed resources at the same rate as does a typical resident of Los Angeles, it would require at least three planet Earths to provide all the energy they required.

Second because environmental processes are much modified as a result of urbanization: the modified climate of London was first recognized by L. Howard in 1833 and it is now known that, compared to surrounding rural locations, cities of the world disrupt the climatic properties of the surface and of the atmosphere, altering the exchanges and budgets of heat, mass and momentum with a range of effects (Table 22). Localized effects of urban climate tend to be merged above roof level, forming the urban boundary layer (UBL), and below roof level is the urban canopy layer (UCL). As more heat is used to warm the air and the ground in urban environments, the relative warmth of the city provides its urban heat island and a large city is typically 1.3 °C warmer annually than the surrounding area, although the heat island varies diurnally, with up to 10 °C difference near midnight. Hydrologically the extensive impervious areas of urban environments reduce surface storage so that infiltration is not possible and evapotranspiration is much less than in rural areas. Increased amounts of surface runoff are complemented by the surface water runoff system which collects water from roads and roofs. Stream discharges from urban areas tend to have higher peak flows and lower base flows, and the flood frequencies of rivers draining urban areas will be significantly changed from the time before the urban area existed. Urban areas also generate characteristic water quality with water temperatures often higher than those of rural areas, higher solute concentrations reflecting additional sources including pollutants, and suspended sediment concentrations high during building activity but lower after urbanization when sources are no longer exposed (Table 22). Urban ecology is changed as a result of exterminations and introductions: although once thought of as ecological deserts, urban environments are now known to support a variety of plant and animal species, including songbirds, deer and fox, and an increasing number of North American metropolitan areas now have to contend with large predators including

Table 22 *Effects of urban areas on some aspects of environment*

Atmosphere	Hydrology	Morphology and soils	Ecology
Increased:	*Increased*:	*Increased*:	*Increased*:
Temperature (0.5° to 4.0°C giving heat island);	Magnitude and frequency of floods; annual surface runoff volume; stream velocities;	Impervious areas; Soil compaction; Subsidence (draining of aquifers shrinks building foundations);	New habitats (e.g., parks, backyards, reservoirs); Introduced species (ornamental and exotic plants, pets,
Rainfall (5–15% more especially in downwind areas, due to localized pollutants acting as condensation nuclei);	Sediment pulses; pollutant runoff; nutrient enrichment and bacterial	Soil pollution (with waste material and industrial pollutants);	street trees, e.g., Norway maple) 'Weed' species
Cloudiness (+5–10%); atmospheric instability (by 10-20%, caused by surface and near-surface heating; increased turbulence from rougher city surface);	contamination; toxins, trace metals, hydrocarbons; Water temperature; Debris and trash dams;	Soil erosion (during building construction); *Decreased*: Soil erosion (in built up urban areas)	Some birds become more dominant (pigeons, starlings, sparrows) *Decreased*: Natural vegetation;
Thunderstorms (10–15% more frequent) PE and transpiration rates; air pollution (by 10 fold);	*Decreased*: Baseflow; Infiltration;	*Changes in*: Topography and landforms (including accumulation of	Native animal species; Wetlands; riparian buffers and springs
Dust (+1000% relative to rural, SO$_2$ + 500%, CO + 2500);	Bank erosion (when sediment sources protected) pool riffle structure (removed in channelized reaches);	materials, extraction of building materials); Erosion rates (greatly increased during building constriction,	*Changes in*: Plant growth (light levels reduced, dust greater, pollutants greater); Animal behaviour (cease
Decreased: Solar radiation (up to 20% less);	Ground water reserves (pumping for water supply); Aquatic life in rivers	but may be decreased when urbanized);	hibernation); Diversity of aquatic insects;
Relative humidity (5–10% lower); Wind speeds (20–30% lower);	*Changes in*: Channel capacity; Sedimentation		Diversity and abundance of fish; Remaining natural areas
Changes in: Thermal circulation (analogous to sea/land breeze with 'country' breezes converging on the city centre, then rising and diverging to form counter flow);	(aggradation if sediment supply large; scour if sediment transport lower)		(become refuges for plant and animals)
Fog (+30 – +100%)			

alligators, coyotes, pumas and bears. Such changes are summarized in Table 22 and you may add more.

A third reason for interest is the opportunity to determine future urban environments. Cities have been described as a karst topography, with sewers performing precisely the function of limestone caves in Yugoslavia, which causes a parched physical environment, especially in city

Table 23   *Some methods for shaping urban environments*

	Urban climate	Hydrology	Morphology and soils	Ecology
Mitigation and minimization of impacts	Control programmes for atmospheric emissions;	Flow velocity reduction; Channelization; Floodplain levees; Land use zoning; Diversion channels for water from construction sites; Recharge of aquifers; Land use regulation; Flood insurance;	Minimize loading effects of buildings; Minimize exposure of bare soil (temporary ground cover – geotextiles, mulches, plastic sheeting); Control water and sediment on building sites; Landslide protection measures; Insulating procedures in permafrost areas; Protection against salt weathering in drylands	Reduce pollution (to encourage organisms to return); Remove undesirable species; Reduce use of fertilizers and pesticides;
Design aspects	Air conditioning; Tree planting (to augment tree biomass and diversity and influence atmospheric environment)	Stream restoration (e.g., Urban Streams Restoration Act in California 1984); SUDS (sustainable drainage systems – militate against flooding and pollution); Minimize connections between impervious surfaces; Local storage such as rain gardens in each garden/backyard to reduce runoff	Slope stabilization and design; Integrated planning for river corridors; Include soil as integral component of park planning and management;	Street tree planting; Create urban nature reserves; Establish conservation areas and natural reserves; Botanical gardens and parks; Golf courses, Public gardens, Backyards; Window boxes

centres, or as ecosystems which embrace population ecology, system ecology, the city as a habitat, and energy and material transfer within cities. The European Commission (18 February 2004) revealed plans to improve environmental aspects of towns and cities with a new EU-wide strategy which aims to provide a 'best practice' style approach with successful projects implemented on a widespread basis across the Union.

People created the urban environment but for many a more natural setting is preferred – reflected in window boxes, botanical gardens, natural reserves, golf courses and public parks. Typically at least 30% of urban areas remain vegetated. Planning urban environments can involve:

1   mitigating or at least minimizing some of the problems – of climate, hydrology, ecology (Table 23). Maintaining natural areas within urban environments is essential for both survival of resident plants and animals and for well-being of inhabitants. Wise urban planning is based on knowledge of dynamics and functioning of urban ecosystems; ideally urban natural areas should be not only diverse functionally and aesthetically appealing, but also self-perpetuating ecosystems that require minimal maintenance and are sustainable (Table 23).

2   creating environment characteristics that people want, by taking initiatives (e.g., Table 23). Whereas nature used to be seen as a machine which could be engineered to provide the maximum output of desired products, this was challenged in the mid–late 20th century by a new attitude to environment, with scientific and political concern about ecological, economic and cultural sustainability of monocultural resource systems. This is exemplified by rivers and in 1994 D.M. Bolling concluded that 'Where cities once exploited, abused and then ignored rivers in their midst, they are now coming to recognise, restore and appreciate them'. In attempting to make urban environments more natural do we fake nature or endeavour to recreate it? Although inappropriate for the Australian urban environment, many gardens in Australia originally imitated the British style and not until a new Nationalism was born in the 1950s, were lawns replaced by bush-floor effect with promotion of indigenous plants and maintenance of the bush garden ethos. In some cases densification of urban environments is said to provide a solution to local scale environmental problems such as traffic and air pollution but it may intensify problems in others. Planning should be specific to the particular urban environment.

See also: ATMOSPHERIC SCIENCES, CULTURES, ENVIRONMENTAL MANAGEMENT

Further reading

Bolling, D.M., 1994. *How to save a river. A handbook for citizen action*. Island Press, Washington, DC.

Burch, W.R. Jr., 1999. Introduction. In Aley, J., Burch, W.R., Conover, B. and Field, D. *Ecosystem Management: Adaptive strategies for Natural Resources Organisations in the Twenty First Century*. Taylor and Francis, Philadelphia.

Douglas, I., 1983. *The urban environment*. London, Arnold.

Emery, M., 1986. *Promoting nature in cities and towns.* Croom Helm, Dover, NH.

Ford, G., and Ford, G., 1999. *The natural Australian garden.* Bloomings Books, Hawthorn, Victoria.

Gordon, D. (ed), 1990. *Green cities: ecologically sound approaches to urban space.* Black Rose, New York.

Grimmond, S., 2006. Urbanization and global environmental change: local effects of urban warming. *Geographical Journal* 173, 83–88.

Howard, L., 1833. *The climate of London.* Hutchinson, London.

Knowles, R.L., 1974. *An ecological approach to urban growth.* MIT Pres, Cambridge, Mass.

Landsberg, H.E., 1981. *The urban climate.* Academic Press, New York.

Oke, T.R., 1987. *Boundary layer climates.* Routledge, London. 2nd edn.

Wackernagel, M. and Rees, W., 1995. *Our ecological footprint: reducing human impact on the Earth.* New Society Publishers, Gabriola Island, BC.

White, R., 1994. *Urban environmental management: environmental change and urban design,* Wiley, Chichester.

VEGETATION TYPES

The word 'vegetation' is not a precise term. It is more inclusive than 'flora', which is confined to the taxonomic varieties found at a particular place, but does not take in the data about climate and soils which are found in the definitions of biome. The term is spatially undefined and can legitimately be applied to e.g., the whole of the tropical forests, a patch of relict woodland in an agricultural landscape, a field of soy beans or even a lawn. It is no surprise that there is no one system of classification of vegetation types. In Europe, for instance, the dominant species tends to give its name to the vegetation type, without any mention of climate: oak woodland/beech woodland/birch woodland might be found in any topographical guide to the types. Classifications which recognize the role of domestic escapes or of weed species are also common in lands without great areas of wild terrain such as the British Isles. In North America, a classification which is truly hierarchical (as with soil classification) is being adopted in official circles: this goes down from climatic zone, through plant habit and growth form (tree, shrub, herb etc.) to the dominant species. Both types of classification can be used of 'natural' vegetation as well as of wild or semi-wild areas with distinct human influence.

The architecture of vegetation is another key to its description and classification. Intuitive classes such as forest, grassland, cropland, desert or even waste land are used and these depend very often on plant habit, which is the form of the mature plant (e.g., herb, low shrub, cushion cactus) and its foliage (e.g., deciduous tree, evergreen shrub).

All descriptions and classifications have to acknowledge that they may be snapshots in time of a dynamic system. Vegetation changes through two main influences. The first is endogenous change, when a biotic community creates the conditions for its own replacement with a later flora and fauna (Fig. 32). Thus on bare rock after glaciation, bacteria, mosses and lichens gradually create enough loose rock and humic material for flowering plants, which in turn provide a habitat for shrubs and then trees. This is called succession. It contrasts with exogenous change which is forced upon an ecosystem by external events. The blanketing of a forest by volcanic ash is one example, or the drowning of a reclaimed area by the breakdown of a sea wall. Many human influences exert exogenous pressures on plant communities and indeed may provide a bare surface from which endogenous succession may start: mining waste-heaps provide an example.

Figure 32   *A forest in southern Bohemia (Czech Republic) showing a dominant layer of long-lived forest trees (pine and beech) and a ground layer of herbaceous plants. Because this forest has been unmanaged for some time, fallen trees are also part of the ecosytem. A forest with dead trees and small clearings is probably like those areas which long ago had little or no human impact (photograph, I.G. Simmons)*

See also: BIOMES, ECOSYSTEMS, SOIL CLASSIFICATION, VEGETATION CLASSIFICATION

Further reading

Impacts on vegetation at http://ww.maweb.org/en/framework.aspx

Collinson, A.S., 1988. *Introduction to world vegetation*. 2nd edn. Unwin Hyman, London.

Begon, M., Townsend, C.R. and Harper, J.L., 2006. *Ecology from individuals to ecosystems*. Blackwell, Oxford.

PART III

PARADIGMS/CONCEPTS

INTRODUCTION

The science fiction writer Isaac Asimov (1920–1992) is often cited as saying that 'nature is not divided into departments the way universities are'. Sometimes the analogy of a cake is used: the cosmos is a whole and it is we who decide how it is to be divided up in order to try to understand it or to manipulate it. Some divisions appear to be obvious and 'natural': rocks are clearly different from the atmosphere, for example; living organisms like plants, animals and bacteria are different from non-living entities like water. But when we get away from a simple classification based on appearance and want to understand how things work, their relationships turn out to be very important: the atmosphere helps to break rocks down into smaller particles and hence is important in soil formation; an excess of rainfall makes for landslides; without taking in water, living things would die. So the dynamics of the planet (and beyond) depend upon us finding categories that make sense of the fluxes and flows of the world around us.

The ways in which environmental sciences approach these tasks are by no means uniform but have certain common features, which lead us to use words like 'concept' and 'paradigm'. A concept is a general notion or idea, which therefore applies to more than one instance of a phenomenon. It must therefore move away at least one step from an initial observation: 'I see a bird from my window and it flies' is an observation; 'all birds fly' is a concept, and in this case a mistaken one which needs more observation to establish its veracity – or lack of it. Behind most, if not all, science there is the concept that the cosmos is not a random and disorganized place. Rather, as the original Greek implies, it is a place with order and therefore can be understood, whereas if it were completely chaotic we would have no chance of understanding it. The idea that it hangs together causes some people to think it must have a designer, but since Charles Darwin (1809–1882) the concept of the cosmos evolving in a self-organizing way has been available for scientific consideration.

One of the characteristics of the sciences is that they are cumulative: their knowledge builds up as time progresses, whereas in the humanities each generation can rework its understandings anew. So while Europeans knew about the giraffe in mediaeval times, they now know immensely more about that species' physiology, habits, habitats and relationships with other species. That knowledge is stored in print and in electronic form and is available to the next generation of mammalogists and wildlife researchers. This availability means that knowledge is continually being tested and that progress consists not only of adding more observations but of improving an understanding of the way in which the various components of any natural system interact, both now, in the past and, possibly, in any future scenario to be imagined. We write about science as if it were a kind of abstract practice or even ritual, in which 'scientists' move about their work clad in white coats. But they work within a social framework, since they are human. They exercise choice about what to investigate in the first place; they are constrained by what it costs to carry out their work or by the time they have available from other aspects of their work (like teaching, in universities, for example), by problems of access (both upper atmosphere and deep ocean floors present difficulties) or by security considerations. Some will work on secret military projects whereas others are prevented by their conscience from doing so. Some science lies unappreciated in drawers or on disks because the investigator never got round to writing it up or because it was rejected for publication. The whole bundle of practices within which a science happens is called a paradigm and it is not forever fixed. At any one time there may be an accepted wisdom and an agreed set of procedures but at some moment these will change: a paradigm shift. So the categories of sciences and classifications of findings in this book are representative of current concepts and paradigms but do not expect that they will stay that way for long into the future (Fig. 33).

A means of understanding the land surface of the earth, probably influenced by Darwin's ideas on evolution, was provided by the proposal by William Morris Davis of the cycle of

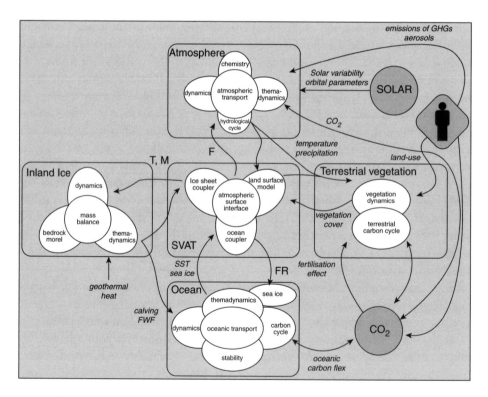

Figure 33 *A model of intermediate complexity of the interaction between the main Earth systems and human activity. Based on the CLIMBER model developed by the International Geosphere-Biosphere Programme (IGBP), it is one of a class of models where sufficient reliable data are being steadily garnered to allow realistic simulations to be made of Earth conditions under various scenarios. From IGBP Science No. 4 Global Change and the Earth System, p. 26. Published with permission from Prof. Dr. Vladimir Petoukhov*

erosion in 1895. This dominated thinking for more than half a century, but was then succeeded by investigations of processes, later complemented by realization of the significance of human impacts, of clearer documentation of stages of evolution of environments in the past and then realization of how knowledge could be applied to contemporary environmental problems. Although the Davisian approach is now largely of only historical interest, the environmental sciences would not have advanced as well without its formulation. George Bernard Shaw's statement in *The Doctor's Dilemma* that 'Science becomes dangerous only when it imagines it has reached its goal' remains as apposite now as in 1911 when it was written, reminding us that knowledge of concepts and paradigms that have prevailed in the past provide the building blocks for sciences of the future and that we should strive to fashion new concepts and paradigms. Furthermore it is the contribution by individuals, originating paradigms and concepts, that have been instrumental in advancing the environmental sciences. As Mark Twain (1835–1910) in *Life on the Mississippi* (1883) observed 'There is something fascinating about science. One gets such wholesale returns of conjecture out of such a trifling investment of fact'. Long may conjecture exist to stimulate further progress of the environmental sciences.

BIOCOMPLEXITY

Natural and environmental phenomena that are multi-scale and inclusive of human dimensions are considered under the umbrella of biocomplexity. Biocomplexity refers to the complex interactions of organisms with their environments and the complex nature of responses, which are often mediated through biological processes and interactions, and the expression on biodiversity. Biodiversity can refer to the diversity of species, populations, ecosystems, life histories, habitats, links in food webs, or the diverse pathways of energy flow and nutrient cycles. Biodiversity components refers as an ecological reference to these hierarchical levels, and the idea is appropriate to describe biocomplexity of natural ecosystems. Biocomplexity provides a framework or perspective of the consequences of natural and human-induced impacts on ecosystems. An example of this is the effect of climate change for coastal wetlands. The diverse landforms of coastal regions can be considered as a biodiversity component of coastal wetland ecosystems, and they are susceptible to a changing climate. Sea-level rise, storms, and changing freshwater input can directly influence coastal wetlands.

However, some physical outcomes will depend heavily on biological processes such as above ground baffling of currents by plant communities and subsequent sedimentation, below ground plant productivity, organic soil formation by wetland plants, and activities of resident communities. Thus, feedbacks between the biotic and abiotic components of ecosystems will affect the response of coastal systems to climate change. Furthermore, geomorphologic changes to coastal and estuarine ecosystems can induce complex outcomes for the biota that are not intuitive due to biological interactions. Human impact can greatly increase the impact of climate change. For example, freshwater diverted from reservoirs and not delivered to the coast can reduce the ability of coastal systems to survive sea-level rise.

See also: BIODIVERSITY

Further reading

Michner, W.K., *et al.*, 2001. Defining and unraveling biocomplexity. *BioScience* 51, 1018–1023.

Morris, J.T., *et al.*, 2002. Responses of coastal wetlands to rising sea level. *Ecology* 83, 2869–2877.

Ray, G., and McCormick, M.G., 1992. Functional coastal-marine biodiversity. *Transactions of the 57th North. American Wildlife and Natural Resources Conference*, 384–397.

BIODIVERSITY (FUNCTIONAL BIODIVERSITY OF TROPICAL MARINE SYSTEMS)

Biodiversity is usually defined at three levels, i.e., genetic, species/population, and ecosystem. The biodiversity of coastal environments is generally high in all respects; however information on tropical lagoon-estuarine ecosystems is usually not as detailed as required, given the variety of responses to environmental changes, the diverse space and timescales with which organisms interact with their habitats, and the general lack of information on the areas as compared with temperate systems. The diverse biological, ecological and physical interactions which occur within tropical estuaries and the adjacent ocean produce a highly dynamic and variable mosaic of ecosystems.

From an ecological standpoint, the term biodiversity can have several meanings when applied to specific ecosystems, i.e., tropical lagoon-estuarine ecosystems. It can mean that there is a high diversity of species, or that there is a high diversity of environmental factors, habitats, connections in the food webs, and a high diversity of couplings, both internally and with neighbouring systems. Also the diversity of forcing functions effectively modulating ecosystem functioning is high, and includes wind, tide, river flows, littoral currents, sediment input, and others. Moreover there is a high diversity among primary producers and consumers, many with different types of life history.

Grassle and Lasserre have both pointed out that large-scale experiments are needed on whole system responses and on carefully selected tropical sites in order to identify a new paradigm for understanding coastal-marine functional biodiversity. More information is needed on tropical coastal zones for their complexity and biodiversity to be understood. In tropical coastal demersal communities the structure and function of multi-species fish stock are characterized by 'functional ecological groups'. In a multistock context, functional ecological groups have been defined by: (1) species redundancy; (2) appearance frequency; (3) dominance in biomass and number of individuals; and (4) persistence in multi-annual cycles. They correspond to fish population assemblages with similar behaviour that play an analogous role or determined function in the ecosystem. They are key in the structure and the dynamics of the community, controlling the biomass flows and the functional role of the biodiversity in the ecosystem, but also by representing themselves important fishery resources in the continental shelf. The number of functional ecological groups in a ecological system depends principally on: (1) the spatial heterogeneity; (2) the environmental variability; (3) the total number of species in the community; (4) the ability of the populations to recover from a decreased density in the population, resulting from natural mortality or from fishing; (5) environmental disturbance; (6) fluctuations of population size; (7) primary productivity levels.

See also: ECOSYSTEM, PRODUCTIVITY

Further reading

Day, J.W. and Yáñez-Arancibia, A., 1982. Coastal lagoons and estuaries: ecosystem approach. *Ciencia Interamericana* 22 (1–2), 11–25.

Grassle, J.F., Lasserre, P., McIntyre, A.D. and Ray, G.C., 1991. Marine biodiversity and ecosystem function: A proposal for an international program of research. *Biology International, Special Issue* 23. IUBS, SCOPE, UNESCO, 19 pp.

Lasserre, P., 1992. The role of biodiversity in marine ecosystems. In Solbrig, O.T., van Emden, H.M. and van Oordt P.G. (eds), *Biodiversity and global change*, International Union of Biological Science, Monograph 8, Paris, 105–130.

Longhurst, A. and Pauly, D., 1987. *Ecology of tropical oceans*. Academic Press Inc., San Diego, California.

Mitsch, W. J. and Gosselink, J. G., 2000. *Wetlands*. Wiley, New York, 3rd edn.

Pauly, D. 1986. Problems of tropical inshore fisheries: Fishery research on tropical soft bottom communities and the evolution of its conceptual base. In Mann Borgesse, E. and Ginsburg, N. (eds), *Ocean year book 6*. The University of Chicago Press, London, 29–54.

Ray, G. C., 1991. Coastal-zone biodiversity patterns. *BioScience* 41, 490–498.

Rojas Galaviz, J.L., Yáñez-Arancibia, A., Day, J.W. and Vera, F., 1992. Estuarine primary producers: Laguna de Terminos a study case. In Seeliger, U. (ed), *Coastal plant communities of Latin America*. Academic Press, New York, 141–154.

Yáñez-Arancibia, A., Sánchez-Gil, P. and Lara-Domínguez, A.L., 1999. Functional groups and ecological biodiversity in Terminos Lagoon, Mexico. *Revista Sociedad Mexicana Historia Natural*, 49, 163–172.

BIOGEOCHEMICAL CYCLES

Biogeochemical cycling is the cycling of elements that is affected by a combination of biological, chemical, and geological processes. This cycling is a result of inputs to an ecosystem and the internal cycling within the ecosystem. Inputs can come from rain, runoff, dust, and other materials. A major input to the biogeochemistry of the ocean, and thus of the earth, is the addition of materials to ocean water at the mid-ocean ridges. Chemicals cycle internally in ecosystems, passing through air, water, soil, and organisms. Research on biogeochemical cycling often focuses on the 'big six' elements that are the basic building blocks of life. These are carbon, nitrogen, phosphorous, hydrogen, oxygen, and sulphur. Biological activity greatly affects biogeochemical cycling of carbon and other materials through the uptake and release of materials and by converting inorganic chemicals into organic chemicals. Upon death and decomposition, partially decomposed organic chemicals cycle, sometimes accumulating over the long term, as organic fossil material such as oil and coal. The nitrogen cycle is complex and important because of the critical role that nitrogen plays in living organisms (i.e., proteins and DNA). Phosphorus is also important because it is critical to life and is an important component of many rocks. Like nitrogen, excessive concentrations of phosphorus can lead to water quality problems. Factors such as rainfall, rates of biological activity, and soil type affect biogeochemical cycles. There are a number of important biogeochemical cycles that are important ecologically and to humans. These include carbon, sulphur, nitrogen, phosphorus, and silicon. Humans have strongly affected this cycling by extraction, addition of materials to ecosystems (such as nitrogen fertilizer), and by the addition of toxins such as heavy metals and pesticides. More detail on biogeochemical cycling is found in the entries for individual cycles.

See also: CARBON CYCLE, CYCLES, NITROGEN CYCLE

Further reading

Bianchi, T., Pennock, J. and Twilley, R., 1999. *Biogeochemistry of the Gulf of Mexico estuaries*. Wiley, New York.

Botkin, D.B. and Keller, E.A., 2007. *Environmental science – Earth as a living planet*. Wiley, New York.

Day, J.W, Hall, C.A.S., Kemp, W.M. and Yáñez-Arancibia, A., 1989. *Estuarine ecology*. Wiley, New York.

Mitsch, W.J. and Gosselink, J.G., 2000. *Wetlands*. Wiley, New York, 3rd edn.

Nriagu, J. (ed). 1976. *Environmental biogeochemistry*. Ann Arbor Science, Ann Arbor MI.

BIOMASS

Biomass is a measure of the amount of organic matter on Earth or some part of it. Biomass can be expressed as the total amount of organic matter of a resource, or more commonly on a unit area basis. For example, one can talk of the total forest biomass of the Earth or the Brazilian rainforest in metric tonnes. But biomass are normally given per unit area, such as gm^{-2}, $Kgha^{-1}$, or tonnes km^{-2}. Biomass is dramatically different for different ecosystems (Table 24). For example, the biomass of tropical rainforests averages about 45 kgm^{-2}, while that of a temperate grassland is about 1.6 kgm^{-2}, that of a salt marsh is about 2.5 kgm^{-2}, and that of open ocean surface waters is 0.003 kgm^{-2}. This great difference is because there is a great deal of structural organic matter in a tree, less in grass, and very little in oceanic phytoplankton. Biomass is generally expressed on a dry weight basis to correct for varying water content in different kinds of organic matter. Biomass is increased through biological production and decreased through consumption and decomposition.

It is clear that the relationship between biomass and biological production is not constant. For example, the rainforest and salt marshes mentioned above have about the same net production of organic matter even though the biomass of the forest is about 20 times higher than the marsh. The relationship between production (P) and biomass (B) is called the P/B ratio. This ratio varies for different ecosystems depending on productivity and biomass values. For a forest, the P/B ratio decreases over time because in a young forest the biomass increases rapidly. This is because as the forest matures, total biomass reaches a steady state maximum. A biomass pyramid normally results if the biomass of organisms in an ecosystem is arranged by trophic or feeding level. For example, in a healthy grassland, there may be 1000 $g.m^{-2}$ during the summer. Herbivores grazing on the grass might be 100 $g.m^{-2}$, while primary carnivores eating the herbivores may average about 10 $g.m^{-2}$ and top carnivores 1 $g.m^{-2}$. These numbers are for illustrative purposes, but are generally representative of many ecosystems. Thus, most terrestrial ecosystems have a biomass pyramid where biomass is highest for the plants. In many

Table 24 *Estimates of Net Primary Production (NPP) and Biomass for major ecosystems*

Ecosystem	NPP in G m^{2}yr^{-1}	Average biomass in Kg m^{-2}
Tropical rainforest	2200	45
Temperate evergreen forest	1300	35
Temperate deciduous forest	1200	30
Boreal forest	800	20
Savanna	900	4
Temperate grassland	600	1.6
Tundra	140	0.6
Desert	90	0.7
Swamp and marsh	2000	15
Lake and stream	250	0.02
Open ocean	125	0.003
Upwelling zone	500	0.02
Estuaries	1500	1

lakes and oceanic systems, the biomass of phytoplankton, the major primary producer, is often low even if there is relatively high productivity. In this case the base of the biomass pyramid can be smaller than that of the biomass of herbivores. Biomass production by plants is called primary production while the production by animals is called secondary production.

See also: ECOSYSTEM, NATURAL PRODUCTION MECHANISM, PRODUCTIVITY

Further reading

Botkin, D.B. and Keller, E.A., 2007. *Environmental science – earth as a living planet*. Wiley, New York.

Brewer, R., 1994. *The science of ecology*. Saunders, Forth Worth, TX and London, 3rd edn.

Whittaker, R.H., 1975. *Communities and ecosystems*. Macmillan, New York, 2nd edn.

CATASTROPHE THEORY

A series of mathematical forumulations which can be used to account for sudden shifts of a system from one state to another as a result of the system being moved across a threshold condition. In mathematics, catastrophe theory was introduced in the 1960s by Rene Thom (1923–) to signify those sudden changes, or jumps, which occur after relatively smooth progress. Such sudden shifts in behaviour arising from small changes in circumstances, constitute a special branch of dynamical systems theory. Catastrophe theory is generally considered as a branch of geometry because the variables and resultant behaviour are usefully depicted as curves or surfaces, with seven fundamental types, with the names that Thom suggested (fold catastrophe, cusp catastrophe, swallowtail catastrophe, butterfly catastrophe, hyperbolic umbilic catastrophe, elliptic umbilic catastrophe, parabolic umbilic catastrophe). Applications have been found in ecology, biology, physics, climate change, psychology and economics.

Catastrophe theory is useful in the environmental sciences because small changes in certain parameters affecting a non-linear system can cause equilibria to appear or disappear, or to switch from repelling to attracting, giving rise to large changes in the behaviour of the system. Catastrophes are bifurcations between different equilibria, or fixed point attractors. Due to their restricted nature, catastrophes can be classified according to how many control parameters are being simultaneously varied. For example, if there are two controls, then one finds the most common type, called a 'cusp' catastrophe, which is the one that has been most frequently used in the environmental sciences (Fig. 34). In fluvial geomorphology, emphasis upon change provides opportunity for the use of catastrophe theory which can be especially useful for expressing theories related to change in geomorphological systems and for analysis in terms of force and resistance. Appropriate problems include issues related to channel patterns, arroyo development, sedimentation at channel junctions, ephemeral stream processes and sediment

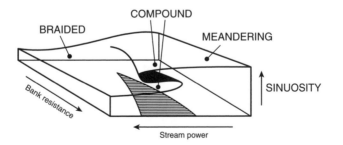

Figure 34 *An example of the cusp catastrophe applied to various channel patterns by relating the control factors of stream power and resistance with the responding variable of sinuosity (After Graf, 1988, Copyright John Wiley & Sons Limited. Reproduced with permission.)*

dynamics at the basin scale. The concept of equilibrium has pervaded geomorphic research, but because some geomorphic systems rarely achieve equilibrium catastrophe theory affords a basis for theory building. It also provides an alternative to thresholds defined as dividing lines by describing zones of transition where two states of equilibrium are possible.

See also: EQUILIBRIUM, THRESHOLDS

Further reading

Graf, W.L., 1988. Applications of catastrophe theory in fluvial geomorphology. In Anderson, M.G. (ed), *Modelling geomorphological systems*. Wiley, Chichester, 33–47.

Thom, R., 1975. *Structural stability and morphogenesis: An outline of a general theory of models.* W.A. Benjamin, Reading, MA.

CATASTROPHISM

Catastrophism is the theory that the Earth's geological and biological processes have been affected by sudden and violent events thus complementing uniformitarianism and gradualism that had succeeded more catastrophic views of the 19th century. Such catastrophic views, prompted by the French naturalist Baron Georges Cuvier (1769–1832), prevailed until the mid-19th century, assuming that catastrophic events had shaped the Earth but were succeeded by more uniformitarian ideas and by gradualism (that geologic change occurs slowly over long periods of time). Catatrophism was revived in the late 20th century because it was realized that events of great magnitude and low frequency can be very influential. It was originally developed, as neocatastrophism, in the mid-20th century, to account for sudden and massive extinction of life forms in palaeontology such as the effects of a 10 km diameter asteroid that struck the Earth at the end of the Cretaceous period, possibly creating a crater in Yucatan, Mexico, wiping out some 70% of all species including the dinosaurs and leaving behind the so called K-T boundary (The Cretaceous-Tertiary extinction which occurred about 65.5 million years ago, saw major changes in both land and marine ecosystems). This is one of the 'big five' extinction events (others occurred at the end of the Ordovician, in the late Devonian, the end of the Permian and the end of the Triassic).

Extinction is a natural process which occurs when all individuals of a species die without producing progeny. Mass extinction episodes have been identified from the fossil record, and

Figure 35   *Meteor Crater, Arizona. (Photograph K.J. Gregory)*

a mass extinction event is when there is a sharp decrease in the number of species in a relatively short period of time. Other catastrophic events such as widespread volcanism could have been responsible for the dinosaur extinction. On the Colorado plateau in north-eastern Arizona, Meteor Crater (Fig. 35) is 180 m deep and 1.2 km in diameter and has a rim which rises 30–60 m above the surrounding plateau. It was realized that the meteorite would have impacted the surface at an angle, that it would have exploded when it struck the ground and that temperatures of 1000°C or higher would have occurred according to the way in which sandstone has been changed into a high pressure form of silica.

In addition to extraterrestrial events, due to asteroids and comets, as well as earthquakes and volcanic eruptions, it has been suggested that truly catastrophic floods have occurred in the past and have been responsible for shaping some landscapes. Thus the channelled scabland of Washington was suggested to have been the result of catastrophic flooding after the drainage of Late Pleistocene proglacial lakes. Although the features such as giant current ripples up to 15 m high were not initially accepted as results of catastrophic floods, the advent of remote sensing showed how the features produced could be interpreted and in 1981 Baker demonstrated that discharges could have been as great as $21.3 \times 10^6 \, m^3 s^{-1}$. In addition there is growing evidence for superwaves, originated by submarine landslides or by the impact of asteroids or comets in the ocean, flooding continental lowlands. In the earth sciences the catastropic view is reinforced by catastrophe theory and aided by non-linear dynamics and systems.

Although uniformitaranism and gradualism still prevail it is now appreciated that they require qualification because significant catastrophic changes have also occurred.

See also: UNIFORMITARIANISM

Further reading

Albritton, C.C., 1989. *Catastrophic episodes in Earth history.* Chapman and Hall, New York.

Baker, V.R., 1981. *Catastrophic flooding: the origin of the Channeled Scabland.* Dowden, Hutchinson and Ross, Stroudsburg, PA.

Huggett, R.J., 1990. *Catastrophism: systems of earth history.* Arnold, London.

Huggett, R.J., 1997. *Catastrophism: asteroids, comets, and other dynamic events in earth history.* Verso, London.

Society for Interdisciplinary Studies (http://www.knowledge.co.uk/sis/index.htm) formed in 1974 to consider the role global cosmic catastrophes may have played in our history.

CHAOS THEORY

Chaos theory attempts to explain the fact that complex and unpredictable results can and will occur in systems that are sensitive to their initial conditions, and it describes the behaviour of non-linear dynamical systems that under certain conditions can exhibit chaos. Most environmental systems are non-linear, they do not have an easily derived solution and give rise to responses which do not settle down to a fixed equilbrium condition or value and so are described as chaotic. However many systems are thought of as linear when viewed over a restricted range of action.

Chaos theory describes systems which are apparently disordered, but endeavours to find underlying order in apparently random data. Investigation of meteorological systems led Lorenz in 1963 to identify chaos theory but also to appreciate that it is impossible to predict the weather accurately. He showed that when running programmes to simulate weather systems a very slight change in the initial conditions could lead to a very large difference in the final outcome. This led to the butterfly effect, also known as sensitive dependence on initial conditions, whereby just a small change in the initial conditions can drastically change the long-term behaviour of a system, so that in theory, the flutter of a butterfly's wings in China could, in fact, actually affect weather patterns in New York City, thousands of miles away. However chaotic systems are deterministic in that they are well defined and do not have any random parameters so that they have well-defined statistics, hence the weather is chaotic, but its climate statistics are not. In ecology, rates of the growth and regulation of animal and plant populations (resulting from the individual processes of birth, death, immigration and emigration in natural, managed or artificial environments) in the field of ecological demography involved the concept of 'carrying capacity' whereby no population can increase indefinitely. However non-linear models are needed for density dependence and Robert May in 1974 showed that for higher growth rates different populations could appear; in effect the line would bifurcate, and subsequently bifurcations came faster and faster until suddenly, chaos appeared. This meant that past a certain growth rate, it becomes impossible to predict the behaviour of the growth rate equation in population dynamics.

Chaos has become part of modern science, so that some have suggested that relativity, quantum mechanics, and chaos may be the three major scientific theories of the 20th century, with physics being the study of chaotic systems and how they work. In the environmental sciences in addition to applications to weather prediction and ecology it has been suggested that chaos theory can be applied to the flow of ocean currents, the branches of trees, the effects of turbulence in fluids, plate tectonics, the ENSO circulation, and global change consequent upon global warming. It seems that chaos theory will continue to be important throughout the environmental sciences and it can apply at a range of temporal scales. Big effects can have very small causes according to proximity to thresholds, and can be expressed in catastrophe theory.

See also: CATASTROPHE THEORY, ECOSYSTEM APPROACH, EQUILIBRIUM, SYSTEMS: THEORY APPROACH AND ANALYSIS, THRESHOLDS

Further reading

Gleick, J., 1998. Chaos: *The amazing science of the unpredictable.* Vintage, London.

May, R.M., 1974. Biological populations with nonoverlapping generalizations: stable points, stable cycles, and chaos. *Science* 186, 645–647.

CULTURES

This phenomenon is normally a topic for the human sciences rather than the natural sciences which underpin the study of physical geography. But in a world with so much human impact and influence, the way in which this one species, *Homo sapiens,* models the world in its head is clearly of great importance. It is likely that other great apes have something approaching culture but none has developed the richness, subtlety and ability to adapt that is characteristic of the human's own species.

What comprises culture? The human sciences (such as anthropology, sociology, archaeology and cultural studies) do not always agree, but would probably converge on three main components: values, norms and artefacts. Values determine what is important in a culture and may be derived from the ecological surroundings, such as the appreciation of scenery or the supply of resources; equally they may be non-material, as in with altruism or individualism. A purely rational motivation may be espoused or a metaphysical entity such as a god taken as the ultimate source of the values. Norms deal with how people behave in different situations, with a strong suggestion that there is a 'right' way to act. So not setting fire to forests is usually regarded as a good norm to adopt, as is buying fair trade foods. But in times of rapid change, norms are difficult to establish: knowledge of the extent of damage to the environment of the possession of SUVs (sport utility vehicles) is quite recent and has yet to establish itself among those whose values include a large dose of individualism. The SUV is an artefact, and one example among millions of material objects which have meaning in many cultures. Artefacts often express values and norms and are thus often called symbolic, with an extra layer of meaning beyond any utilitarian purpose. The quantity of meat eaten in a western industrial culture, for instance, goes well beyond the needs of nutrition: it is symbolic of wealth, success at environmental manipulation and indeed of power over nature.

One human quality which is contiguous with culture is language: apart from a few spontaneous movements such as gesture, dance, flight, most experiences are communicated via language. Thus a particular language may frame reality just as much as describe it. Languages which go beyond the inherent cultural biases of English, Chinese or Arabic are especially valuable in the natural sciences: mathematics is generally reckoned to be the best culture-free language. Language gains a particular authority when it is written, especially for the guardians of the writing in a largely illiterate society, be they priests in Neolithic times or theoretical physicists. Within the concept of culture as a whole there are many sub-divisions: some along national lines ('American culture'), by age ('youth culture') or by gender ('machismo culture'). Included in such classifications there is the notion that the natural sciences have their own subculture, with its values (a commitment to following the truth and sharing findings with others), norms (publication in agreed outlets, deference to senior figures) and artefacts (books and journals, lab and field equipment) which can be transgressed at times of new discoveries.

In physical geography, culture is of importance in terms of the way models of the physical world are made. These may be very positive: a model may look back into history and deduce that technological developments have solved many of the social and economic problems confronted at any one era in the past. Thus as wood became scarce in Europe, ways to use coal for smelting iron became more efficient and so industrial development became possible; this included the fixing of nitrogen from the atmosphere in fertilizers which made food production more efficient. A model may predict that as the price of oil rises, it is profitable for exploration companies who then 'discover' new resources. Other models may not be so optimistic. In the 1960s and 1970s there was a substantial scientific movement based on the notions that world population growth (then running at about 2.8% per annum) would outrun resources and cause many kinds of environmental degradation. Their hypotheses have not yet been disproved, though world growth rates have fallen back to 2.0% and below: either from technological advance in family spacing, or from cultural feedback about the negative effects of growth. Cultural disparities however mean that there is a loss of population in some countries and over 3.0% per annum growth in others. Other models run across any simple positive-negative axis: contrarians in science may not believe that climatic warming is due to human activity, attributing it instead to natural fluctuations in the flux of solar energy.

Cultural change is an important interface with the natural world. The role of animals in human societies has clearly changed through time: hunter-gatherers treated many species (especially those they ate) with reverence, propitiating their remains so that future abundance was assured. With the coming of agriculture, a more instrumental set of attitudes was often widespread though tempered with care when the beast was basically a unit of currency or essential to human survival. In 21st-century western cultures, animals are increasingly objects of moral concern in ways formerly restricted to humans (and not all of them). The New Zealand Parliament, for instance, debated a motion extending the right of legal representation to other primates, suggesting that their representation in court was little different from humans of tender ages or those adjudged to be insane. But differences between human cultures differ on a regional basis: the exhibition of caged bears in Hokkaido, Japan, would never be acceptable in western Europe, and the Japanese insistence on killing whales 'for scientific research' is outside the values and norms of most western countries, though not those of Iceland and some parts of Norway.

The world is at present dominated by 'western' capitalist values, norms and artefacts. Most of these facets are underlain by the plentiful use of fossil fuels, notably oil and coal. Even though technology 'stretches' these natural resources, they are not renewable and not infinite; coupled with the effects of the emissions after their use, adaptation to a different dominant culture may well be one of the greatest challenges ever faced by human beings.

See also: NATURAL RESOURCES

Further reading

Goody, J. 2000. *The power of the written tradition*. Smithsonian Institution, Washington, DC.

Ziman, J., 2000. *Real science. what it is and what it means*. Cambridge University Press, Cambridge.

Gellner, E., 2003. *Cause and meaning in the social sciences*. Routledge, London and New York.

Early, J., 1997. *Transforming human cultures: social evolution and the planetary crisis*. SUNY Press, Albany, NY.

UNESCO 2002 *Universal Declaration on Cultural Diversity* at www.unesco.org/education/imld_2002/universal_decla.shtml

CYCLES

Cycles are extremely important in the biosphere and lithosphere. Since material essentially cannot be created or destroyed on the Earth's surface, materials must cycle. And these cycles are involved in geochemical and biogeochemical processes. These cycles occur on timescales of 100s of millions of years to a day or less. Botkin and Keller discussed the cycles of nature and they are summarized them below. Several of these cycles are considered in more detail in other entries.

Geologic cycle. Over the several billion year history of the Earth, the surface of the Earth has constantly changed due to physical and biogeochemical forces. Taken together these changes are called the geologic cycle. The tectonic cycle involves the movement of large plates in the lithosphere, the outer surface of the Earth. The lithosphere is about 100 km thick. The movement of these plates is called continental drift or plate tectonics. Although the lithosphere is solid, it is not fixed and can be deformed. The force that drives plate tectonics comes from deep within the Earth due to heat released from these deeper layers. Material is added to the bottom of the ocean at mid-oceanic ridges as volcanic activity occurs. This is called a divergent plate boundary. As material moves at about 2–15 cm/yr, it pushes the plates around. Where an oceanic plate collides with a continent plate, the heavier ocean rock flows under a continental plate in a subduction zone. If two continental plates collide, a mountain range can form. Active earthquake zones form where two plates slide past one another. Moving plates change the size of continents and the oceans and isolate organisms from one another. This has been an important factor in evolution.

Hydrologic cycle. The hydrologic cycle is the movement of water on the surface of the Earth. Water evaporates from the ocean and some falls on the land. Water falling on land can evaporate or runoff as surface or ground water. About 97% of water on Earth is in the ocean and only about 1% is stored as liquid water on land. About 550,000 km³ of water evaporates each year and then falls back to the surface as precipitation. On land, the basic unit in hydrology is the drainage basin watershed. This is the surface area that contributes to a particular river or stream. Drainage basins can vary greatly in size from a few hectares or less to very large. The Mississippi River drainage basin is about 3.2 million km². Humans are altering the hydrologic at many different levels. Global climate change is altering precipitation patterns. Much surface and ground water is now used for agriculture, industry and human consumption. Subsurface aquifers are being drawn down as water is pumped out. Much surface water is held in reservoirs behind dams.

Rock cycle. This cycle consists of the processes that produce, change, and weather rocks and soils. The rock cycle is strongly affected by the tectonic and hydrologic cycles. Rocks can be classified as igneous, sedimentary, and metamorphic. Igneous rocks solidify from molten material associated with volcanic activity such as lava. Rocks weather both chemically and physically due to water flowing over them and from breaking up as water freezes and thaws. This produces finer and fine particles from boulders, to rocks, gravel, sand and silt. These materials accumulate in sediments to form sedimentary rocks due to compaction, consolidation, and dewatering. As sedimentary rocks are buried to great depths, they are transformed by heat and pressure, and chemical processes to form metamorphic rocks. These rocks may be transported to the surface by uplift and then undergo weathering. Living organisms produce organic carbon which can be incorporated into rocks, forming coal for instance.

Biogeochemical cycles. Biogeochemical cycling is the cycling of elements that is affected by a combination of biological, chemical, and geological processes. Thus, all the other cycles discussed affect biogeochemical cycling. This cycling is a result of inputs to an ecosystem and the internal cycling within the ecosystem. Inputs can come from rain, runoff, dust, and other materials. A major input to the biogeochemistry of the ocean, and thus of the Earth, is the addition of materials to ocean water at the mid-ocean ridges. Factors such as rainfall, rates of biological activity, and soil type affect biogeochemistry. Humans have strongly affected this cycling by extraction, addition of materials to ecosystems (such as nitrogen fertilizer), and by the addition of toxins such as heavy metals and pesticides. More detail on biogeochemical cycling is found in the entry and in the entries for individual cycles.

Cycle of life. Living organisms undergo a grand cycle due to production and consumption. As plants and algae photosynthesize, they take CO_2, water and other materials from the environment and produce living organic matter and water. As this organic matter is consumed and dies, it decomposes and CO_2, water and inorganic compounds are released back into the environment. These two processes are called production and respiration. Production and respiration do not occur at the same rate over the surface of the Earth. Areas of the land surface that received abundant rainfall have much higher rates than much of the ocean. The highest rates of production and respiration occur in tropical rainforests, temperate and tropical flood-plains, and in coastal ecosystems. Deserts, high mountains, and tundra have low rates. In the ocean, high productivity occurs in the coastal ocean, upwelling zones, and in high latitudes during the spring plankton bloom. The centres of the great ocean gyres (see Ocean Circulation) and most of the tropical ocean have low productivity. Living processes strongly affect other cycles. Over time much organic material is not rapidly decomposed and ends up in rocks or in underground storage as coal, oil, or natural gas. Humans have greatly affected the cycle of life by changing the distribution of living organisms (i.e., growing crops in the desert, cutting forests) and by drawing down the great reservoirs of fossil fuels.

See also: BIOGEOCHEMICAL CYCLES, CARBON CYCLE, NITROGEN CYCLE, OCEAN CIRCULATION

Further reading

Botkin, D.B. and Keller, E.A., 2007. *Environmental science – earth as a living planet.* Wiley, New York.

Brewer, R., 1994. *The science of ecology.* Saunders College Publishing, Orlando, FA, 2nd edn.

Cotgreave, P. and Forseth, I., 2002. *Introductory ecology*. Blackwell Scientific, Oxford.

Kormondy, E., 1995. *Concepts of ecology*. Prentice Hall, NJ.

Mitsch, W.J. and Gosselink, J.G., 2000. *Wetlands*. John Wiley, New York, 3rd edn.

Odum, E.P., 1971. *Fundamentals of ecology*. W.B. Saunders, Philadelphia, 3rd edn. (the most influential ecology text of the mid-20th century).

Stiling, P., 2001. *Ecology: theory and applications*. Prentice Hall, New Jersey.

Townsend, C., Begon, M. and Harper, J.L., 2008. *Essentials of ecology*. Blackwell Publishing, Oxford.

DIASPORA

'Diaspora' is a term generally used in demographic studies to indicate people of a particular culture who have settled outside their homeland and formed a distinct cultural group. In an environmental context, its relevance is to practices involving the use of resources. For example, a group may take a favoured crop plant or animal with them or endeavour to cultivate it in a new and possibly hostile environment. Examples might be the strains of rice taken to Hawaii and California by Japanese immigrants so as to be able to eat their preferred sticky varieties; the introduction of citrus fruits and of sugar into Iberia by the Muslim communities which settled there in the middle ages and perhaps most hegemonic of all, the introduction of the pizza and thus the need to produce industrial quantities of tomatoes and a stringy cheese like mozzarella — first into the USA and then more or less worldwide. With climatic change and the likely movement of large numbers of humans, more examples are likely to occur in the future.

See also: POPULATION GROWTH, CULTURE

Further reading

McMichael, A. 2002. Population, environment, disease, and survival: past patterns, uncertain futures. *The Lancet* 359, 1145–1148.

Dunlap, T.R., 1999. *Nature and the English diaspora: environment and history in the United States, Canada, Australia, and New Zealand*. Cambridge University Press, New York.

ENVIRONMENT

Environment is made up of the set of characteristics or conditions that surround an individual human being, an organism or a group of organisms or a community. Such environmental conditions will influence the growth and development of organisms. It therefore constitutes the environment of a particular organism so that arguably without organisms there is no environment to be occupied. In the early twentieth century it was thought that human activities and cultures were strongly influenced and constrained by natural environment – a view styled environmental determinism. By the mid-20th century it was realized that such a rigid deterministic view was not consistent with the way that people could modify their environments so that environmental possibilism became the notion that environments provided possibilities for human choice. However this was realized to be insufficient in view of the extent to which human impact has affected environments of the world.

Several types of environment have therefore been defined (Table 25) including the natural environment which may be a theoretical concept, the physical environment which excludes the influence of human activity, the cultural and built environments used for environments created by humankind. Specific types of environment have been recognized by distinguishing between

Table 25 *Types of environment*

Type of environment	Useage
Natural	Includes all living and non-living things that occur naturally on Earth. Although this could exclude human activity it is generally thought that the natural environment includes all living things including human beings
Continental	When used for types of aquatic sedimentary environment includes fluvial, lacustrine, limnic, paludal and paralic, and can also refer to environments of particular processes such as glacial environments, aeolian environments
Marine	Used particularly for types of sedimentary deposition including zones described as abyssal, aphotic (and photic), bathyal, littoral, neritic
Human	At a UNEP conference in 1972 it was asserted that both aspects of man's environment, the natural and the man-made, are essential to his well-being and to the enjoyment of basic human rights the right to life itself. The Conference called upon governments and peoples to exert common efforts for the preservation and improvement of the human environment, for the benefit of all the people and for their posterity
Cultural	The complex of social, cultural and ecological conditions that affect the nature of an individual or community
Built	Man-made surroundings that provide the setting for human activity, ranging from the large-scale civic surroundings to the personal places. A concern of many organizations such as the Environmental Protection Agency (EPA) in the US

fluvial and glacial environments for example, and in ecology habitat and ecosystem have been employed to refer to interpretations of environment.

In view of the increasing importance accorded to environment a number of organizations have been established, including the United Nations Environment Programme (UNEP) created in 1972 to provide leadership and encourage partnership in caring for the environment by inspiring, informing, and enabling nations and peoples to improve their quality of life without compromising that of future generations. Subsequent milestones have included creation of the Intergovernmental Panel on Climate Change (1988–), publication in 1992 of Agenda 21 as a blueprint for sustainable development, and world summit meetings held in 2002 and 2005. In addition, other international organizations have established environment strategies such as that of the World Bank, established in 2001 to improve the quality of life, the quality of growth, and to protect the quality of the regional and global commons. Regional policies have been developed such as that by the European community (EU) with an action programme entitled 'Environment 2010' focused on climate change and global warming; the natural habitat and wildlife; environment and health issues; and natural resources and managing waste. Legislation has been increased since the greater environmental awareness in the 1960s as exemplified by the National Environmental Policy Act (NEPA) in the USA in 1969 which encouraged '… productive and enjoyable harmony between man and his environment; to promote efforts which will prevent or eliminate damage to the environment and biosphere and stimulate the health and welfare of man; to enrich the understanding of ecological systems and natural resources to the Nation …'. It is now widely accepted that protecting the environment is essential for the quality of life of current and future generations but that an outstanding challenge is to combine protection with continuing economic growth in a sustainable way.

See also: CONSERVATION, CULTURES, ECOSYSTEM, ENVIRONMENTAL SCIENCES, HUMAN IMPACTS ON ENVIRONMENT, SUSTAINABILITY

Further reading

Environments is a journal published three times a year, *Environments* addresses people in their social, natural and built environments. The intent is to promote scholarship and discussion in a multidisciplinary and civic way, providing ideas and information that people might use to think effectively about the future.

O'Riordan, T. (ed), 1994. *Environmental science for environmental management.* Longman, Harlow.

UNEP www.unep.org

ENVIRONMENTALISM

A set of ideas and a citizens' movement to protect the quality and continuity of life through advocacy for the preservation or improvement of the natural environment. Environmentalism effectively arose with the conservation movement in the late 19th century, when there were initiatives towards establishment of state and national parks and forests, wildlife refuges, and national monuments all intended to preserve noteworthy natural features. Early conservationists included President Theodore Roosevelt, Gifford Pinchot, and the Izaak Walton League and the philosophical foundations for environmentalism in the USA were established by Thomas Jefferson, Ralph Waldo Emerson, and Henry David Thoreau.

In the social sciences, the term has been used to refer to any theory that emphasizes the importance of environmental factors in the development of culture and society. Greater public awareness in the 1950s and 1960s that conservation of wilderness and wildlife had to be complemented by other concerns relating to pollution (air, water and noise), waste disposal, energy resources, radiation, and human impacts including application of pesticides prompted what came to be known as the 'new environmentalism'. This has been reinforced leading up to the 21st century by concerns for the implications of global warming. Public concern has seen the formation of activist organizations such as Greenpeace which is an independent global campaigning organization that acts to change attitudes and behaviour, to protect and conserve the environment and to promote peace with particular reference to addressing the climate change threat; challenging wasteful and destructive fishing, and creating a global network of marine reserves; protecting the world's ancient forests and the animals, plants and people that depend on them; urging safer alternatives to hazardous chemicals in products and manufacturing; and campaining for sustainable agriculture.

O'Riordan suggested in 1981 that there are three views of the world at the heart of environmentalism. The *technocentric* views humanity as manipulative and capable of transforming the Earth for the benefit of both people and nature. An *ecocentric* view aims to incorporate the costs of altering the natural world and can embrace sustainable development, the precautionary principle (accepting that scientific knowledge may never be available or sufficient to enable action to be taken), ecological economics, environmental impact assessment, and ecoauditing. Thirdly is the *deep ecology or steady state economics* view which emphasizes the management of small-scale self-reliant communities according to local resources and local needs.

See also: CONSERVATION OF NATURAL RESOURCES, ENVIRONMENT, ENVIRONMENTAL SCIENCES

Further reading

O'Riordan, T., 1981. *Environmentalism*. Pion, London.

Simmons, I.G., 1993. *Interpreting nature. Cultural constructions of the environment*. Routledge, London and New York.

EPISTEMOLOGY

This term literally means the study of (Greek *logos*) knowledge or science (Greek *episteme*), it is concerned with how humans think, so that it has become known as the theory of knowledge. Humans need to understand how knowledge is acquired and how concepts are developed. It has developed as a core area of philosophy concerned with the nature, origins and limits of knowledge. It has been employed in the environmental sciences as they severally became concerned with closure (the limits which separate different academic disciplines), with objectives and methods of approach for academic subjects. In the 1960s as the environmental sciences became more aware of different scientific approaches, there was a greater awareness of the way in which epistemology focuses on the means for acquiring knowledge, how it may be possible to differentiate between truth and falsehood, and to make the distinction between rationalism and empiricism, or understand whether knowledge can be acquired *a priori* or *a posteriori*. Whereas rationalism is knowledge acquired through the use of reason, empiricism is knowledge obtained through experience. Thus two main approaches were identified in environmental sciences – a theoretical or modelling approach (rational) which is dependent upon an assumed theoretical model, and the other an empirical approach, based upon field, laboratory or analytical data. The methodology of any discipline covers the methods and techniques that are employed and so embraces the ontology (concerned with nature of being and phenomena) and epistemology (concerned with obtaining knowledge about being and phenomena) which are together concerned with concepts and the methods collectively constituting the philosophy of the discipline (Fig. 36). As the way in which knowledge is obtained can affect human

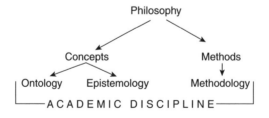

Figure 36   *Relationships between terms used in the philosophy of an academic discipline (after Bird, 1989, p. 231)*

Table 26 *Relationships between philosophies and aspects of scientific method. (Developed from Richards, 2003)*

Scientific method	Methods involved
Positivism	Representing: observation, measurement
Logical empiricism	Intervening: experimentation, laws of constant conjunction
Critical rationalism	Theorizing: hypoethesize, experiment, test, falsify, refine
Realism	Uncovering hidden structures and mechanisms
(Post)modern science	Perception reflects the perspective of the observer; the uncertainty principle; at the limits of observation and intervention; chaotic behaviour of non-linear dynamical systems; sensitivity to initial conditions

views of environment it is inevitable that methods of study and the scientific method (Table 26) will change with time and may alter according to scale of attention. To understand how the environmental sciences have evolved it is therefore necessary to appreciate the epistemology of those sciences.

See also: PHILOSOPHY OF SCIENCE, POST-MODERNISM

Further reading

Bird, J.H., 1989. *The changing worlds of geography. A critical guide to concepts and methods.* Clarendon Press, Oxford.

Richards, K.S., 2003. Geography and the physical sciences tradition. In Holloway, S.L., Rice, S.P. and Valentine, G. (eds), *Key Concepts in Geography.* Sage Publications, London, 23–50.

EQUIFINALITY

When a particular characteristic of environment can arise as a result of different processes so that different initial conditions can lead to similar effects. It was defined by the originator of general systems theory (Ludwig von Bertalanffy, 1901–1972) and it has been argued that equifinality can characterize the behaviour of biological organisms, but it has also been applied in the study of landforms and is sometimes referred to as 'convergence'.

For example, hillslopes can be similar in shape but be the product of very different assemblages of processes. Similarly tors on hilltops and slopes can arise as a result of either a sequence of processes involving deep tropical chemical weathering or as a result of frost action in periglacial conditions. However in this and other cases it may be that there is insufficient precision in describing the tors, because there may be subtle differences between periglacial tors and subtropical tors, so that in effect equifinality may not be as applicable as has been thought. Also, as many environmental features have developed as a result of past processes there may be insufficiently detailed knowledge to know exactly what processes were significant. Thus dry valleys, defined as valleys without a stream channel along the valley axis, can arise in different ways but there are subtle differences between several types of dry valley. Hence it has been proposed that equifinality should be reserved for those cases where the same final state is produced by the same processes from a range of initial conditions.

See also: EARTH SURFACE PROCESSES, GEOMORPHOLOGY, SYSTEMS: THEORY, APPROACH AND ANALYSIS

Further reading

Gerrard, A.J., 1984. Multiple working hypotheses and equifinality in geomorphology: comments on a recent article by Haines-Young and Petch. *Transactions Institute of British Geographers* 9, 364–366.

Richards, K.S., Brooks, S., Clifford. N., Hams, T. and Lane, S., 1997. Theory, measurement and testing in 'real' geomorphology and physical geography. In Stoddart, D.R. (ed), *Process and form in geomorphology*. Routledge, London, 265–292.

EQUILIBRIUM

In a system, if inputs are equal to output, then the system is said to be in a steady state or in equilibrium. But this is a dynamic equilibrium because material is entering the system and leaving the system, but at equal rates. For example, over a day or two, a person's weight will generally change very little, if at all. But food is being consumed, digested, and turned into carbon dioxide as oxygen is consumed. This is a dynamic steady state. Such a steady state can occur at many different levels of organization, from the whole earth to a single organism. Larger scales can be considered, the solar system for example, and smaller scales such as a single cell.

For example, over a period of a few days to a few weeks the surface of the Earth does not gain or loose much temperature. Seasonally, the Earth heats up and cools, but over an annual period there is generally little net change. This is changing, however, due to global warming. If the outputs to a system are greater than the inputs, then the reservoir of the system one is considering declines. For example, many aquifers are being drawn down faster than they are replenished. If the inputs to a system are greater than the losses, then the reservoir increases. This is happening for CO_2 in the atmosphere and leading to global warming. In the latter two cases, the system is not in equilibrium.

In general, in the absence of strong human pressure, the biosphere tends towards an overall dynamic equilibrium or self-regulating homeostasis. James Lovelock has called this the Gaia hypothesis. This is also called the balance of nature. This incorporates the idea that feedbacks affect dynamic equilibriums and that biological feedbacks help keep the environment in a range that is suitable for life. A concept related to equilibrium or steady state is the idea of residence time. In systems with a short residence time, the inputs are large relative to the size of the reservoir. Such systems are generally more susceptible to human impacts. For example, it is easier to overfish a small pond than it is to overfish the ocean. People have been overfishing small ponds for long periods of time, but it is only with the development of large modern fishing fleets that oceanic stocks of some species, such as cod, have been depleted. Human impacts can rapidly affect systems near steady state because the impacts are often non-linear. Human population growth and per capita use of resources are having dramatic impacts on the environment because the growth rates are exponential. The increase in CO_2 in the atmosphere is not linear and projections for increases in temperature and sea level in the 21st century are also non-linear. This means that systems can quickly be driven out of steady state and that it is not possible always to look to the past for ideas about future rates of change.

See also: GAIA HYPOTHESIS, SYSTEMS: THEORY, APPROACH AND ANALYSIS

Further reading

Botkin, D.B. and Keller, E.A., 2007. *Environmental science – earth as a living planet*. Wiley, New York.

Brewer, R., 1994. *The science of ecology*. Saunders, Forth Worth, TX and London. 3rd edn.

Cotgreave, P. and Forseth, I., 2002. *Introductory ecology*. Blackwell Scientific, Oxford.

Kormondy, E., 1995. *Concepts of ecology*. Prentice Hall, New Jersey.

Lovelock, J., 1995. *The ages of Gaia: a biography of our living planet*. Norton, New York.

Townsend, C.R., Harper, J.L. and Begon, M., 2002. *Essentials of ecology*. Blackwell Science, Oxford.

ERGODICITY

In the environmental sciences ergodicity arises from the substitution of spatial variations for temporal ones. Because it is often difficult to determine how environment changes over time then arranging present spatial sequences as equivalent to temporal change can be helpful. It has been employed, especially for relatively small-scale systems, because changes occur over much longer time periods than the time of empirical observations and because until recently dating techniques were unable to give precise dating for stages of environmental change. Space-time substitution underlay attempts to deduce stages in drainage network development. Slope profiles from coastal cliffs where there has been protection from coastal erosion were used to construct sequences of possible slope evolution. The ergodic hypothesis was applied by using the distribution of thermokarst alas features in Siberia as a basis for producing a six stage model showing the progress of a thermokarst surface from syngenetic ice wedges to a thermokarst valley. Difficulties with substitution of space for time arise in geomorphology because landforms may be assembled into time sequences to accord with preconceived theories; only one variable is assumed to be an ergodic indicator whereas other variables alter as well; and a temporal sequence may be assumed to exist when variations are in fact chance fluctuations around an equilibrium state.

In mathematics early work in chaos theory was preceded by ergodic theory, which has also been used in statistical physics. Applications in the environmental sciences other than in geomorphology are possible and are implicit in the construction of some models. An example of an application is provided by a study of Wheeler Ridge, an anticlinal fold in California, where drainage networks are younger towards the eastern tip of the anticline, a gradient of between 4.8 and 10° is needed to initiate the channel networks, and near-constant valley densities are reached 40–80 ka after channel network initiation.

See also: CHAOS THEORY, ENVIRONMENTAL CHANGE

Further reading

Chorley, R.J., Schumm, S.A. and Sugden, D.A., 1984. *Geomorphology*. London and New York, Methuen.

Czudek, T. and Demek, K.J., 1970. Thermokarst in Siberia and its influence on the development of lowland relief. *Quaternary Research* 1, 103–120.

Talling, P.J. and Sowter, M.J., 1999. Drainage density on progressively tilted surfaces with different gradients, Wheeler ridge, California. *Earth Surface Processes and Landforms* 24, 809–824.

ESTUARINE-DEPENDENT OR -RELATED SPECIES

In defining estuarine-dependent or -related species, Chesney and his co-workers. used the distribution and abundance of a species or life-history stage over environmental gradients to define its habitat. Taking this view, a species may be characterized as estuarine-dependent if a life history stage requires some combination of environmental conditions typically found in estuaries. Environmental conditions that control recruitment success for a species may vary across its range, so a species may be dependent on estuaries (low salinity shallows, high productivity, rich in prey) throughout most of its range but rely heavily on sea-grass covered coastal sand flats (low salinity shallows, rich in prey) in another location. As the identification of nursery habitat types should require at least a comparison of the densities of early life-history stages in other nearby habitat types but a better characterization would derive from enhanced survival, growth, and recruitment. It is well documented that changes in estuarine conditions when the juveniles are in the estuary are related to annual variations in future adult stock sizes.

Some important points have emerged from these studies: 'if the "recruitment" both biological and fishery, utilize estuaries and coastal lagoons, those resources are "estuarine-dependent"', but if that resources utilize regularly the option of the estuarine plume on the continental shelf, they are 'estuarine-related' or estuarine opportunistic. For estuarine-dependent species, seasonal and inter-annual variation in environmental conditions (e.g., temperature and salinity) may serve as ecological filters that define the quantity of suitable habitat for an early life history stage. Utilization of coastal habitat types, particularly with regard to estuarine dependence, the occurrence of a life-history stage in more than one habitat type is insufficient to refute the nursery function of a given habitat type. A species may be estuarine-dependent on one end of its range and rely on similar, estuarine-like conditions at another extreme: it is the controlling variables that are important, not the biologist's definition and characterization of a habitat type.

See also: ESTUARINE ENVIRONMENTS

Further reading

Baltz, D.M. and Yáñez-Arancibia, A., 2008. Ecosystem-based management of coastal fisheries in the Gulf of Mexico: environmental and anthropogenic impacts and essential habitat protection. In Day J. W. and Yáñez-Arancibia, A. (eds), *The Gulf of Mexico: ecosystem-based management*, The Harte Research Institute for Gulf of Mexico Studies, Texas A&M University Press, College Station, TX, Chap 19.

Boesch, D.F. and Turner, R.E., 1984. Dependence of fisheries species on salt marshes: a question of food or refuge. *Estuaries* 7, 460–468.

Chesney, E.J., Baltz, D.M. and Thomas, R.G., 2000. Louisiana estuarine and coastal fisheries and habitats: perspective from a fish's eye view. *Ecological Applications* 10, 350–366.

Pearcy, W.G. and Myers, S.S., 1974. Larval fishes of Yaquina Bay, Oregon: a nursery ground for marine fishes? *U. S. Fisheries Bulletin* 72, 201–213.

Miller, J.M., 1994. An overview of the Second Flatfish Symposium: recruitment in flatfish. *Netherlands Journal of Sea Research* 32, 103–106.

Pauly, D. and Yáñez-Arancibia, A., 1994. Fisheries in coastal lagoons In Kjerfve, B. (ed), *Coastal lagoon processes*. Elsevier, Amsterdam, 377–399.

Potter, I.C., Beckley, L.E., Whitfield, A.K. and Lenanton, R.C.J., 1990. Comparisons between the roles played by estuaries in the life cycles of fishes in temperate Western Australia and Southern Africa. *Environmental Biology of Fishes* 28, 143–178.

Pritchard, D.W., 1967. What is an estuary: physical standpoint. In Lauff, G.H. (ed), *Estuaries*. American Association for the Advancement of Science Publication 83, Washington, DC, 3–5.

Walsh, H.J., Peters, D.S. and Cyrus, D.P., 1999. Habitat utilization by small flatfishes in a North Carolina estuary. *Estuaries* 22, 803–813.

EVOLUTION

E volution is the central unifying theory of biology. It explains the vast variety of life that exists on the planet today. Evolution can be defined as the interaction of all the processes that have transformed biodiversity over long periods of time from the earliest life to the present. Charles Darwin is perhaps the most famous biologist of all times. In 1859, he published *On the Origin of Species by Means of Natural Selection*. With this publication, biology was unified into a comprehensive science because Darwin presented an explanation of how biodiversity came about. His theory addressed the great diversity of life, geographic distributions of animals and plants, how species originate and adapt, and why there are similarities and differences. Darwin argued that species evolved over time from earlier forms and he proposed a mechanism for evolution, natural selection. Random changes or mutations take place in species over time and some of these changes confer advantages or disadvantages that are inherited. Thus, organisms change over time.

Darwin, of course, did not know about the molecular basis for life, but later work on DNA has provided the molecular mechanisms for evolution. Explanations in the field of geology about the origins of sedimentary rock layers and the fossils they contained paved the way for Darwin's theory. Darwin collected data that formed much of his theory during his work on the voyage of the *Beagle*. One of the most famous of this information is the description of different species of finches on the Galapagos Islands off the west coast of South America.

Evidence for evolution is very strong and includes biogeography (the geographical distribution of species), the fossil record, comparative anatomy, comparative embryology, and molecular biology. A common genetic heritage is very strong evidence that all life is related, and species that are close to each other have more similar genetic codes. The origin of species is central to the theory of evolution. A species is a population of organisms that can interbreed in nature and produce viable, fertile offspring. Members of this species cannot interbreed with members of another species to produce viable, fertile offspring. The process of speciation or the formation of new species is based on reproductive isolation that isolates gene pools of species. Isolation can be due to habitat isolation, behavioural isolation, temporal isolation, mechanical isolation (individuals are physically unable to mate), and various problems that cause genetic incompatibility even if organisms do mate. The mule, the sterile offspring of a horse and a donkey, is an example of this.

There have been discussions of the rate of evolutionary change; whether evolution occurs gradually or in a discontinuous manner. The latter has resulted in the so-called punctuated equilibrium model, which describes the sudden appearance of new species. An important point to consider in evolution is the extraordinary long period of time of the geological record that allowed time for evolution to take place. Since its beginning, evolution has sparked debate between scientists and some religious persons who consider evolution a threat to religious teaching. But many religious people do not find a contradiction between the scientific and spiritual views.

See also: BIOLOGY, GENETICS

Further reading

Allmon, W.D. and Bottjer, D.J. (eds), 2001. *Evolutionary palaeoecology*. Columbia University Press, New York.

Darwin, Charles, 1859. *Origin of Species*. John Murray, London.

Grant, V., 1991. *The evolutionary process – a critical study of evolutionary theory*. Columbia University Press, New York, 2nd edn.

Larson, E.J., 2004. *Evolution – the remarkable history of a scientific theory*. The Modern Library, New York.

Lewin, R. and Foley, R., 2003. *Principles of human evolution*. Blackwell Publishing, Oxford.

Ridley, M., 2003. *Evolution* Blackwell Publishing, Oxford, 3rd edn.

GAIA HYPOTHESIS

At its simplest, this model reverses long-held ideas about the relations of the living and non-living components of this planet. Since Darwin's time in the 19th century, the overarching view was that life had adapted to the changing conditions of a cooling planet. The Gaia model, on the other hand, holds that life itself has created many of the physical conditions which are now found. The model is named after the Greek goddess of the Earth and was first formulated by the analytical chemist James Lovelock in the 1960s, though largely ignored until the more favourable intellectual climate of the 1970s. His 1979 book launched the idea into both scientific and humanistic communities and was much derided by the former. His arguments showed the whole planet to be acting like one self-organizing organism which moved towards optimizing the conditions for life. Thus, the composition of the atmosphere and of the oceans and the global surface temperature of the Earth all have values which would not pertain if they were simply those of a cooling fragment of the cosmos.

Lovelock argued that it was the development and continuing presence of life which regulated the system so as to produce unlikely values for e.g., the proportions of oxygen, nitrogen and carbon dioxide; likewise the continuing salinity of the oceans at about 3.4% is best explained by the role of life forms as regulators. The nature of the mechanisms controlling this homeostasis was for many years little understood, but more recent work on the role of phytoplankton controlling the production of aerosols near the ocean surfaces, which in turn act as cloud-forming nuclei that then determine rain over continents downwind, is one example of the complex sequences of biogeochemical interactions that tend to confirm Gaian views. Elaboration of the basic idea has spawned two variants: the 'weak Gaia' hypothesis that life and non-life have co-evolved and one result has been stable conditions for the evolution of many life forms, and the 'strong Gaia' model that asserts that the developments were purposive, i.e., there was an end result to be attained, which was that life should flourish. The strong Gaia notions were anathema to most scientists, who argued that teleological approach (i.e., one with a definite end in view) was contrary to Darwinian ideas and therefore more in the sphere of religion than science; equally they were skeptical that there was no way of rejecting the whole Gaian set of ideas through controlled experiment. In general, the argument over teleology has faded, with pro-Gaians saying that it was never part of the original hypothesis anyway. The integrating discoveries of many General Circulation Models (GCMs) have on the whole tended to argue towards a Gaian position and Lovelock has suggested that the 'enhanced greenhouse effect' will produce unforeseeable feedbacks, which will not necessarily favour the continuance of human life, especially if the output of carbon dioxide and other greenhouse gases continues. In the early 2000s, Lovelock shocked the environmentalist community (normally his strong supporters) by advocating the increased use of civil nuclear power.

See also: GREENHOUSE EFFECT, EVOLUTION

Further reading

Lovelock, J. 2005. *Gaia: a new look at life on earth*. Oxford University Press, Oxford. [Revised and corrected version of first edition of 1979].

Lovelock, J. 2006. *The revenge of Gaia: earth's climate in crisis and the fate of humanity*. Basic Books, New York.

GREENHOUSE EFFECT

The greenhouse effect occurs due to the absorption of infrared radiation emitted from the Earth by the atmosphere and then warming by re-radiating heat back to the surface (see Table 39, P. 245 under ENERGY AND MASS BALANCE). It was first discovered by Joseph Fourier in 1824, and first investigated quantitatively by Svante Arrhenius in 1896. The term 'greenhouse effect' may be used to refer either to the natural greenhouse effect, due to naturally occurring greenhouse gases, or to the enhanced (anthropogenic) greenhouse effect, which results from gases emitted as a result of human activities, such as carbon dioxide. Without these greenhouse gases, the Earth's surface would be about 30°C cooler. The name comes from an analogy with the way in which greenhouses are heated by the sun in order to facilitate plant growth. It is not an exact analogue, since a real greenhouse does not allow heat movement away from the floor and windows by convection. In the real atmosphere, convection away from Earth's surface is a substantial heat loss, particularly from warm land areas. Mars and Venus also have greenhouse effects. To the extent that the Earth is in energy equilibrium, the energy stored in the atmosphere and ocean does not change in time, so energy equal to the incident solar radiation must be radiated back to space. Radiation leaving the Earth takes two forms: reflected solar radiation and emitted thermal infrared radiation. The Earth reflects about 30% of the incident solar flux; the remaining 70% is absorbed, warms the land, atmosphere and oceans. Eventually this energy is reradiated to space as infrared radiation. This thermal infrared radiation increases with increasing temperature.

See also: ANTHROPOGENIC IMPACT, ENERGY AND MASS BALANCE, ENERGY FLUX

Further reading

Aguado, W. and Burt, J.E., 2006. *Understanding weather and climate.* Pearson Prentice Hall, Upper Saddle River NJ, 4th edn.

White, R.M. 1990. The great climate debate. *Scientific American*, 263, 36–43.

HABITAT OF COASTAL AND MARINE FISHES (ESSENTIAL AND CRITICAL)

M anaging habitat has long been recognized in wildlife management as key to main-taining healthy populations (Leopold, 1933). Recognizing that habitat destruction, as in the case of wetland destruction, was responsible for as much as half of the depleted coastal fisheries in United States waters, the US Congress added Essential Fish Habitat provisions to the revision of the Magnuson Act of 1976 (Magnuson-Stevens Act, 1996). Defined by Congress as 'those waters and substrate necessary to fish for spawning, breeding, feeding or growth to maturity', the designation and conservation of essential fish habitats seeks to minimize adverse effects on habitat caused by fishing and non-fishing activities.

'Habitat' is a loosely used ecological term that can be applied at the individual, population, and community levels and is often entangled with so many other ecological concepts that it can mean everything and therefore nothing. The term 'habitat' has almost been relegated to the status of a pseudocognate (*sensu* G Salt) in that it is a term in common use and each individual who uses it feels that all others share his own intuitive definition. Nevertheless, it is used and can be useful. Habitat can be regarded as the range of environmental conditions in which a species/population/life-history stage can live. It is a general term that broadly defines where a species lives without specifying patterns of resource use (*sensu* realized niche = resources used: energy, materials, and sites). From a fish's eye view, its distribution over environmental gradients describes its habitat. In defining habitat, the basic problem is that there are a number of points of view. From a biologist's point of view, strata in the environment can be arbitrarily described as habitats, but more properly as 'habitat types'. At the community level, the environments dominated by a single species (e.g., smooth cordgrass, *Spartina alterniflora*) may be characterized as Spartina habitat, but more properly as a 'Spartina community'.

Essential fish habitat can best be described by means of a microhabitat approach which yields a fish's eye view of its habitat requirements. At the finest scale, the microhabitat of an individual is the site it occupies at a given point in time. Sites are presumably selected to optimize an indvidual's net energy gain while avoiding predators (i.e., tradeoffs of growth vs. mortality) Since similarly sized individuals of a species select similar microhabitats, many care-ful measurements of individuals and associated physical, chemical, and biological variables should define the population's responses to environmental gradients. As defined here, the micro-habitat is an occupied site, not a very small habitat type. Fine-scale measurements of environmental conditions at a site occupied by one or more individuals constitute an observation, and many independent observations characterize the population's response to complex gradients.

From the point of view of estuarine fish, critical and essential habitats have the following characteristics: (1) The utilization of estuaries, coastal lagoons, and tidal wetland ecosystems is an integral part of the life cycle of numerous finfish and shellfish, particularly in areas like the Gulf of Mexico, (2) These coastal ecosystems are mainly utilized by juveniles and young adults, (3) There is a greater number of fish species in the coastal zone of the Gulf of Mexico than in comparable temperate or boreal systems in North America, (4) Second-order consumers are more abundant and diverse than first-order consumers or top carnivores. Second-order consumers are the most common commercial fish stocks, in contrast with the top carnivores, which are the most common sport fishing stocks. (5) Functional components in the fishery-ecosystem of the Gulf of Mexico are: (a) freshwater spawners occurring in waters <10 ppt salinity, (b) brackish water group limited to 10–34 ppt salinity, (c) marine spawners occurring in waters ≥35 ppt salinity. (6) Three temporal groupings of fish occur in coastal ecosystems in the Gulf of Mexico (a) resident species, those which spend their entire life cycle within estuaries, lagoons, or coastal wetlands, (b) seasonal migrants, those which enter the estuary during a more or less well-defined season from either the marine or the freshwater side and leave it during another season, (c) occasional visitors, those which enter and leave the estuary and associated coastal wetlands without a clear pattern within and among years. To these, two other groups may be added: (d) marine, estuarine-related species, those which spend their entire life cycle on the inner sea shelf under the estuarine plume influence, and (e) freshwater, estuarine-related species, those which spend their entire life cycle in the fluvial–deltaic river zone, associated with the upper zone of the estuarine system.

So, the essential habitat protection is a major concern in the sustainable fish stocks in areas like the Gulf of Mexico. Coastal fish resources are an expression of the ecosystem functioning and to assure the persistence of fish resources, the protection and conservation of essential habitats is the keystone for sustainable management of coastal fisheries.

See also: ECOLOGY, ECOSYSTEMS, WETLANDS

Further reading

Baltz, D.M., 1990. Autecology. In Schreck, C.B. and Moyle, P.B. (eds), *Methods for fish biology*. The American Fisheries Society, Bethesda, MD, 585–607.

Baltz, D.M. and Jones, R.F., 2003. Temporal and spatial patterns of microhabitats use by fishes and decapods crustaceans in a Louisiana estuary. *Transactions of the American Fisheries Society* 132, 662–678.

Baltz, D.M., Fleeger, J.W., Rakocinski, C.F. and McCall, J.N., 1998. Food, density, and microhabitat: factors affecting growth and recruitment potential of juvenile saltmarsh fishes. *Environmental Biology of Fishes* 53, 89–103.

Chesney, E.J., Baltz, D.M. and Thomas, R.G., 2000. Louisiana estuarine and coastal fisheries and habitats: perspective from a fish's eye view. *Ecological Applications* 10, 350–366.

Hurlbert, S.H. 1981. A gentile depilation of the niche: Dice and resource sets in resource hyperspace. *Evolutionary Theory* 5,177–184.

Leopold, A., 1933. *Game management*. Charles Scribner and Sons, New York and London.

Magnuson-Stevens Act, 1996. *Essential Fish Habitat*, Office of Habitat Conservation, NOAA Fisheries www.nmfs.gov

Salt, G.W., 1979. Letters to the Editor: A comment on the use of the term emergent properties. *The American Naturalist* 113, 145–148.

Yáñez-Arancibia, A., Lara-Domínguez, A.L. and Pauly, D., 1994. Coastal lagoons as fish habitat. In Kjerfve, B. (ed), *Coastal lagoon processes*. Elsevier, Amsterdam, Oceanography Series, 60, 363–376.

HAZARDS

The term 'hazard' refers to those natural processes that have the potential to damage human property and take human lives. In a broader sense, hazards also have the potential to damage ecosystems. Hazards are often natural processes that are hazardous because people live and work in places where natural processes that produce hazards occur. These processes include: flood, earthquake, volcanic eruption, landslide, coastal erosion, hurricane, tornado, lightning, drought, and frost among others. The occurrence of these processes may be influenced by human use of the land and in particular land-use changes, such as urbanization, agriculture, and deforestation, which affect natural processes and influence the magnitude and frequency of events such as flooding and landsliding.

An important concept related to natural hazards involves their potential to produce a catastrophe. When a property is damaged and human lives are lost, then the event is labelled as a disaster. The difference between a disaster and a catastrophe is that a catastrophe is a hazardous event in which the damage to human property and people is sufficiently severe that recovery is a long and involved process. Table 27 lists some of the natural hazards faced by people in terms of whether the event may be influenced by human use and its potential to produce a catastrophe. Flooding is the most universally experienced hazard primarily because there are so many streams and rivers across humanized landscapes that periodically flood. While a high magnitude earthquake or tsunami (giant sea wave) may take thousands and even hundreds of thousands of lives in a single event, it is the weather-related hazards that, over a period of many years, take

Table 27 *Effects of selected hazards in the United States (*Estimate based on recent or predicted loss over 150-year period. Actual life and /or property could be much greater. Modified from White, G.F. and Haas, J.E., 1975. Assessment of research on natural hazards. MIT Press, Cambridge, MA)*

Hazard	Deaths per year	Occurrence influenced by human use	Catastrophe potential
Flood	86	Yes	High
Earthquake*	50+?	Yes	High
Landslide	25	Yes	Medium
Volcano*	<1	No	High
Coastal erosion	0	Yes	Low
Expansive soils	0	No	Low
Hurricane	55	Perhaps	High
Tornado and windstorm	218	Perhaps	High
Lightning	120	Perhaps	Low
Drought	0	Perhaps	Medium
Frost and freeze	0	Yes	Low

the most lives. The events that are particularly hazardous from year to year include windstorms, tornado, and lightening strike.

There are several fundamental principles concerning hazards that are important in environmental science.

» Scientific evaluation of natural processes has led to a better understanding of the hazards. In particular, hazardous processes such as earthquakes, volcanic eruption, landslide, floods, and coastal erosion have been sufficiently studied utilizing scientific methods that these events, with their future activity, may be predicted. When the term prediction is used here, it is not about specifying a magnitude of an event and a time when it will occur, but where it is likely to occur based on frequency of past events, types of precursor activity and patterns of activity of a particular process.

» Hazards that previously produced disasters are more likely now to produce catastrophes. This results because as human population has increased and poor land-use choices have been made, the effects of hazardous processes have been amplified. For example, more people are living in low-lying areas vulnerable to hurricanes and tsunami. In some cases, they are living there because that is the only land available. In other cases, more cities and towns are located in coastal areas where hazards occur because of the desire of people to live and spend vacations in coastal areas. In respect of land use, some of the effects of hurricanes and intense rainstorms have increased as a result of urbanization and deforestation, which renders the land more vulnerable to rapid runoff and flooding, as well as to landslides on steep slopes.

» Analysis of risk associated with hazards has greatly improved in recent years. The risk from a particular event is the product of the probability of that event occurring times the consequences should it occur. For example, there is sufficient experience of earthquakes that it is possible to predict the probability (in terms of a percentage) of a moderate or large earthquake occurring in a given area over a given time and coupling that with potential damages allows the risk to be better evaluated.

» Hazards may be minimized. With improved scientific understanding of hazardous processes in terms of where and when they occur coupled to risk analysis, societies are better prepared to minimize the effects of hazards. Specific processes that may help minimize damages include land-use planning to avoid hazardous areas, engineering structures to render homes and other buildings better able to withstand hazardous events, and social and political resolution to better prepare for hazardous events and how they may be proactively mitigated. The effects of a specific disaster or catastrophe can be potentially reduced if it is feasible to predict or forecast an event and issue a warning. Minimizing the effects of hazardous events involves:

  ○ Identification of locations where hazardous events are more likely to occur
  ○ Calculating the probability that a given event will occur
  ○ Identifying precursor events that are likely to occur prior to a hazardous event
  ○ Predicting or forecasting the event
  ○ Issuing a warning to the public

Of particular importance is to become proactive with respect to hazards. Hazardous events will continue to occur and property will be damaged and human lives lost. What is of particular importance is to be prepared for hazardous events by proper planning, which is generally known as disaster preparedness. Such preparation will involve programmes such as plans for evacuation, as well as providing insurance and other relief for people who suffer damages.

As part of disaster preparedness, an understanding of what happens during and following a disaster is essential. Fig. 37 shows the generalized model for recovery following a disaster or catastrophe. Understanding the phases of recovery is a positive step in taking a proactive response to hazards. Another valuable approach is to anticipate and avoid hazards whenever possible. The most environmentally sound adjustment to a number of hazards is land use planning. This recognizes that it is possible to avoid building on floodplains or in areas where active landslides have occurred in the past or are likely to occur in the future. In a similar manner, an authority can provide for setbacks from coastal areas to reduce or put off into the more distant future consequences of coastal erosion. Taking a proactive stance in terms of natural processes that produce hazardous events is superior to taking a reactive stance, which simply involves dealing with the event once it has occurred.

Environmental science offers a number of opportunities to better evaluate and prepare societies for future disasters and catastrophes. In particular, environmental planning linked to the nature and extent of natural hazardous processes is a proactive stance to help minimize effects of events such as flooding, earthquake, hurricane, and volcanic eruption. As human population continues to increase in coming decades, more challenges will become apparent in trying to minimize effects of hazardous events. Working toward that goal now will result in reduced losses of human life as property damage may continue to increase because there is simply so much more infrastructure and building in potentially hazardous areas. Natural hazardous events occur almost everywhere where people live, but through careful planning, adverse effects of those processes may be minimized.

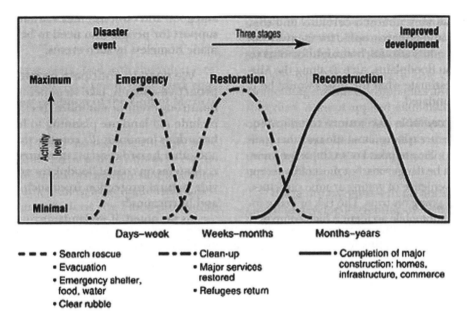

Figure 37 *Recovery following disasters my take years. Shown here is a highly generalized idea of what happens after a disaster, from emergency to restoration and reconstruction. The period of emergency lasts from a few days to a few weeks, and is followed by restoration activities, which can take a number of months or longer. The longest part is the reconstruction, which may take several years or more. (Source: Modified from R.W. Kates and D. Pijawka. From rubble to monument: The pace of reconstruction. Cambridge, MA: MIT Pess, 1977.)*

See also: CATASTROPHISM, DISASTERS, EARTH SURFACE PROCESSES

Further reading

Abramovitz, J.N., 2001. Averting unnatural disasters. In Brown, L.R. *et al.*, (eds), *State of the world 2001*. World Watch Institute. W.W. Norton, New York, 123–142.

Abramovitz, J.N., and Dunn, S. 1998. *Record year for weather-related disasters*. World Watch Institute, Vital Signs Brief 98–5.

Kates, R.W. and Pijawka, D., 1977. Reconstruction following disaster. In Haas, J.E. Kates, R.W. and Bowden, M.J. (eds), *From rubble to monument: the pace of reconstruction*, MIT Press, Cambridge, MA.

Keller, E.A. and Blodgett, R.H., 2006, *Natural hazards: earth's processes as hazards, disasters, and catastrophes*, Pearson Prentice Hall, Upper Saddle River, NJ.

White, G.F. and Haas, J.E., 1975. *Assessment of research on natural hazards*. Cambridge, MA: MIT Press.

HOLISM

Literally this term means that the whole is more than the sum of the parts, a general principle attributed to Aristotle. Applied to the environmental sciences all the properties of an environmental system or ecosystem, including biological, chemical, social and economic ones, cannot be explained entirely by the sum of the component parts because the system as a whole determines how the system operates. In just the same way human personality is more than can be predicted from biochemistry. In the development of the environmental sciences use of a reductionist approach involved increasingly detailed investigation of the component parts of systems. This has been characterized as 'knowing more and more about less and less', and although very necessary needs to be complemented by a more holistic view which focuses upon the complete system and the interaction between the components. An excellent example is provided by land management and by integrated basin management. Whereas early attempts tended to deal with problems as they arose and treat the symptoms, Aldo Leopold developed a biotic view of land which was in effect a holistic approach to land management so that his *A Sand County Almanac* is often cited as a foundation for environmental ethics. In the case of integrated basin management the usual reactive approach has been succeeded by reliance upon a much deeper understanding of natural phenomena and processes together with a larger-scale, more interdisciplinary, integrated and holistic methodology necessitating a paradigm shift in approaching land use change and watershed issues.

The implications of a holistic approach are therefore that a particular problem should not be approached exclusively from a single discipline such as biology, ecology, geology or geomorphology; the investigation should involve studies of surrounding areas that may influence the subject under consideration; and the research should be undertaken at different spatial scales as well as over different time scales. Systems theory, the ecosystem approach and the Gaia hypothesis are all essentially holistic in character. Because a holistic approach means looking at the wider context and relationship to the whole, it is considered by environmental ethics and by deep ecology, providing an antidote to scientific reductionist accounts of nature which focus on its constitutive parts.

See also: ECOSYSTEM, ENVIRONMENTAL ETHICS, GAIA HYPOTHESIS, INTEGRATED BASIN MANAGEMENT, NATURE, REDUCTIONISM, SYSTEMS: THEORY, APPROACH AND ANALYSIS

Further reading

Diplas, P., 2002. Integrated decision making for watershed management. *Journal of the American Water Resources Association*, 38, 337–340.

Gardiner, J.L. (ed), 1991. *River projects and conservation: a manual for holistic appraisal*. Wiley, Chichester.

Leopold, A., 1949. *A Sand County Almanac*. Oxford University Press, Oxford.

Newson, M.D., 1997. *Land, water and development: sustainable management of river basin systems*. Routledge, London, 2nd edn.

LANDSCAPE

Though in fact used more widely, a scientific description of a landscape comprises the visible features of an area of land, including physical elements such as landforms, soils, plants and animals, the weather conditions, and it also includes any human elements, such as the presence of agriculture or the built environment. Thus although often used primarily of rural areas, it is equally applicable to urban zones. There are a number of specific uses in the context of the environmental sciences. These include landscape ecology which brings ecological ideas to the causes and consequences of pattern and process in landscapes. Likewise, landscape history looks into the ways in which landscapes have evolved through time periods (mostly within the last 10,000 years). Landscape management is the deliberate management of landscapes and is a main aim of landscape architecture, which is specifically about the creation of humanized landscapes for particular purposes, of which gardens and other amenity sites are a chief element. It may also be used to meliorate the impact of large and unsightly buildings or industrial plant. In the 17th and 18th centuries, large-scale landscape gardening was carried on most western countries where the owners of estates wished to follow fashion and create particular combinations of grass, trees and water to form a landscape which often had Classical allusions. If the hand of humans is visible (or detectable by the techniques of palaeoecology or historical ecology) then the term 'cultural landscape' may properly be used. These usually demonstrate the power of human societies to overcome constraints and utilize opportunities presented by the natural (or at any rate the pre-existing) conditions of the region concerned.

See also: PALAEOECOLOGY, HUMAN ECOLOGY

Further reading

Head, L. 2000. *Cultural landscapes and environmental change*. Arnold, London.

LANDSYSTEMS APPROACH

A landsystem is a sub-division of a region in which there is a recurring pattern of topography, soils, vegetation reflecting the underlying rock types and climate of the area and affecting the erosional and depositional surface processes. Early attempts at land classification were developed from national surveys, for example from soil maps. In the USA, after the dust bowl era, the Soil Conservation Service was established and developed methods of soil capability classification as a basis for determining the potential uses of soil types. However this approach, and others developed from a single characteristic of environment such as slope or climate as well as soil, did not take sufficient account of all other environmental characteristics. Extensive resource surveys in undeveloped parts of Australia and Papua, New Guinea were initiated in 1946 by the Australian Commonwealth Scientific Industrial Research Organization (CSIRO). These surveys acquired information on the geology, climate, geomorphology, soils, vegetation and land use of areas and, by collating the information, designated land systems as areas or groups of areas with recurring patterns of topography, soils and vegetation with a relatively uniform climate.

A landsystem could be divided into units or land facets which in turn are composed of individual slopes or land elements. The landsystem approach has been modified for application to the problems of urban and suburban areas, it can be the basis for an information system that can be used to describe extensive regions, and structured methods of land classification can provide a database to give a practical summary of the distribution and management of resources in specific localities. The approach can be compiled using GIS for example with numerical land classification based on 1 km grid squares. Such quantitative approaches have the advantage of combining many aspects of environmental character and are not use-specific; in that sense they are an example of landscape ecology rather than land evaluation. It was criticized because it does not readily relate to a particular use; because it is not easily related to results obtained from field survey and because it emphasizes a static view, according to evolution of environment rather than upon the dynamics of the environmental system. An alternative approach can be based upon how the environment works and what it does rather than merely what it is.

See also: ECOSYSTEM-BASED MANAGEMENT, GIS, LANDSCAPE ECOLOGY, LAND EVALUATION

Further reading

Cocks, K.D. and Walker, P.A. 1987. Using the Australian Resources Information System to describe extensive regions. *Applied Geography* 7, 17–27.

Young, A. 1998. *Land resources: now and for the future.* Cambridge University Press, Cambridge.

MODERNISM

This term is mostly used in the humanities, social sciences, architecture and design. It refers to developments with their origins in the late 19th century but which came to fruition in the first half of the 20th century, being superseded by post-modernism. Thus the relevance to environmental science may seem tangential at best. However, the economies spawned in that period, and encouraged by the spread of electricity, the move to consumerism and the speed of communications all have had considerable environmental consequences. The mass production of motor vehicles, for example, using the assembly-line techniques developed by Henry Ford in Detroit, brought in their wake huge alterations of the land surface for roads, parking lots and oil refineries; the extension of the technology into agriculture in the form of the tractor revolutionized many agro-ecosystems. Potatoes, said one American ecologist, were now partly made of diesel oil. Thus modernism stands for the underpinning of many economic activities with fossil fuels in the form of refined oil and thus the penetration of a new source of energy into human–environment relations. The process is largely complete in the richer countries of the world but is still under way in the poorer places, where it is usually labelled as 'development'. Another change was the drop in family size as affluence spread, made possible by high-quality rubber, a product also underlying the spread of the motor vehicle; yet another was the shift towards enhanced consumption of meat which could be produced cheaply in e.g., Argentina, the western USA and Canada, and New Zealand but shipped to the eastern USA and to Europe in refrigerated ships at low cost. Modernism also implanted the idea of science and technology as fundamentally beneficent forces in the world. To extend e.g., western technology to everybody was a solution to their problems of resource supply and any environmental consequences could be mopped up by more technology. This attitude is still prevalent in many 'western' countries.

See also: POST-MODERNISM, ENERGY

Further reading

Lash, S., Szerszynsky, B. and Wynne, B. (eds), 1996. *Risk, environment and modernity*. Sage, London.

Lash, S., 1999. *Another modernity, a different rationality*. Blackwell, Oxford.

NATURAL RESOURCES

The idea of a 'natural resource' draws upon the valuation of the Earth as a store of materials and habitats for human use. Therefore the definition of a natural resource depends on being identified as such by a human society which depends in turn upon its cultural attitudes towards it and its technological ability to access it. A mineral ore will not be identified as a useful resource by a society unable to raise the temperatures necessary to smelt it; a perfectly nutritious animal will not be eaten by a society for whom it is taboo or in whom it invokes nausea. Thus the phrase 'natural resources are cultural appraisals' is a staple of examination questions in this area of study.

Most classifications of natural resources divide them into two or three types. There are first those which are called *renewable resources*, materials whose supply is, under optimum conditions, renewed by the very processes that create them. The most obvious is that of living organisms, which reproduce themselves; renewability depends upon the rate of reproduction so that living off elephant steak is less sustainable than eating yoghurt, since elephants breed every two years whereas bacteria can divide in two every few seconds. A cow normally has only one offspring whereas a head of wheat turns one seed-corn into hundreds, so beefsteak is a scarcer resource than bread. Another example is water, which moves perpetually through the hydrological cycle. Its renewability depends largely upon the time it spends in the different 'stores' of that cycle although its quality may also affect its utility for humans. The second category comprises *non-renewable resources*, whose supply is finite. Minerals such as coal, oil and metal ores are outstanding examples: fuel sources, especially, once used are converted to gases, heat and particulate wastes and the energy cannot be recovered. With for example metals there are more chances of recovering the materials for re-use; the same is true of materials like plastics though less is done re-processing most plastics compared to metals. In industrialized countries, these two categories now require high levels of fossil fuel input to produce: beef production and deep sea fishing alike are subsidized by oil in various forms. The third group can be termed *non-material resources* and are not altered by being used: scenery is one example and several other recreation resources fall into the same category. However, very heavy use can convert these into a non-renewable state, as when footpaths in popular regions are so heavily trodden that they erode away.

The availability of these resources to human societies has varied through time. Hunter-gatherers relied on fire as a major manipulator of environment and resource-producer; pre-industrial agriculturalists do not use fossil fuels in the shape of tractors, chemical fertilizers and biocides. The full panoply of solar energy, fossil fuel sources and nuclear energy has been available to humans (and then only to those in 'western' economies) only in the last 50 or so years. Consequently the quantity and variety of resources supplied to some people has grown enormously: the intensity of resource use, which is the amount per unit area per unit time (think of oil usage per square kilometre per week in Los Angeles) has risen since the speed of throughput of material resources in industrialized economies is now so high. All such use normally produces waste and, while a low density of population and a slow rate of use mean that the surrounding ecosystems can process the waste, this is not likely with industrial levels of production and with large cities. Furthermore, technology has made it possible to synthesize substances for which there are no evolutionary breakdown pathways in nature enabling them to accumulate for example living tissue or estuarine sediments. Waste can then in turn have an impact upon resource availability: untreated sewage in rivers will often lead to fish-kills and to the diminution of catch either for food or the anglers' pleasure.

The garnering of resources and the harnessing of nature are very complex processes beyond the obvious technological and scientific skills needed. Societies usually have strong attitudes toward resource use and expect that the future will hold greater availability rather than less. Thus they formulate models of the resource–population–environment interactions, partly at least in order to assuage their anxieties about future supply. Some of these models have a strong coercive element in them, as when a powerful nation decides to assure its future oil supply should its own resources start to decline in quantity. In a world ruled by neoliberalism, however, the main model is that provided by the mind-world of the discipline of economics in which 'the market' is the dominant mechanism. The discipline of ecology is also noticeable on the fringes but usually as a *post-hoc* amelioration rather than an initial determinant of possibility. It operates on a much longer timescale than economics and finds the use of money as the common currency more difficult to use than does economics. Attempts to develop branches of economics, such as 'environmental' and 'evolutionary' economics, aim to bring the two models of the world and its regions towards a fruitful interface. Much interest has also been generated by the unequal access to resource use by rich and poor people. All these models are only part of a cultural nexus which provides definitions of, and demands for, resources and so other less immediate factors may intervene. Horse-meat is not culturally acceptable in Britain whereas in France it appears on most meat counters; the Amish people of rural Pennsylvania may still eschew automobiles in favour of the horse, which also they do not eat. In past times, much of the world was seen as sacred and thus there were sanctions of what might be done to nature; these have largely been broken down in the search for a higher material standard of resource consumption whose benefits few question.

See also: ECOLOGY, ECOSYSTEM, HYDROLOGICAL CYCLE

Further reading

Daly, H.E. and Farley, J., 2004. *Ecological economics: principles and applications.* Island Press, Washington, DC.

Hodgson, G.M., 1993. *Economics and evolution: bringing life back into economics.* University of Michigan Press, Ann Arbor, MI.

Jones, G. and Hollier, G., 1991. *Resources, society and environmental management*. Chapman, London.

Peet, R. and Watts, M. (eds), 2004. *Liberation ecologies: environment, development, social movements*. Routledge, London and New York, 2nd edn.

Rees, J., 1990. *Natural resources. allocation, economics and policy*. Routledge, London and New York, 2nd edn.

Smil, V., 1993. *Global ecology: environmental change and social flexibility*. Routledge, London and New York.

Turner, R.K., Pearce, D. and Bateman, I., 1994. *Environmental economics*. Harvester Wheatsheaf, Hemel Hempstead.

World Resources Institute at http://www.wri.org/

World Resources Institute, 2000–2001. *People and ecosystems*. WRI, Washington, DC.

World Resources Institute, 2005. *The wealth of the poor. managing ecosystems to fight poverty*. WRI, Washington, DC.

NATURE

'Nature' is a word for the essential qualities of environment. P.J. O'Rourke (1991) said that 'Worship of nature may be ancient, but seeing nature as cuddlesome, hug-a-bear and too cute for words is strictly a modern fashion'. Natural history as the study of the physical and biological environment was popular in the 19th century but was displaced as the environmental sciences became more specialized. It can be argued that nature is a human construct and that it exists only when there are individuals to perceive it and that it is perceived differently by different cultures. For this reason nature has been studied by philosophers, drawing attention to the fact that there are alternative views of nature: as a number of discrete elements including wildlife and endangered species; as a living organism, as a mechanism, as a self-restoring organism or Gaia; as a bottomless mine of resources; as a bottomless sink for the absorption of pollution; or as a sanctuary or temple such as the Grand Canyon. These various views explain why it is perceived in different ways and also why different cultures have their own ways of recognizing nature.

However perceived, it is agreed that nature is threatened by human activity so that various methods have been devised to preserve or to conserve nature, often by saving or setting aside portions of the natural environment. Early conservation efforts focused on the protection of wildlife and/or the establishment of parks. Yellowstone became the first national park when it was established in 1872, was followed by others including Yosemite in 1890 (Fig. 38) in the

Figure 38 *Yosemite National Park. (Photo K.J. Gregory)*

USA and in other parts of the world. Nature conservation was reinforced in the second part of the 20th century with the advent of environmental legislation including the National Environmental Policy Act (NEPA) in 1969 in the USA. A contrast developed between a preservationist view which focused upon setting aside a portion of the natural environment in a relatively pristine state and minimizing the extent to which its environment is degraded, exploited or changed; and a conservationist alternative using the resources of an area in a sustainable way that does not degrade the environment. Most recently this more dynamic conservationist agenda has become more holistic by seeking ways of more sustainable management ensuring that the environment benefits from a wiser use of resources. Aspects of nature can also be preserved and displayed in museums and zoos and the Eden project (P. 429) in Cornwall, UK and Biosphere 2, Arizona are examples of this type of nature conservation. Individuals may express their ideas about nature and environment through gardening and the English landscape garden was a way of creating nature or environment.

More interventionist approaches have been developed when attempts have been made to restore environments to some preceding condition. In areas of environmental degradation, for example where wetlands have been drained or rivers have been straightened and channelized, this has prompted attempts to restore nature. Although offering greater respect for nature it raises issues including whether restoration is possible physically, whether a restored system can ever have the same value as a natural system, and whether 'faked nature' can ever approach a natural, undisturbed, system. A design-with-nature movement can be traced to the concerns of fishermen and hunters for loss of wildlife habitat in mid-19th century Europe but such concerns were largely disregarded in favour of hard engineering design solutions until the mid-1960s when a movement began towards a softer management approach that mimicked the landscape's 'natural' characteristics. Rather than stemming from the earth and environmental sciences as might have been expected, it was the disciplines of landscape architecture and ecology that became most significantly involved. A book *Design with Nature* by Ian McHarg, developed ideas initially applied to the city, subsequently providing a method whereby environmental sciences, especially ecology, could be used to inform the planning process.

See also: CONSERVATION OF NATURAL RESOURCES, ENVIRONMENT, ENVIRONMENTAL SCIENCES, PERCEPTION

Further reading

Attfield, R., 1999. *The ethics of the global environment.* Edinburgh University Press, Edinburgh.

Elliott, R., 1997. *Faking nature: the ethics of environmental restoration.* Routledge, London.

McHarg, I.L., 1969. *Design with nature.* 25th Anniversary edition. Wiley, New York. 1992.

McKibben, B., 1989. *The end of nature.* Random House, New York.

O'Rourke, P.J., 1991. *Parliament of whores.* Picador, London.

Simmons, I.G., 1993. *Interpreting Nature. Cultural constructions of the environment.* Routledge, London and New York.

OCEAN CIRCULATION

The circulation of the oceans is a complicated process affected by the rotation of the Earth, winds, salinity and temperature gradients, and the position of the continents. Ocean circulation is characterized by a six great surface circuits or geostrophic gyres in the north and south Atlantic, north and south Pacific, and in the Indian ocean. Water flows westerly along the equator (the north and south equatorial currents), then towards higher latitudes along the western edges of the oceans, then easterly towards the eastern boundaries of the ocean, and finally back towards the equator to join the equatorial current. In the north Atlantic, the current along the western boundary is the Gulf Stream, this joins the easterly North Atlantic current, and the southerly flowing water mass along western Europe and northwest Africa is the Canary current. There are also two smaller, counterclockwise currents in the North Pacific and North Atlantic. The Antarctic circumpolar current flows in an easterly direction around Antarctica driven by almost continuous westerly winds. If there were no continents, ocean circulation would likely be more like the circumpolar current with bands of east and west moving water masses. These surface currents are driven by winds. Near the equator, the trade winds blow from east to west causing the westerly equatorial currents. The continents deflect the currents north and south. At mid-latitudes, westerly winds drive the currents towards the eastern margins of the oceans. Another important force affecting ocean circulation is the Coriolis effect. This is an apparent force due to the turning of the Earth. The Coriolis effect causes moving water masses to curve to the right in the Northern Hemisphere and to the left in the Southern Hemisphere. The net surface ocean circulation is mainly a result of the interaction of winds and the Coriolis effect.

See also: THERMOHALINE CIRCULATION

Further reading

Garrison, T., 2007. *Oceanography an invitation to marine science.* Thomson Brooks/Cole, Belmont, CA. 6th edn.

Pinet, P., 2006. *Invitation to oceanography.* Jones and Bartlette Publishers, Boston.

Stowe, K., 1996. *Exploring ocean science.* Wiley, New York, 2nd edn.

PARADIGMS

A paradigm is a philosophical and theoretical framework of a scientific school or discipline in which theories, laws, and generalizations and the experiments performed in support of them are formulated. Thomas Kuhn (1922–1996), a philosopher of science, presented the word 'Paradigms' in the sense of its modern meaning when he referred to the set of practices that define a scientific discipline during a particular period of time. However, in his book *The Structure of Scientific Revolutions* Kuhn defines a scientific paradigm as that which is observed and scrutinized and the sort of questions that are supposed to be asked and answered in relation to a subject, how these questions are to be structured, and how the results of scientific investigations should be interpreted. An additional component included in his definition is how should an experiment be conducted, and what equipment is available to conduct the experiment? Within normal science, the paradigm is the set of exemplar of experiments that can be copied or mimicked. 'Paradigm shifts' tend to be most dramatic where they are least expected, as in Physics. At the end of the 19th century, physics seemed to be a discipline filling in the last few details of a largely worked-out system.

In 1900, Lord Kelvin famously stated, 'There is nothing new to be discovered in physics now. All that remains is more and more precise measurement'.[1] Five years later, Albert Einstein published his paper on special relativity, which challenged the very simple set of rules laid down by Newtonian mechanics, which had been used to describe force and motion for over three hundred years. In this case, the new paradigm reduces the old to a special case (Newtonian mechanics is an excellent approximation for speeds that are slow compared to the speed of light). In *The Structure of Scientific Revolutions,* Kuhn stated that 'Successive transition from one paradigm to another via revolution is the usual developmental pattern of mature science' (p. 12). Kuhn's idea was itself revolutionary in its time, as it caused a major change in the way that academics talk about science. Thus, it caused or was itself part of a 'paradigm shift' in the history and philosophy of science. Kuhn's original model is now generally seen as too limited. In the social sciences, the term is used to describe the set of experiences, beliefs, and values that affect the way an individual perceives reality and responds to that perception. Social scientists have adopted the Kuhn phrase 'paradigm shift' to denote a change in how a given society goes about organizing and understanding reality. Paradigms are shaped by both the community's cultural background and by the context of the historical moment. Kuhn defines a paradigm as an entire constellation of beliefs, values, and techniques shared by scientists and philosophers.

See also: EPISTEMOLOGY, PHILOSOPHY OF SCIENCE

Further reading

Kuhn, T.S., 1970. *The structure of scientific revolutions*. University of Chicago Press, Chicago and London, 2nd edn.

Note

[1] Address to the British Association for the Advancement of Science (1900), as quoted in *Superstring: a theory of everything?* (1988) by Paul Davies and Julian Brown; also in *Rebuilding the matrix: science and faith in the 21st century* (2003) by Denis Alexander.

PERCEPTION

Perception is the subjective way in which people image the natural and built environment. The perceived environment is converted according to an individual's beliefs, knowledge and experience to a cognized environment which forms the basis for action. Whereas perception relates to the neurophysiological processes of the reception of stimuli from a person's surroundings through sight, smell and hearing, cognition involving memory, experience, values, evaluation and judgement is also necessary to achieve environmental construction by an individual. Perception in psychology is the process of acquiring and organizing sensory information. Information is stored in the brain in the form of cognitive maps and it is in neural networks that individuals know and think about environment.

In addition to understanding how people individually and collectively achieve understanding of environment, recent examples of the importance of perception include:

» In responding to natural hazards the actual incidence of hazards may differ from the way in which hazards are perceived.
» When determining what is natural as the basis for ecological restoration the environment perceived to be natural may differ from reconstructed natural conditions.
» Wilderness conditions have been shown to be very differently perceived by different people.
» Surface reclamation after surface coal mining can depend upon the perception of what was the original environmental condition rather than the original environment that actually existed.
» Downstream of a dam landowners believe that bank erosion has been instigated since dam construction whereas field measurements show that bank erosion and bed degradation have declined since dam construction with the channel now approaching dynamic equilibrium.

In many environmental decisions it is now required that public views are obtained and that there is public participation in decision-making. However it has to be remembered that perception by individuals and by groups of individuals as well as by decision-makers can influence the way in which decisions are reached. There is therefore ample scope for environmental education to ensure that environment is fully understood by as many people as possible.

See also: ENVIRONMENTAL PERCEPTION, PLACE, HAZARDS

Further reading

Garling, T. and Colledge, R. (eds), 1993. *Behaviour and environment: Psychological and geographical approaches.* North Holland, Amsterdam.

Tuan, Yi-Fu, 1990. *Topophilia. A study of environmental perception, attitudes and values*. Columbia University Press, New York, Revised edn.

White, G.F. (ed), 1974. *Natural hazards local, national, global*. Oxford University Press, New York.

PHILOSOPHY OF SCIENCE

The branch of philosophy concerned with the foundations, presumptions and implications of science, including the physical, environmental and social sciences. The main purposes are critical (to expose confused or mistaken interpretations); overview (the rationale of scientific change, how specific beliefs develop and change, how scientific concepts and terms are produced and used, how specific beliefs are considered knowledge, and how concepts of scientific reasoning and knowledge are formulated), and detailing (through detailed case studies showing how methods and beliefs have developed, including how science explains nature). A journal, *Philosophy of Science*, sponsored by The Philosophy of Science Association, was established in 1934.

Although the environmental sciences were not concerned with the philosophy of their disciplines up to the middle of the 20th century, it became necessary to identify processes of thinking, ways in which data was collected and analyzed, and ways in which conclusions were reached. This led to distinction between inductive and deductive methods and coincided with a time when the environmental sciences were becoming more generalist and seeking general models rather than exclusively based upon detailed field investigations of specific areas. Aided by the quantitative revolution and developments in computing, this meant that general quantitative models could be produced and so it was imperative that scientific methods were understood as a basis for comprehending alternative modes of theory construction and testing. In addition to terms associated with epistemology, Kuhn had suggested that science develops through a series of phases in which a pre-paradigm phase with conflicts focused around individuals, is succeeded by professionalization when definition of the subject is acute, and then succeeded by a paradigm phase, each characterized by a school of thought and each separated by a crisis phase when revolution occurs because problems cannot be solved by the prevailing paradigm. Paradigms have not been accepted by all scientists so that exemplars (concrete problem solutions accepted by the discipline group) were an alternative. A distinction is sometimes made between basic sciences such as physics, chemistry and biology and the composite sciences such as geology, ecology and geomorphology.

Greater concern by the environmental sciences for the philosophy of science has raised awareness of approaches and methods used in each of the constituent sciences, in the course of which particular approaches have been considered (Table 28) additional to those associated with epistemology.

Table 28 *Some terms associated with concern for philosophy of science in the environmental sciences*

Term or approach	Brief explanation
Logical positivism	Scientific theories evaluated solely on the basis of observational data in accordance with a set of formal rules. Tends to employ sampling theory, statistical methods and empirical generalization
Critical rationalism	A rational basis for scientific knowledge is provided by deducing the consequences of theories and then attempting to expose their falsity by critical testing
Methodology of scientific research programmes	Research programmes are the units of scientific achievement and each is supported by a heuristic, problem-solving machinery using mathematical techniques can digest anomalies and convert them into positive evidence
Post-modernism	A reaction to modernism, which abandons the idea of master narratives advocates pluralism and is place-specific.
Realism	Using abstraction to identify the causes and conditionalities of the structures under specific conditions. Occupies a position between empiricism (belief that sense experiences provide the only true basis for knowledge) and idealism (tendency to represent things in an ideal form)

See also: EPISTEMOLOGY, PARADIGMS, POST-MODERNISM

Further reading

Kuhn, T.S., 1962, 1970. *The structure of scientific revolutions*. University of Chicago Press, Chicago. (1970 edn. was revised and enlarged)

Osterkamp, W.R. and Hupp, C.R., 1996. The evolution of geomorphology, ecology and other composite sciences. In Rhoads, B.L. and Thorn, C.E. (eds), *The scientific nature of geomorphology*. Wiley, Chichester, 415–441.

Philosophy of Science. Five issues per year. Published January, April, July, October and December.

Shapere, D., 1987. Method in the philosophy of science and epistomology. In Nersessian, N.J. (ed), *The process of science*. Martinus Nijhoff, Dordrecht, 1–39.

PLACE, SENSE OF

The word is used to refer to the particular part of space occupied by organisms or possessing physical environmental characteristics. Places are located within the spheres in the envelope from about 200 km above the Earth's surface to the centre of the Earth and are not exclusive to any one discipline. Ecology, geology, physical geography, other environmental sciences and landscape architecture focus upon place, each using their own terms and approach. Each discipline has terms for the basic atoms of the environment studied, often for the units which are basic to mapping schemes. In ecology, habitat has come to signify description of where an organism is found, whereas niche is a complete description of how the organism relates to its physical and biological environment. In ecology and biogeography, a habitat plus the community it contains forms a single working system, the ecosystem, thus embodying the community in a place together with the environmental characteristics, relief, soil, rock type (the habitat) that influence that community. In soil investigations the site was a primary feature because it was the place over which soil profiles were investigated. Soil profiles and sites are then grouped, or classified, into the fundamental soil mapping unit which might be a series, defined as groups of soils with similar profiles formed on lithologically similar parent materials.

Ways in which places are associated, arranged and interrelated in the environment is the basis for system models which can be identified for climate (climosequences); relief (toposequences); lithology (lithosequences); ecology (biosequences); and time (chronosequences). Recurrent patterns of spatial variation have included the catena concept, expressing the way in which a topographic sequence of soils of the same age and usually on the same parent material, can occur in landscape usually reflecting differences in relief/slope and drainage, an arrangement which others have described as a toposequence. Investigations have progressed from an initial focus on unique places, to generalization in models, to concern with the individuality of places necessary for environmental management.

See also: EARTH SPHERES, ECOSYSTEM, LANDSCAPE ECOLOGY

Further reading

Gregory, K.J., 2003. Place: The management of sustainable physical environments. In Holloway, S.L., Rice, S.P. and Valentine, G. (eds), *Key concepts in Geography*. Sage, London, 187–208.

Phillips, J.D., 2001. Human impacts on the environment: unpredictability and the primacy of place. *Physical Geography* 22, 321–332

POST-MODERNISM

Post-modernism is a reaction to modernism in the sphere of culture, which abandons the idea of master narratives. Especially relevant here is the rejection of the objectivity of the natural sciences in favour of localized findings specific to the people and places involved in them. The sociology of science for example is much concerned with the cultural background and motivations of individual researchers, with the culture of their field (what to publish, where and with whom?), and the institutional constraints they may find. Knowledge is therefore of its time and place and not necessarily part of a master narrative of the whole of the scientific endeavor. This notion extends itself into the area of values, where any pretence to universal values (such as justice) is abandoned: there are only local, time-specific justices. In the sphere of the environment, a number of developments are in accord with the post-modern movement. There is to some extent a rejection of consumerism, brought about by the realization that a high-carbon economy creates as many costs as benefits; thus the goods brought by mass production are valued less than those made in small batches by craft workers. Foods produced by 'organic' farmers are sought above those in which industrial production methods are dominant. Needless to say, these are all found among those who are already affluent: the poor are less impressed. There is however one other major environmental linkage of importance. Post-modernism has largely rejected the idea of an external world (i.e., 'nature' as usually defined) and instead adopted the notion that 'environment' is a cultural construct which is subject to the relativism inherent in the abandonment of master narratives. This contrasts strongly with the picture given by science where nature is an integrated system with its own feedback loops and which has limits which humans transgress only at high costs.

See also: CULTURES, MODERNISM

Further reading

Soulé, M. and Lease, G. (eds), 1995. *Reinventing nature?: responses to postmodern deconstruction.* Island Press, Washington, DC.

PROCESS–FORM RELATIONSHIPS

Process–form relationships refer to those links between physical and biochemical processes that cause change at or near the surface of the Earth to the form of the land. For example, runoff from precipitation over the land is concentrated in channels that eventually form rivers that flow to the sea. The running water is a process that interacts with materials such as sand and gravel that self-organize to form the bed and banks of a stream system. The flowing water modifies and forms the river channel, but the river channel itself organizes to control the flow that causes deposition and erosion of materials at varying locations along the river. When the relationships between process and form are investigated, the chicken-and-egg paradox soon appears. Does the process produce the form or does the form somehow produce the process? In reality, the Earth works in harmonious ways through feedback mechanisms where process and form develop together to produce the surface of the Earth that are observed by humans.

In environmental science, the linkages and relationships between process and form produce the major ecosystems that allow life to persist on Earth. For example, wildfire is a natural rapid biochemical oxidation process that helps balance the global carbon cycle. The wildfire also helps maintain the grasslands of the world and rejuvenates many other ecosystems. Wildfires in southern California greatly increase erosion rates and production of sediments from landslides and debris flows. This sediment enters channel systems and as it works its way through river systems, providing important resources for aquatic organisms such as fish. Were it not for the delivery of coarse sediment to rivers, spawning gravels for fish would likely be depleted.

Another area where process and form are tightly linked is coastal environments. Linkages between processes in the ocean that deliver waves to the shoreline help build and maintain beaches while driving sediment transport along the coast. In some areas, wave energy is feeble and the coastline is much different from those that receive high-energy waves capable of causing rapid change in a coastline.

See also: ENVIRONMENTAL SCIENCES, FIRE, PHILOSOPHY OF SCIENCE

Further reading

Keller, E.A. and Pinter, N., 2002. *Active tectonics*. Prentice Hall, Upper Saddle River, NJ.

Bull, W.B., 1991. *Geomorphic response to climate change*. Oxford University Press, New York.

PULSING

The most productive ecosystems are generally those that receive high levels of energy subsidies (e.g., tides, river floods, upwelling) in addition to solar energy. Often these extra sources are seasonal or regular 'pulses' that not only enhance productivity but also contribute to the stability of the ecosystem. At the same time, pulses make the system more 'open' to exchange of elements with adjacent systems. This concept of stability with pulses was termed 'pulse-stability' by E.P. Odum. Examples of pulsing events that contribute to the structure and productivity of ecosystems include ice storms, fires, large river floods, hurricanes, droughts, etc. Much of the southeast of the United States is dominated by pine fire-climax ecosystems.

Pulsing is important in river and deltaic ecosystems. Understanding river ecosystems has evolved from the River Continuum Concept of Vannote *et al.* where the longitudinal pattern changes from upstream to downstream are emphasized, to the Flood Pulse Concept of Junk *et al.* where exchange between a river and its floodplain is emphasized as the main factor determining the function of both the river and its adjacent riparian floodplains.

Deltas provide a good example of the importance of ecological and energetic pulsing events. These energetic events range from waves and daily tides to switching of river channels in deltas that occur on the order of every 500–1000 years, and include frontal passages and other

Table 29 *A hierarchy of forcings or pulsing events affecting the formation and sustainability of deltas (modified from Day et al. 1997)*

Event	Timescale	Impact
Major changes in river channels	500–1000 yrs.	New delta lobe formation; Major sediment deposition
Major river floods	50–100 yrs.	Avulsion enhancement; Major sediment deposition; Enhancement of crevasse formation and growth
Major storms	20–25 yrs.	Major sediment deposition; Enhanced production
Average river floods	Annual	Enhanced sediment deposition; Freshening (lower salinity); Nutrient input; Enhanced 1° and 2° production
Normal storm events (frontal passage)	Weekly	Enhanced sediment deposition; Enhanced organism transport; Higher net materials transport
Tides	Daily	Marsh drainage; Stimulated marsh production; Low net transport of water and materials

frequent storms, normal river floods, strong but infrequent storms such as hurricanes, and great river floods (Table 29). A primary importance of the infrequent events such as channel switching, great river floods and very strong storms (e.g., hurricanes) is in sediment delivery to coastal systems and in major spatial changes in geomorphology. The more frequent events such as annual river floods, seasonal storms such as frontal passages and tidal exchange are important in maintaining salinity gradients, delivering nutrients and regulating biological processes. To understand the impacts of human activities and climate change on fluvial systems in general, and coastal systems specifically, and to effectively manage these systems, it is necessary to understand how human and climate change impact coastal systems on all these spatial and temporal scales.

See also: ECOSYSTEM, HAZARDS, SENSITIVITY

Further reading

Day, J.W., Martin, J.F., Cardoch, L. and Templet, P.H., 1997. System functioning as a basis for sustainable management of deltaic ecosystems. *Coastal Management* 25, 115–153.

Day, J.W., Psuty, N.P. and Perez, B.C., 2000. The role of pulsing events in the functioning of coastal barriers and wetlands: Implications for human impact, management and the response to sea level rise. In Weinstein, M. and Dreeger, D. (eds), *Concepts and controversies in salt marsh ecology*. Kluwer, Dordrecht, 633–660.

Junk, W.J., Bayley, P.B. and Sparks, R.E., 1989. The flood pulse concept in river-floodplain systems. In D.P. Dodge (ed), Proceedings of the international large river symposium. Special Issue of *Journal of Canadian Fisheries and Aquatic Sciences* 106, 11–127.

Junk, W.J., 1999. The flood pulse concept of large rivers: learning from the tropics. *Archiv fur Hydrobiologie* Suppl. 115, 261–280.

Odum, E.P., 1980. The status of three ecosystem-level hypotheses regarding salt marsh estuaries: tidal subsidy, outwelling, and detrital-based food chains. In V. Kennedy (ed), *Estuarine perspectives*. Academic Press, New York, 485–495.

Odum W.E., Odum E.P. and Odum H.T., 1995. Nature's pulsing paradigm. *Estuaries* 18, 547–555.

Tockner, K., Malard, F. and Ward, J.V., 2000. An extension of the flood pulse concept. *Hydrologic Processes* 14, 2861–2883.

Vannote, R.L., Minshall G.W., Cummins, K.W., Sedell, J.R. and Cushing, C.E., 1980. The river continuum concept. *Canadian Journal of Fisheries and Aquatic Sciences* 37, 130–137.

REDUCTIONISM

This term refers to the view that a system can be understood by the operation of its ultimate component parts or sub-systems. It has come to be applied to the approach adopted in environmental sciences whereby increasingly detailed study and investigation leads to more sub-divisions of disciplines into increasingly specialized branches of research. Thus explanation may be sought at the cell level or successively at the molecular, atomic or sub-atomic level. A danger of such increasingly detailed investigation is the loss of the overall perspective which is why holism offers an alternative and holistic thinking has been encouraged by systems approaches. Whereas reductionism focuses on increasingly detailed levels of investigation the antonym is emergence whereby progression is from the detailed investigation of physics and chemistry to a higher level in biology or in other environmental sciences. A long-standing view (dating from Descartes in the 17th century) is that understanding the world can be achieved by taking environment apart, just like the components of a machine, so that potentially there are no undiscoverable scientific facts. However this is challenged by the fact that the level to which it is possible to investigate environmental systems depends upon technology available and questions whether there are finite components beyond which investigation is not feasible and also whether the whole is greater than the sum of the individual parts as in holism. A reductionist approach therefore assumes that all biology might ultimately be explained in terms of physics and chemistry. It is possible to see disciplines according to the level of investigation that they employ so that physics is the most fundamental followed by chemistry, biology, geology, physical geography, psychology, sociology, anthropology and economics.

See also: HOLISM, SYSTEMS: THEORY, APPROACH AND ANALYSIS

Further reading

Dawkins, R., 2006. *The blind watchmaker*. New edn. Norton, London and New York.

SELF-DESIGN

S elf-design is probably the most important 'design concept' incorporated into ecological engineering. It acknowledges that nature, through seed dispersal, bird migration, sediment dispersal, etc., is the chief contractor. Humans simply create a system that will allow nature to manifest itself. The following quotation from Mitsch and Jorgensen in 2003 illustrates these points:

> Self-design and the related concept of self-organization must be understood as important properties of ecosystems in the context of their creation and restoration. Self-organization is the property of systems in general to reorganize themselves, given an environment that is inherently unstable and non-homogeneous. Self-organization manifests itself in microcosms and newly created ecosystems 'showing that after the first period of competitive colonization, the species prevailing are those that reinforce other species through nutrient cycles, aids to reproduction, control of spatial diversity, population regulation, and other means' (p. 28)

Ecological engineering, defined by Mitsch and Jorgensen as 'the design of sustainable ecosystems that integrate human society with its natural environment for the benefit of both' (p. 23), involves creating and restoring sustainable ecosystems that have value to both humans and nature (Table 30). This approach combines basic and applied science for the restoration, design, and construction of aquatic and terrestrial ecosystems. Ecological engineering relies primarily on the energies of nature, with human energy used in design and control of key processes. Ecological engineering combines basic and applied science for the restoration, design, and construction of aquatic and terrestrial ecosystems. The goals of ecological engineering are to restore ecosystems that have been substantially disturbed by human activities

Table 30 *Systems categorized by types of organization (modified from Pahl-Wostl, 1995)*

Characteristic	Imposed organization	Self-organization
Control imposed	Externally	Endogenously
Control type	Centralized	Distributed
Rigidity	Rigid networks	Flexible networks
Potential for adaptation	Little potential	High potential
Application	Conventional engineering	Ecological engineering
Examples	Machine	Organism
	Fascist or socialist society	Democratic society
	Agriculture	Natural ecosystem

such as environmental pollution or land disturbance; and to develop new sustainable ecosystems that have both human and ecological value.

This is engineering in the sense that it involves the design of this natural environment using quantitative approaches grounded in the basic science of ecology. It is a technology where the primary tools are self-designing ecosystems. It is biology and ecology in the sense that the components are all of the biological species of the world, and is often involved in ecological restoration.

See also: ECOLOGICAL ENGINEERING, ECOTECHNOLOGY

Further reading

Botkin, D.B. and E.A. Keller, 2007. *Environmental science – Earth as a living planet.* Wiley, New York.

Kangas, P.C., 2004. *Ecological engineering – principles and practice.* Lewis Publishers. Boca Raton, FL.

Mitsch, W.J. and Jørgensen, S.E., 2003. *Ecological engineering and ecosystem restoration.* Wiley, New York.

Pahl-Wostl, C., 1995. *The dynamic nature of ecosystems: chaos and order entwined.* Wiley, New York.

SYSTEMS: THEORY, APPROACH, AND ANALYSIS

Systems have been defined in a number of ways. A system is a set of components that function together to act as a whole. According to Hall and Day in 1977 any phenomenon, either structural or functional, having at least two separable components and some interaction between these components may be considered a system. Another more general definition is any object whose behaviour is of interest. Depending on ones point of view, systems may be the actual building blocks of nature or an effort by humans to impose order on a seemingly chaotic natural world. Systems analysis is the formalized study of systems and of the general properties of systems. Holism is the approach to studying the total behaviour, or other total attributes, of complicated systems. Lugwig von Bertalanffy (1901–1972) was a pioneer and founder of general systems theory. His 1968 text is one of the classic books in the field. He wrote that:

systems theory is a broad view which far transcends technological problems and demands, a reorientation that has become necessary in science in general and in the gamut of disciplines from physics and biology to the behavioural and social sciences and to philosophy. It is operative with varying degrees of success and exactitude in various realms, and heralds a new world-view of considerable impact.

Systems are universally part of everyday lives. Humans live with these systems, often hardly ever thinking of them. For example, human bodies have a digestive system and their houses have an electrical system. Yet, each of these has components that can be considered systems (the stomach is an organization of muscles, acid-producing cells, and other tissues), and in turn each is part of larger system, for example a house or a body. Thus, any system studied is part of a hierarchy of systems and hierarchy is a characteristic of systems in general. Systems may be open or closed. Practically speaking almost all systems of nature are open systems. For all practical purposes, the Earth can be considered a closed system in terms of matter, although energy enters and leaves. Systems are characterized by feedbacks, that is, a stimulus to the system most often elicits a reaction that affects how the system responds to the stimulus. Feedbacks can be positive or negative. Positive feedback is when an increase in output results in more increase in output. For example, if an animal population is allowed to grow with little restraint, this results in exponential growth. This has been the case for the human population for several centuries. However, ultimately for any population, forces come into play that reduce the rate of growth, such as disease or food scarcity. This is negative feedback in which an increase in output leads to a later decrease in output.

Systems analysis came about because of the limitations of analytical solutions to many problems of science and technology. For many of these problems that can be expressed mathematically, an analytical solution is impossible and interactive simulation solutions are necessary. In common language, this reflects the fact that generally speaking, the whole is more than the sum of its parts. Systems theory and systems analysis deals with problems that are more of a general nature, such as gaining an understanding of the systems of nature. Because the systems approach deals with complicated problems that almost always have non-linear interactions, computers and simulation models are commonly used. Often these models have many non-linear equations. A number of theoretical approaches and theories have grown out of systems theory and the systems approach. Some of these are set theory, network theory, cybernetics, information theory, game theory, and decision theory.

In the natural world, the ecosystem is the basic systems unit. Science can consider cells, individuals, populations, and communities as systems. But it is at the ecosystem level, that an integrated system that includes both biotic and abiotic elements emerges. An ecosystem is an organized system of land, water, mineral cycles, living organisms and their programmatic behavioural control mechanisms. It includes all the organisms living in a community as well as the abiotic factors with which they interact. Forests, lakes, estuaries, tundra, coral reefs are all different types of ecosystems. Ecosystems can be characterized in terms of trophic (feeding) relationships, energy flow, structure of community of organisms living in the ecosystem, and cycling of chemical elements. Human impacts are often described at the ecosystem level. Human impacts often affect entire ecosystems (see pollution of air and water). Increases in inorganic nutrients, such as runoff from agricultural fields, lead to enrichment in aquatic ecosystems and cause eutrophication (algal blooms, fish kills, and low oxygen). Humans often physically change ecosystems such as the construction of dams and cutting of forests. Toxic materials are released into the environment leading to both immediate, acute impacts (such as poisoning) and long-term, cumulative impacts (such as concentration of pesticides in the food chain). Systems analysis and the systems approach have proven very valuable in studying and solving problems of human impacts. A formalized approach has been developed to understand and manage in a sustainable manner the system of humans and nature. This is called the ecosystem approach and is defined as a strategy for management of land, water and living resources that promotes conservation and sustainable use in an equitable way. Capturing and optimizing the functional benefits of ecosystems is emphasized. The importance of biodiversity management beyond the limits of protected areas is emphasized, while protected areas are recognized as being vitally important for conservation. The flexibility of the approach with respect to scale and purpose makes it a versatile framework for biodiversity management. Trans-boundary biodiversity problems can be addressed using the ecosystem approach and regional political structures.

See also: ECOSYSTEM, HOLISM, MODELLING, PARADIGMS, PHILOSOPHY OF SCIENCE

Further reading

Botkin, D.B. and Keller, E.A., 2007. *Environmental science – earth as a living planet.* Wiley, New York.

Haefner, J.W., 1996. *Modeling ecological systems – Principles and applications.* Chapman and Hall, New York.

Hall, C.A. and Day, J.W., (eds). 1977. *Ecosystem modeling in theory and practice: an introduction with case histories.* Wiley, New York.

Kormondy, E., 1995. *Concepts of ecology.* Prentice Hall, New Jersey.

Von Bertalanffy, L. 1968. *General systems theory.* George Braziller, New York.

THRESHOLDS

Thresholds are stages, or tipping points, for an environmental system at which the essential characteristics of the natural system's state change dramatically or where there is a significant impact on socioeconomic systems. Some environmental thresholds are well known, including the wilting point which occurs when plants can no longer extract water from the soil, the plant then starts to wilt and may die. In hydraulics there are values (Froude and Reynolds numbers) which define the conditions at which flow becomes supercritical or turbulent, and there also are threshold velocities that are necessary to move sediment particles of a particular size. As velocity increases the threshold velocity is when movement begins and later as velocity decreases there will be a threshold velocity at which movement ceases. Soil erosion by gullying may occur when a certain amount of vegetation is removed from the surface – the threshold at which erosion begins.

An intrinsic threshold is when changes take place within the system without change of an external variable. Thus long-term progressive weathering on a slope can lead to reduced strength which at a particular threshold level will fail by landsliding or other type of mass movement. The build-up of snow and ice on a glacier may occur up to a threshold level at which a glacier surge occurs, such surges occurring periodically even though the precipitation as snow remains consistent over decades. A particular type of intrinsic threshold is a geomorphic threshold whereby a change within the system leads to incipient instability and failure as when sediment storage increases in a valley of a semi-arid area until the point at which failure occurs by gullying. An extrinsic threshold is when an abrupt change occurs as a response to an external variable, illustrated by the flow velocity necessary to instigate particle movement or the removal of vegetation cover triggering surface erosion.

Thresholds are important for environmental management but are not always easy to define and isolate. Linear modelling has often been used to identify threshold conditions but non-linear methods such as catastrophe theory may be more appropriate. Threshold conditions may relate to global change, for example in relation to circulation of the north Atlantic where the southward flow of Arctic meltwater could deflect the Gulf Stream and then affect the climate of the British Isles.

See also: CATASTROPHE THEORY

Further reading

Schumm, S.A., 1979. Geomorphic thresholds: the concept and its applications. *Transactions of the Institute of British Geographers* 4, 485–515.

UNIFORMITARIANISM

Uniformitarianism is a major principle of the earth sciences and in particular geological sciences. Most simply stated, uniformitarianism states that the present is the key to the past. The idea is that basic physical, chemical and biological processes have operated for much of Earth's history. As a result, studying present-day processes can help interpret the geologic history. For example, if stream and river gravels are studied, along with how they are transported and what the deposits look like, the rock record allows an inference that similar-looking deposits were deposited by a river sometime in the past. This is true even if the deposits are found presently at the top of a mountain. For example, the top of Mount Everest is limestone, which forms in ocean basins and shallow bays and other marine environments today. Therefore, uniformitarianism postulates that the limestone at the top of Mount Everest began its history at a very low-lying place and has been uplifted during the past millions of years as a result of tectonic activity related to the collision of India with Eurasia.

Uniformitarianism does not demand that processes operating today have operated throughout the entire history of the planet or that the processes have operated at the same rate. However, so long as these processes have operated, inferences can be made about the past from study of present processes. In environmental science, it is important to turn uniformitarianism around and say that the present is a key to the future. That is, by studying past and historic activity of floods, landslides, and earthquakes, inferences are made about where these events are likely to occur in the future. By studying the return period of past events, it may be possible to predict the probability of future earthquakes. For example, the way the probability of a flood of a particular magnitude is made involves the examination of the occurrence of past floods from a record of flows from a gauging station.

When uniformitarianism is used to make predictions about the future, there is no demand that there must be gradual conditions throughout long periods of time. Occasionally, very large events may occur and these are often followed by a sequence of similarly large events. For example, large earthquakes on a long fault are often clustered in time. Thus, in the study of Earth history and environmental science in order to develop plans to minimize damage from natural processes or hazards, uniformitarianism is used in a variety of ways, but generally it involves studying of the past records to make predictions about the future.

See also: EARTH SURFACE PROCESSES, GEOLOGY, GEOMORPHOLOGY, PHYSICAL GEOGRAPHY

Further reading

Keller, E.A., 2005. *Environmental geology*. Prentice Hall, Upper Saddle River, NJ.

Shea, J., 1982. Twelve fallacies of uniformitarianism. *Geology* 10, 455–460.

PART IV

PROCESSES AND DYNAMICS

INTRODUCTION

Lord Kelvin (1824–1907), a founder of modern physics, contended 'when you can measure what you are speaking about and express it in numbers, you know something about it, but when you cannot express it in numbers your knowledge is of a meagre and unsatisfactory kind'. Finding out in science has been much associated with measurement but is not so easy for the environment: it was necessary to observe and classify, to organize and then to measure shapes before later moving on to measure processes, dynamics and change. Although collecting, whether it be plants, butterflies, fossils or rocks, was one response of the earliest environmental scientists, this had to give way to serious classification systems of plant and animal taxonomy, with more data eventually needed to improve understanding of how the environment works.

Exploration was vital for obtaining more information about other parts of the world and was rather like collecting; explorers communicated their experiences and trophies. National surveys and mapping agencies were established to document the characteristics of environment. In Britain the Ordnance Survey, founded in 1795, and the Geological Survey, established in 1801, both initiated national mapping. In the USA the USGS (United States Geological Survey) took responsibility for mapping the country in 1879, when the first topographic map was published, continues as the primary civilian mapping agency, and from 1899 to 1935 about half of the USA was provided with soil maps. World-wide, the UN stated that in 1976 well in excess of £2000 million was spent each year on surveying and mapping. However how often does an environmental scientist find that the information required is not provided by these national surveys?

Measurement of processes and dynamics is not as easy in environmental sciences as are controlled experiments in laboratory sciences. As there seem to be an infinite number of points in space and time, some form of sampling is needed. This is undertaken in the context of a frame of reference, often in the form of a hypothesis based upon a specific context or paradigm. There is scope for laboratory experiments, for example using a hardware model of a stretch of coastline, but the difficulty of scaling down sand or pebbles on a beach to smaller material in the laboratory gives fine clay which will not behave in the same way. Theoretical equations can be used to deduce mathematically how certain processes operate but in most cases so many variables have to be 'controlled' that the theoretical method is rarely sufficient on its own. Sometimes historical data can tell us about processes, in the way that maps of vegetation and land use at different dates show how processes of change have occurred and at what rate. However, such cases inform us about certain times so that deductions made for intervening periods may or may not be correct; there is no substitute for continuously measuring change in environmental systems although this is often expensive in terms of cost of equipment and people. Happily, national data collection schemes afford much basic data: an automatic weather station was shown to the Royal Society by Robert Hooke in 1679, rainfall records have been kept at Burnley, England since 1677 and by 1950 the US Weather Bureau had 10,000 regular and cooperative stations measuring precipitation. The beginning of oceanography as a science is usually ascribed to the five year voyage of HMS Challenger (1872–1876) which explored ocean depths, analyzed sea water samples, studied fauna and collected sediment samples from all major world oceans. The first records of river stage were made for the Elbe at Magdeburg, Germany from 1727–1869, and systematic and continuous measurements of streamflow began in the USA in 1900, with a basic network established from 1910 to 1940 so that, by 1950, observations were being made regularly at 6000 points. Around the coast of the UK, permanent records at tide gauges have been made since 1860.

Such process measurements are vital but it is often important to have other continuous records. With the expansion of the environmental sciences, the cost of acquiring the necessary continuous process data from sufficient locations could have been a stumbling block, limited by cost and persons needed to collect and analyze the data. However this potential crisis for the environmental sciences was largely solved by several developments including the computer,

Figure 39   *Flooded Lake Forest area of New Orleans, Louisiana after Hurricane Katrina. Source: ISTOCK © Joseph Nickischer. Hurricane Katrina was one of the strongest storms ever to hit the coast of the United States bringing intense winds (offshore speeds up to 250 km per hour), high rainfall (up to 34cm/13.6 inches fell in some areas over the 24-hour period), waves, and storm surge. Winds in category 2 and a tidal surge of 7m equivalent to a category 3 hurricane resulted as the centre of the hurricane passed to the east of New Orleans on 29 August. More than 50 breaches in drainage canal levees contributed to what has been described as one of the worst disasters in US history. More than 75% of New Orleans was flooded by 31st August, some areas were under 4m of water and the flooding was largely the result of breaches in the levees in the city. More than 1100 people died, thousands of homes were destroyed, direct damage to residential and non-residential property was estimated at $21 billion and damage to public infrastructure a further $6.7billion. A report produced by the American Society of Civil Engineers, Hurricane Katrina External Review Panel considers What Went Wrong and Why (HYPERLINK ""http://www.asce.org/files/pdf/ERPreport.pdf) and plans are in progress to guard against similar disasters in the future. Such an event demonstrates the power and dynamics of environmental processes.*

the satellite and the microchip together with inspired technological advances which have generated a revolution in environmental process monitoring. By 1980 virtually the whole world was monitored on an 18-day cycle by the Landsat satellite and there is currently a great range of satellite platforms. No longer does hard copy of process measurements need to be brought from the field because it can be accessed remotely and retrieved in an already processed form. A deep-sea submersible was launched in the 1960s allowing direct observation of deep sea environments and drill ships have probed ocean floors since 1968. The Optically Stimulated Luminescence (OSL) method is a dating technique which can be used on individual grains and can indicate the burial time of Quaternary deposits. Radionuclides deriving from the fallout from thermonuclear weapons testing in the late 1950s and 1960s have enabled stunning improvements in dating of erosion rates and of accumulation of sediments, for example, on floodplains. Overall, the great achievements have been the ability to make measurements in the remotest parts of the planet and its atmosphere, together with the development of ever more sensitive instruments, so that ever lower concentrations of contaminants in oceans and atmosphere can be detected.

In 1930 Albert Einstein said 'I never think of the future. It comes soon enough' but we now have to think seriously about the environmental future aided by even more imaginative ways of expanding our knowledge of processes and dynamics. Until our actions are based on accurate, tested and critically evaluated knowledge of both the components and the dynamics of environmental processes then the risks of producing uncontrollable amplitudes of environmental change will get ever higher (Fig. 39).

ACID RAIN

Acid rain, defined as any type of precipitation with a pH that is about 5.6 or less, is produced through sulphur dioxide and nitrogen oxides that enter the atmosphere and are absorbed by water droplets. This process can increase the soil's acidity as well as affect the chemical balance of lakes and streams. Acid rain causes weathering rates in carbonate rocks and structures to increase, in addition to contributing to high acidity in rivers and streams, as well as damages to trees at high elevation (see Fig. 40). The first reporting of such a phenomenon was by Robert Angus Smith in Manchester, England, in 1852 – Manchester was an important city during the Industrial Revolution. Smith studied this phenomenon for a number of years and correlated the relationship between acid rain and atmospheric pollution. This discovery was later (1872) labeled 'acid rain' (Seinfeld and Pandis, 1998). Although acid rain was essentially 'discovered' in 1852, it was not until the late 1960s that scientists began widely acknowledging the phenomenon.

In the 1990s, the New York Times published newsworthy reports from the Hubbard Brook Experimental Forest in New Hampshire, which demonstrated the environmental impact resulting

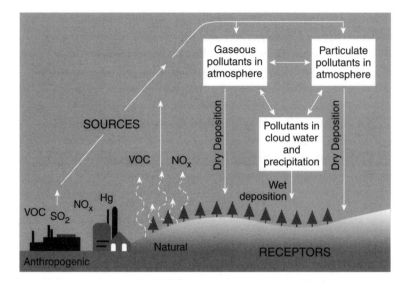

Figure 40   *Sources and receptors involved in the acid rain process. http://www.epa.gov/acidrain/what/index.html*

from acid rain. Industrial acid rain is a substantial problem in China, Russia, and Europe. The same problem also affects areas down-wind from these countries. Acid rain from power plants in the mid-western United States has also harmed the forests of upstate New York and New England. Organisms known as diatoms inhabit ponds. They die and are deposited in layers of sediment on the bottoms of the ponds. They also thrive in certain pHs, so the numbers of diatoms found in layers of increasing depth give an indication of the change in pH over the years.

pH readings of well below 2.4 (the acidity of vinegar) have been reported in some industrialized areas. These sources, in addition to the transportation sector, are the major causes of increased nitrogen oxides. Scientific studies suggest that the principal natural phenomena that contribute acid-producing gases to the atmosphere are emissions from volcanoes as well as emissions from biological processes that occur on the land, in wetlands, and in oceans. The effects of acidic deposits have been detected in glacial ice that is thousands of years old. Decades of enhanced acid input have increased the environmental stress on high elevation forests and on aquatic organisms in Earth's sensitive ecosystems. In some instances, input has altered entire biological communities, at times eliminating fish species from certain lakes and streams. This is especially the case in the northeastern United States and Canada, where rain tends to be most acidic, and often the soil has less capacity to neutralize the acidity.

Changes in ecosystems have been relatively unnoticeable but these changes have reduced the diversity of organisms. In the United States, Congress developed the Acid Rain Program under Title IV (Acid Deposition Control) of the Clean Air Act Amendments of 1995. The idea was to require reductions in NO_x and SO_2. However, the problems of acid rain still remain in the northeast United States. In a response to clean up Europe's air, in 1979 the United Nations

Economic Commission for Europe (UNECE) implemented the Convention on Long Range Transboundary Pollution, with the aim of reducing acidic emissions. Since its implementation, sulphur emissions across Europe have fallen significantly, but with the increase in volume of traffic, nitrogen oxide emissions have only seen slight improvements. Acid rain in Europe will therefore continue to be a problem until these emissions can be dramatically reduced.

See also: HYDROLOGICAL CYCLE, POLLUTION OF AIR AND WATER, PRECIPITATION

Further reading

Some material from Acid Rain (Wikipedia)

Berresheim, H., Wine, P.H. and Davies, D.D., 1995. Sulfur in the Atmosphere. In Singh H.B. (ed) *Composition, chemistry and climate of the atmosphere*. Van Nostrand Reinhold, London and New York, 251–307.

Brimblecombe, P., 1996. *Air composition and chemistry*. CUP, Cambridge

Rodhe, H., Dentener, F. and Schultz, M., 2005. The global distribution of acidifying wet deposition. *Environmental Science & Technology* **36**, 4382–438.

Seinfeld, J.H. and Pandis, S.N., 1998. *Atmospheric chemistry and physics—from air pollution to climate change*. Wiley, Chichester and New York.

ALBEDO

Albedo is typically labeled as the ratio (0 to 1.0) or percentage (0% to 100%) of scattered and reflected to incident sunlight, although the term also refers to longer wavelengths as well. In visible light, albedos of typical materials range from about 4–10% for oceans, water, dark soils to up to 90% for fresh snow. Most land areas have an albedo of 10 to 40%, with the average albedo of the Earth at about 30%. The albedo of a pine forest is around 10%. This is in part due to the dark colour of the pines, and in part due to multiple scattering of sunlight within the trees that lowers the overall reflected light level. Due to light penetration, the ocean's albedo is even lower (at about 3.5%), although it depends on the angle of the incident radiation. The higher the sun in the sky, the lower the albedo, generally. The following are rather low albedoes overall – wet soils, forest, meadows (between 10–15%). A barren field can range from as low as 5% to as high as 40%, depending on the soil's colour – with 15% being about the average

Table 31 *Table of Albedo values*

Material	% Albedo
Oceans	3.5%
Water	8%
Soil wet, dark	10%
Forest	10%
Meadows	15%
Savanna	18%
Crops	20%
Grass	25%
Sand wet	25%
Soil dry	25%
Desert	28%
Ice	35%
Sand dry	40%
Snow	fresh 85–90%; old 40%
Stratus clouds	50%
Altostratus, cirrus	50%
Cumulus clouds	70%

Urban areas (after Brazel and Quatrocchi, 2005)	
Glass	9.0
Blacktop/asphalt	10.3
Tar-gravel roof	13.5
Yard (90% lawn, 10% soil)	24.0
Roofing shingles	25.0
Concrete	27.1
Paint, dark	27.5
Stone	31.7
Brick, red	32.0
Brick, yellow/buff	40.0
Brick, white/cream	60.0
Paint, white	68.7

for farmland and meadows. A desert or large beach is usually around from 25–40% (wet to dry sand), but this could vary due to the colour of the landscape. The exact type of cloud can dictate the albedo of the climate system (see ranges in Table 31). In the northern latitudes, cities are relatively dark, with average albedo at about 5–10%, increasing only slightly during the summer. This is due to the composition of materials such as asphalt, glass, and tar roofs (10–15%). Within a diurnal period for most surfaces the albedo has a marked variation – from lower values at higher sun angles to higher values at lower sun angles.

See also: RADIATION BALANCE OF THE EARTH

Further reading

Brazel, A.J. and Quatrocchi, D., 2005. 'Urban climatology.' In *Encyclopedia of world climatology*, 766–779. Springer Pubs.

Hummel, J.R. and Reck, R.A., 1979. A global surface albedo model. *Journal of Applied Meteorology* 18, 239–253.

Miller, D.H., 1981. *Energy at the surface of the earth*. International Geophysics Series, No. 27, Academic Press, Oxford.

Rees, W.G., 1990. *Physical principles of remote sensing*. CUP, Cambridge.

Weast, R.C. (ed), 1981. *Handbook of chemistry and physics*. CRC Press, Boca Raton FL.

ANIMAL BEHAVIOUR

Behaviour is a fundamental characteristic of animals. Dictionary definitions of behaviour include phrases such as 'way of behaving or acting, to act or react, to respond to some stimulus.' Animal behaviour is the scientific study of what an animal 'does and how it does it.' Behaviour results from both genetic and environmental factors. But it is not an either or situation and there is not a sharp distinction between the two types of behaviour. Environmental factors influence the way that genetically based behaviour is expressed. Innate or intrinsic behaviour is fixed in the genes of an animal. It is said to be developmentally fixed in that almost all individuals exhibit the same response in spite of environmental variations. A new-born infant, if held by its mother, naturally turns its head and begins to suckle. In contrast to innate behaviour, learned behaviour is modified by experience. This is called learning.

Both innate and learned behaviour have their basis in the evolutionary history of a particular species. In the past, specific behaviours developed because they provided an evolutionary advantage; the behaviour increased the potential for survival. A decline in response to a repeated stimulus is called habituation. This saves energy. For example, a person moving to a city is at first likely to be disturbed by traffic and sirens, but then learns to ignore much of it. Trial-and-error learning is a learned association between a stimulus and a response. This is called conditioning. An animal can learn, for example, which foods taste good and which do not.

Modern animal behaviour developed from the field of ethnology where behaviour of animals was studied in their natural habitat. Animal communication is a component of the study of animal behaviour. The communication involves visual, auditory, and chemical cues that provide information on the animal's location, level of aggression, condition, readiness to mate, etc. Communication is defined as 'the production of a signal by one organism that causes another organism to change its behaviour in a way beneficial to one or both'. Most communication occurs between members of the same species. Visual communication is most effective over short distances while sounds are more effective over long distances. Dogs barking at night let other dogs know their territory. Chemical messages last longer such as a dog marking its territory. Behaviour helps animals in the competition for resources. Optimal foraging allows animals to choose foraging behaviours that provide the most food for the least effort. Aggressive behaviour helps secure resources and the development of dominance hierarchies help manage aggressive interactions. Thus, in wolf packs, there is an alpha individual that to whom all others are subordinate. This results in less energy wasted on disputes. One of the most important behaviours is mating and reproduction. Innate and learned behaviours allow animals to recognize members of the opposite sex of its own species, and as being sexually receptive. This is done by a variety of visual, vocal, and chemical signals. Thus, in summary, behaviour has developed over evolution to allow a certain species or individual to be more successful and survive.

See also: BIOGEOGRAPHY, BIOLOGY

Further reading

Audesirk, T., Audesirk, G. and Byers, B.E., 2002. *Biology – life on earth*. Prentice Hall, Upper Saddle River, NJ, 6th edn.

Campbell, N.A., Reece, J.B. and Mitchell, L.G., 1999. *Biology*. Addison Wesley Longman, Inc., Menlo Park, CA, 5th edn.

CARBON CYCLE

Carbon is the most important element for living systems and is involved in geological, biological, and atmospheric cycles. The carbonate system is important in regulating the pH of the world oceans (due to the balance of dissolved CO_2, bicarbonate $-HCO_3^-$, and carbonate $-CO_3^{2-}$) and calcium carbonate is an important compound in the skeletons of many organisms such as coral. Living organisms undergo a grand cycle due to production and consumption. As plants and algae photosynthesize, they take CO_2, water and other materials from the environment and produce living organic matter and water. As this organic matter is consumed and dies, it decomposes and CO_2, water and inorganic compounds are released back into the environment. These two processes are called production and respiration. Production and respiration do not occur at the same rate over the surface of the Earth. Areas of the land surface that received abundant rainfall have much higher rates than much of the ocean. The highest rates of production and respiration occur in tropical rainforests, temperate and tropical floodplains, and in coastal ecosystems. Deserts, high mountains, and tundra have low rates. In the ocean, high productivity occurs in the coastal ocean, upwelling zones, and in high latitudes during the spring plankton bloom. The centres of the great ocean gyres (see Ocean Circulation) and most of the tropical ocean have low productivity. Living processes strongly affect other cycles. Over time some organic matter is not rapidly decomposed and ends up in rocks or in underground storage as coal, oil, or natural gas. Humans have greatly affected the cycle of life by changing the distribution of living organisms (i.e., growing crops in the desert, cutting forests) and by drawing down the great reservoirs of fossil fuels. Humans have also significantly increased the amount of CO_2 in the atmosphere leading to global warming.

See also: CYCLES, NITROGEN CYCLE, OCEAN CIRCULATION

Further reading

Botkin, D.B. and Keller, E.A., 2007. *Environmental science – Earth as a living planet*. Wiley, New York.

Brewer, R., 1994. *The science of ecology*. Saunders College Publishing, Fort. Worth, TX, 2nd edn.

DRAINAGE BASINS

A drainage basin is the area delimited by a topographic divide or watershed as the land area which collects all the surface runoff flowing in a network of channels to exit at a particular point along a river. The subsurface phreatic divide may not correspond exactly to the topographic divide on the surface. In the USA the term 'watershed' is often applied to small and medium-sized drainage basins and 'river basin' to larger areas. In some countries the term catchment can be used to describe small drainage basins.

The importance of the drainage basin as an accounting unit has been understood since the 18th century when Pierre Perrault calculated that river flow of the Seine in France was approximately one-sixth of the precipitation over the basin. As an accounting unit the drainage basin can be employed for water balance equations, for river discharge calculations and for hydrological modelling. The water balance equation relates runoff (Q), to precipitation (P), evapotranspiration (ET) and changes in storage in soil or ground water (ΔS) in the form:

$$Q = P - ET \pm \Delta S$$

At the scale of the drainage basin it is also possible to interpret how the input of precipitation is translated into the streamflow as river discharge. This translation is affected by the losses to evapotranspiration and infiltration to the ground water table, but also depends upon the basin or catchment characteristics which are the geological, topographical, soil and vegetation/land-use characteristics of the basin. These characteristics in turn affect how much precipitation flows over the ground surface, how much is infiltrated into soil and down to the water table, how much runs through the soil, or through the rock above the water table, and how rapidly the concentrated water flow runs through the river channels in the drainage basin. Water draining through drainage basins can follow a variety of routes over the surface (overland flow), through the soil (throughflow or pipeflow), through the rock above the water table (interflow) or into ground water from which it may eventually emerge as a spring providing base flow. These routes are affected by stores of water, on the surface (surface detention, lakes), in the soil, in the aeration zone, or in ground water (Fig. 41).

The hydrograph is the plot of volume of flow per unit time (discharge) against time and for any gauging point along a river can be plotted for any period such as a year, or for the period immediately following a storm. In this case the storm hydrograph (Fig. 42) begin to rise after a specific period after the storm (lag time) which depends upon the size of the drainage basin and the location of the precipitation. Any hydrograph can be thought of as being composed of quickflow and delayed flow components, and analysis of hydrograph characteristics is one basis for hydrological modelling. Water flowing through a drainage basin transports sediment and solutes, the amount transported varying according to the discharge. Sediment may be transported in suspension or as bedload, and solutes are dissolved in the water or attached to particles being transported and can include pollutants. The transport of sediment and solutes

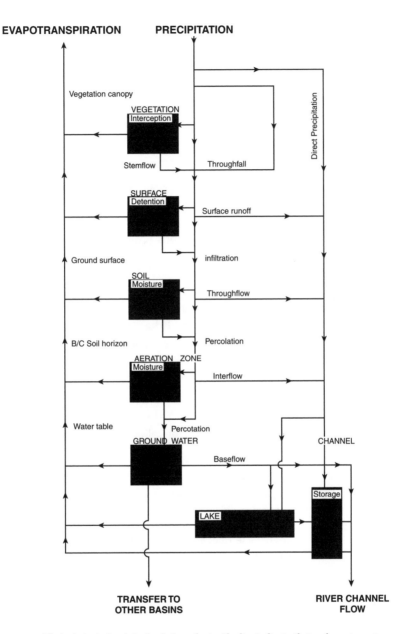

Figure 41 *Components of the hydrological cycle in the drainage basin. Shading indicates that each component can store moisture or transmit moisture to a subsequent component*

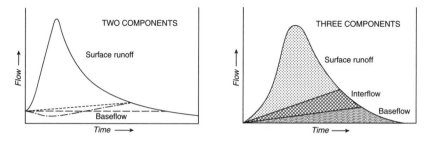

Figure 42   *Hydrograph characteristics. The hydrograph shows the response of river flow, and will rise after a storm event so that individual hydrographs can be analyzed in various ways (after Gregory and Walling, 1973)*

can be expressed as sediment or solute hydrographs (chemographs) which may not always mirror the discharge hydrograph because sources of sediment or solutes may not be the same as the origins of the water in the drainage basin.

The drainage basin is dynamic because the water and sediment-producing areas expand and contract depending upon the catchment characteristics, the antecedent conditions prior to any storm event, and the character of the storm input. It has therefore been necessary to identify the components of the drainage basin nested within each other, (Table 32) which are all affected

Table 32   *Drainage basin components*

Drainage basin component	Explanation
Drainage basin or catchment	Delimited by a topographic divide or watershed as the area which collects all surface runoff flowing in a network of channels to exit at a particular point along a river
Drainage network	The network of stream and river channels that exist within a specific basin. The channels may be perennial, intermittent and ephemeral, the network may not be continuous and connected, and the extent of the network will be affected by storm events, season, basin characteristics and human activity
River corridor	Linear features of the landscape bordering the river channel
Floodplain	The valley floor area adjacent to the river channel. A distinction may be made between the hydraulic floodplain, inundated at least once during a given return period and the genetic floodplain which is the largely horizontally bedded landform composed of alluvial deposits adjacent to a river channel
Channel pattern or planform	The plan of the river channel from the air, may be either single thread or multi-thread and varies according to the level of discharge
River reach	A homogeneous section of the river channel along which the controlling factors do not change significantly
River channel	The linear feature along which surface water may flow, usually clearly differentiated from the adjacent floodplain or valley floor.

by the dynamic character of the basin. Thus the drainage network is composed of some channels which are perennial and always contain flowing water, some which are intermittent and flow during part of the year, and others which are ephemeral and flow after storm events. Thus at any one time the drainage network is at one stage of a range of values of drainage density (total length of stream channel per unit basin area).

The function of the drainage basin is the reason why it has been used as a basis for collection of hydrological information, for the modelling of flows such as flood forecasting, and for the management of physical resources. Although drainage basins may not be the easiest units to use where they do not correspond with political or legal boundaries or with the jurisdiction of environmental management agencies, their use has been restated because drainage basin or watershed management plans should be the starting point of a management scenario.

See also: FLOODS, HYDROLOGICAL CYCLE, INTEGRATED BASIN MANAGEMENT

Further reading

Downs, P.W. and Gregory, K.J., 2004. *River channel management.* Arnold, London.

National Research Council, 1999. *New strategies for America's watersheds.* National Academy Press, Washington, DC.

Newson, M.D., 1997. *Land, water and development: sustainable management of river basin systems.* Routledge. London and New York, 2nd edn.

EARTHQUAKES

An Earthquake is defined as the natural shaking or vibrating of earth in response to the breaking and movement of rocks along faults. A fault is a fracture in rocks along which there has been differential displacement. The displacement may be vertical, horizontal or some combination of both. Most of the earthquake faults on Earth are located close to active plate boundaries, but a few that generate large earthquakes are found in the interior of tectonic plates. The coincidence of earthquakes with plate boundaries is so direct that, the mapping of the plate boundaries has been done largely through the location of earthquakes, as well as volcanoes that are found along plate boundaries (Fig. 43).

When rocks fracture and move along faults, seismic waves or earthquake waves are produced. Some of these waves travel within the Earth and some at the surface. The size of an

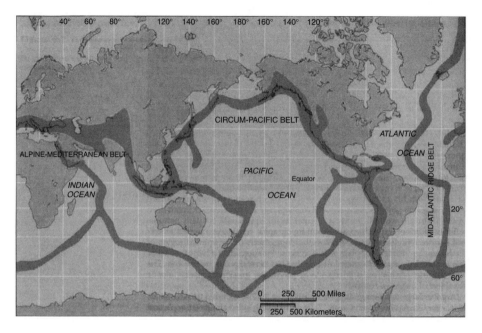

Figure 43   *Earthquake belts map of the world showing the major earthquake belts as shaded areas. (National Oceanic and Atmospheric Administration)*

Earthquake is related to the amount of energy released and is known as the earthquake magnitude. The magnitude of an earthquake is based in part upon physical characteristics of the rocks involved and the total amount of displacement along a fault during the earthquake. The relationship between magnitude and frequency of earthquakes on a global basis is shown on Table 33. The relationship between magnitude, displacement and energy of earthquakes is shown Table 34.

Table 33   *Worldwide magnitude and frequency of earthquakes. (After US Geological Survey, 2000. Earthquakes, facts and statistics. http://neic.usgs.gov)*

Descriptor	Magnitude	Average annual number of events
Great	>8	1
Major	7–7.9	18
Strong	6–6.9	120
Moderate	5–5.9	800
Light	4–4.9	6200 (estimated)
Minor	3–3.9	49,000 (estimated)
Very minor	<3	Magnitude 2–3 about 1000 per day
		Magnitude 1–2 about 8000 per day

Table 34 *Relationships between magnitude, displacement and energy of earthquakes. (After US Geological Survey, 2000. Earthquakes, facts and statistics. http://neic.usgs.gov)*

Magnitude change	Ground displacement change (vertical or horizontal that is recorded on a standard seismograph)	Energy change
1	10 times	About 32 times
0.5	3.2 times	About 5.5 times
0.3	2 times	About 3 times
0.1	1.3 times	About 1.4 times

Great earthquakes with magnitude 8 and higher have the capacity to produce catastrophes. Such events in some parts of the world have killed hundreds of thousands of people in a very short time period as buildings collapse. A large earthquake, when it occurs in the ocean and vertically displaces the water, may produce seismic sea waves known as a tsunami, which can cause widespread destruction and death thousands of miles from the source of the earthquake. For example, in 2004, an earthquake and tsunami in Indonesia spread waves throughout much of the Indian Ocean, killing about 250,000 people and inflicting great damage to property.

The amount of damage from a particular earthquake depends upon a number of factors including the direction of propagation of the rupture, as well as material amplification from weak, water-saturated soils and sediment near the surface. In some instances, the amplification of surface shaking may be increased dramatically as the waves move through materials such as mud and silt. For example, in 1985, a magnitude 8.1 earthquake occurred off the coast of Mexico. The waves moved inland to Mexico City, where buildings were constructed on lake beds that accentuated and increased seismic shaking. The shaking may have increased by a factor of about 4 times and tall buildings collapsed as upper stories pancaked onto lower ones claiming several thousand lives.

The effects of earthquakes are not limited to shaking and tsunami. Large earthquakes also cause liquefaction of the land, particularly in water-saturated, fine-grained sediment, and may induce thousands of landslides while causing fires and spread of disease.

Although we are not able to make short-term predictions of earthquakes on a regular basis, we have done fairly well with long-term prediction of where earthquakes are likely to occur in the future. Perhaps the most significant efforts to reduce effects of earthquakes have come from understanding the effects of seismic shaking on buildings. Engineering designs of buildings and other structures, such as bridges and airports, have led to much improved preparedness for future earthquakes. Where high standards of building construction are required, the buildings seldom fail during large earthquakes. On the other hand, where building codes are lax and people still build with unreinforced walls of adobe, rock or concrete, many more lives are lost. For example, a magnitude 7 earthquake in California may kill a few tens of people while a similar earthquake in other parts of the world, where housing is not designed to withstand shaking, may claim several tens of thousands or even hundreds of thousands of lives.

See also: SEISMOLOGY

Further reading

Bolt, B.A., 2004. *Earthquakes*. W.H. Freeman, San Francisco, 5th edn.

Hanks, T.C., 1985. *The national earthquake hazards reduction program: Scientific status*. US Geological Survey Bulletin 1659.

Scholz, C., 1997. Whatever happened to earthquake prediction? *Geotimes* 42, 16–19.

Scholz, C.H., 1990. *The mechanics of earthquakes and faulting*. Cambridge University Press, New York.

Silver, P.G., and Wakita, H., 1996. A search for earthquake precursors. *Science* 273, 77–78.

Southern California Earthquake Center. 1995. *Putting down roots in earthquake country*. University of Southern California, Los Angeles.

US Geological Survey. 1996. *USGS response to an urban earthquake, Northridge '94*. US Geological Survey Open File Report 96–263.

EARTH SURFACE PROCESSES

Processes that affect and modify the surface of the Earth, including geomorphological, pedological, biological and geological processes. Surface processes that occur on or above the surface are termed exogenetic, whereas those that occur below the surface are termed endogenetic and sometimes geological. Sometimes Earth surface processes are taken to refer to the geomorphological processes as in the journal *Earth Surface Processes and Landforms* published since 1977. Processes affecting the Earth's surface are dependent upon energy, which can be derived from solar radiation that drives the hydrological cycle, from gravity, whereby position above sea level provides potential gravitational energy, and geothermal energy which derives from the store of heat energy inside the Earth. Operation of Earth surface processes involves interaction of several Earth spheres and the dynamic way whereby energy is transferred through cycles in environmental systems including the rock cycle, the hydrological cycle and biogeochemical cycles and food chains. The range of Earth surface processes (Table 35) therefore interact in many ways, although for convenience they are often investigated individually. Thus weathering is interrelated with pedogenic processes of soil formation, mass movements on coastal cliffs interact with coastal processes, and nival processes in periglacial areas operate in combination with fluvial and mass movement processes. Some processes are defined according to the dominance of the materials on which they operate, such

Table 35   *Earth surface processes*

Group of Earth surface processes	Processes included	Summary explanation
Geomorphologic: endogenetic	Plate tectonics	The lithosphere is composed of several large plates that move relative to one another
	Seismic movements	Earthquake activity, folding (orogenesis), uplift of land (epeirogenesis), faults and fault movement
	Vulcanicity	Extrusive volcanic activity where magma or molten rock or associated solids, or gases are extruded on to the Earth's surface through fissures and volcanoes may be produced. Volcanoes vary in shape and character according to nature of erupted material, history of activity, and subsequent erosion
Geomorphologic: exogenetic	Weathering	Changes in rocks and minerals in response to physical, chemical and biological changes resulting in the physical, chemical and biological breakdown of rocks and minerals
	Mass movements and hillslope processes	Downward movement of soil, regolith and bedrock under the influence of gravity as a result of hillslope failure. Hillslope processes can also include water action. Movement can be by fall, creep, slide, flow
	Fluvial processes	Where running water, usually rivers, is the dominant fluid agent and involves water, sediment and solute transport involving erosion such as potholes and deposition including flood plains and terraces
	Coastal processes	Include erosional, transportational and depositional processes arising from wind action possibly combined with mass movement, fluvial and aeolian processes
	Aeolian processes	Where wind is the dominant fluid agent
	Glacial processes	Where glacial ice is the dominant fluid agent; associated with land based masses of moving ice including erosional
	Periglacial and nival processes	Processes operating in environments close to ice sheets or at high altitudes where alternate freezing and thawing may be significant
	Subsidence processes	Sinking of the land surface as a result of withdrawal of ground water or other fluids, compaction of sediments, or thermokarst
	Karst processes	In limestone areas where solution is a major process and cave systems and other distinctive features may result
Pedogenic	Soil profile formation or horizonation: eluviation	Translocation within the soil profile involving downward movement of fine particles and colloids to lower horizons.

(Cont'd)

Group of Earth surface processes	Processes included	Summary explanation
	illuviation	Accumulation of materials as clay particles, organic particles (humus) in lower horizons as a result of translocation of particles and colloids. Calcification is concentration of calcium salts whereas salinization is accumulation of sodium salts.
	Transformation within the profile	Decomposition of inorganic, primary to secondary minerals, and organic compounds by humification transforming plant tissues into humus possibly leading to removal of organic matter as water and carbon by respiration
	Soil erosion	Removal of topsoil by wind or water
	Sheet erosion	Removal of soil from a broad area without clear channels
	Gulleying	Erosion by networks of steep-sided channels called gullies
Ecological	Trophic interactions	Consumption and predation between individual populations or groups of species involved in food webs which are networks of consumer–resource interactions between groups of organisms, populations or groups of species. Energy and matter are transferred

as karst processes exclusive to limestone terrain, and some processes such as subsidence can occur in periglacial environments or be the result of tectonic processes. All processes can be envisaged as involving expenditure of energy acting on materials (rock and solids, water, air, organic materials) to produce products which can include landforms, soil profiles or vegetation types. Several processes may be grouped together, especially when studying them in terms of basic laws, so that the principles of fluid flow and fluid mechanics apply to fluvial, coastal, aeolian and glacial processes.

Throughout environmental sciences, data is required on Earth surface processes and some is collected by national and international organizations such as hydrological and oceanographic data. In addition data is obtained in the course of national surveys for example of soil types or from vegetation surveys. However if data is not already available it can be obtained empirically by field measurements often through carefully designed experiments such as experimental areas or by remote sensing; by laboratory investigations which include scaled down environmental models, of a section of coast for example; and by historical techniques, which use a method of deducing change over time from records or archive materials, such as historical maps of the position of a glacier margin at different dates.

Earth surface processes are combined in particular ways in different environments and so process domains are zones in which particular processes operate. Some processes such as glacial processes are confined to particular zones whereas others such as fluvial, aeolian and mass wasting can occur in most regions of the world. Attempts have been made to identify the world distribution of zones which include particular assemblages of processes and these include morphoclimatic zones, morphogenetic regions, and climato-morphogenetic zones. A difficulty with any consideration of the distribution of Earth surface processes is that there have been variations over time. Over relatively short periods there can be significant changes in the operation of processes, in semi-arid areas for example, but in many Earth zones there have been substantial shifts in the operation of processes over the Holocene and Pleistocene. An understanding of the rate of operation of Earth surface processes is necessary for environmental management and this benefits from understanding how processes have operated in the past and from indications of how such processes, which are responsible for hazards, may occur in the future. One suggestion is that a consequence of global warming may be that Earth surface processes may become characterized by more intense events such as floods and droughts, thus changing the incidence of hazards.

See also: ECOSYSTEM, ENERGETICS, ENERGY, HYDROLOGICAL CYCLE, KARST, SOIL PROFILES

Further reading

Allen, P.A., 1997. *Earth surface processes*. Blackwell, Oxford.

Earth Surface Processes and Landforms. 1977–. Wiley, Chichester.

EL NIÑO-SOUTHERN OSCILLATION

In 1924, Sir Gilbert Thomas Walker, Director General of the Observatory in India, coined the term 'Southern Oscillation' to describe the ups and downs in a 'east–west seesaw' in southern Pacific regional pressure. The Southern Oscillation Index (SOI) is defined as the pressure difference between Tahiti and Darwin, Australia (Fig. 44). El Niño–Southern Oscillation (ENSO) is a global coupled ocean–atmosphere phenomenon.

The Pacific ocean experiences major shifts in sea surface temperature that have become recognized as representative of what is called El Niño and La Niña, which are major temperature fluctuations in surface waters of the tropical Eastern Pacific Ocean. El Niño means 'the child' in Spanish, referring to the Christ child; La Niña means 'the girl child' which has opposite effects to El Niño. El Niño is named because the phenomenon is usually noticed around Christmas time in the Pacific Ocean off the west coast of South America. El Niño and La Niña are defined as sustained sea surface temperature anomalies of magnitude greater than 0.5 °C across the

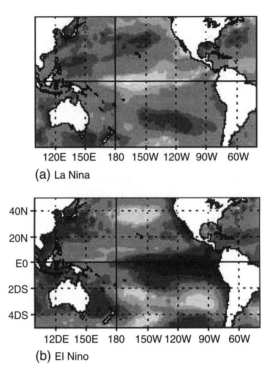

120E 150E   180   150W 120W 90W   60W

(a) La Nina

40N

20N

E0

2DS

4DS

12DE 150E   180   150W 120W 90W   60W

(b) El Nino

Figure 44   *Spatial patterns of La Nina vs El Nino sea-surface temperature departures for December-February. Darker gray is warm anomaly from normal; light areas cooler than normal. From NOAA Climate Prediction Center.*

central tropical Pacific Ocean. El Niño is associated with a positive temperature anomaly, while La Niña is with a negative temperature anomaly. When the condition is met for a period of less than five months, it is classified as El Niño or La Niña condition; if the anomaly persists for five months or longer, it is classified as an El Niño or La Niña episode. Historically, it occurs at irregular intervals of two to eight years and usually lasts one or two years. Coupled ocean–atmosphere climate fluctuations are associated with ENSO. ENSO is the most well-known source of inter-annual oscillations in weather and climate around the world (operating over a time scale of two to eight years), although some areas are more affected than others. ENSO has marked effects in the Pacific, Atlantic, and Indian Oceans. In the Pacific, during major warm events, El Niño extends over much of the tropical Pacific and becomes linked to the SO intensity. While ENSO events are generally in phase between the Pacific and Indian Oceans, events in the Atlantic Ocean tend to lag behind those in the Pacific by 12 to 18 months.

See also: HYDROLOGICAL CYCLE, OCEANOGRAPHY

Further reading

Carredes, C.N., 2001. *El Niño in history: storming through the ages*. University of Florida Press, Gainsville FL.

Glantz, M.H. 2001. *Currents of change: El Niño and La Niña impacts on climate and society*. CUP, Cambridge.

http://www.elnino.noaa.gov/

http://www.cpc.noaa.gov/products/analysis_monitoring/ensocycle/enso_cycle.shtm (a) La Nina relative temperatures of oceans, blue is colder (b) El Nino pattern.

ENERGETICS

The scientific study of energy flows leads to the essence of energetics, and also of a systems approach, which can be expressed by the 'laws' suggested by Barry Commoner in 1972 to be:

» everything is connected to everything else;
» everything must go somewhere;
» nature knows best;
» there's no such thing as a free lunch, because somebody somewhere must foot the bill.

In the environmental system, energy sources include solar radiation and the thermal gradients which arise in the atmosphere and laterally over the Earth's surface; internal energy from the Earth's interior including that from volcanic and seismic sources; rotational energy of the whole and parts of the solar system which can include tides; and human energy. Environmental science disciplines consider sources, fluxes and transformation of energy, expressed in relation to Earth spheres (Table 36), often prompting the development of concepts, many of which are the subject for separate entries. Thus sources, circulation or fluxes and the budget reflecting availability apply to all major spheres (Table 36) but in addition energy expenditure and rate of doing work can be expressed as efficiency. There can also be changes in energy distribution affected by global climate change for example. In ecosystems ecological energetics or bioenergetics refers to the flow, exchange and transformation of energy along trophic pathways. In several ways energetics provides an important framework for the environmental sciences and has prompted the development of concepts, such as power which is the rate of doing work (force × distance) and is expressed in watts which are joules per second ($J\ s^{-1}$). In the case of rivers stream power is the rate at which a stream can do work, especially in the transport of its sediment load, and is usually measured over a specific length of channel. The potential energy

Table 36 *Energetics in major spheres in relation to the environmental sciences. (Developed from Gregory, 2000.)*

Energy	Atmosphere	Geosphere	Hydrosphere	Pedosphere	Biosphere
Sources					
Solar	Internal energy	Solar	Solar radiation	Solar	Solar
Geothermal	Potential	radiation	Atomic energy	radiation	radiation
Gravitational	energy due to	Atomic	Chemical	Chemical	Primary
Rotational	gravity	energy	energy	energy	production
Vital	Kinetic	Chemical	Gravity	Mechanical	
Potential	energy	energy	Earth's	energy	
Kinetic		Gravity	rotation		
Enthalpy					
Circulation, transfers, fluxes					
Conservation	Transfers,	Geochemical	Hydrological	Mineral	Food chains
of energy	transformation,	cycles	cycle	weathering	Nutrient cycles
First law of	storage of			Organic matter	Population
thermo-	energy,			decomposition,	dynamics
dynamics	Exchanges of			Gibbs' free	
Fluxes,	heat, water,			energy	
Energetics	momentum				
	Energy fluxes				
Budget, availability					
Energy balance	Energy balance	Erosion rates	Water balance	Soil moisture	Energy
Thresholds			Sediment	budget	budgets at
Entropy			budget	Soil organic	trophic
				budget	levels,
					Biological
					productivity
Expenditure, rate of doing work					
Equilibrium	Available PE	Force-	Stream power	Soil profile	Ecosystem
Steady state	and KE	resistance		dynamics	dynamics
Power	Energy spectra	relations			Functional
Efficiency	and zonal				approach to
	balance				ecology
Distribution changes					
Maximum	Climatic change	Endogenetic	Palaeo-	Soil profile	Evolution
power		changes	hydrology	erosion	
Minimum					
variance					
Dissipative					
structures					
Fluctuation					

that water possesses at a particular location is proportional to its height above some datum which can be sea level or a lake level; this potential energy is converted into kinetic energy as the water flows downhill under the influence of gravity. Stream power (ω) was first expressed by Bagnold in 1960 as the product of fluid density (ρ), discharge (Q), acceleration due to gravity (g) and slope (s) in the form:

$$\omega = \rho\, Qgs$$

This expression for power can of course be applied to any fluid, and Bagnold used a similar approach in relation to wind movement over the Earth's surface. Odum and Odum in 1976 proposed three principles of energy flows namely the law of conservation of energy, the law of degradation of energy, which introduces entropy as a measure of technical disorder to signify the extent to which energy is unable to do work, and the principle that systems which use energy best survive, which is the maximum power principle or minimum energy expenditure principle.

Energetics, which applies to a great range of scales from the quantum atomic level through ecosystems to world zones and Earth spheres requiring some understanding of thermodynamics, chemistry and biology, is vital for understanding links of environment to the sciences including the social sciences.

See also: EARTH SPHERES, SYSTEMS: THEORY APPROACH AND ANALYSIS

Further reading

Bagnold, R.A., 1960. Sediment discharge and stream power: a preliminary announcement. *US Geological Survey Circular* 421.

Commoner, B., 1972. *The closing circle: confronting the environmental crisis.* Cape, London.

Gregory, K.J. (ed), 1987. *Energetics of physical environment: energetic approaches to physical geography.* Wiley, Chichester.

Gregory, K.J., 2000. *The changing nature of physical geography.* Arnold, London.

Odum, H.T. and Odum, E.C., 1976. *Energy basis for man and nature.* Wiley, New York.

Slaymaker, H.O. and Spencer, T., 1998. *Physical geography and global environmental change.* Longman, Harlow.

Smil, V., 1991. *General energetics. energy in the biosphere and civilisation.* Wiley, Chichester.

ENERGY

E nergy represents the ability to do work and takes on various forms such as potential, kinetic, electromagnetic and chemical. Energy has the same units as work: a force applied through a given distance. For example, the SI unit of energy – the joule – equals one Newton applied through a distance of one metre. Lord Kelvin amalgamated previous laws into Laws of Thermodynamics. Energy is neither created nor destroyed, but only transferred from one place or one form to another. Kinetic energy may represent the amount of work that accelerates a body to a given velocity; gravitational potential energy may constitute the amount of work that elevates or moves a mass against a gravitational pull. Biological chemical processes involve molecular biology and biochemistry, and constitute the making and breaking of certain chemical bonds in the molecules found in biological organisms. The Earth's weather patterns, which include energy-releasing processes such as hurricanes, snow avalanches, floods, and lightning, are all basically powered by the energy of sunlight striking the Earth.

In addition to the atmosphere, other forms of energy are those stored in natural and renewable resources in the Earth's biosphere, hydrosphere, and asthenosphere. The data below represents the world human energy consumption for the period 1980 to 2004 in terawatts (1 TW $=10^{12}$ W) (Table 37). The natural energy budget per year for the Earth's surface is ~400 W/m^2 (and the Earth's land surface area is ~1.5×10^{14} m^2), thus the total natural energy budget is about 60,000 TW. Consumption of the natural and renewable sources shown is about 15 TW or a very little percentage of the natural energy balance of Earth. However, regional consumption of these energy sources causes further anthropogenic heat emission directly into the atmosphere, especially in highly populated regions, where under certain colder climates this heat emission may exceed the natural radiation balance of the surface.

Table 37 *Terawatt or 10^{12} watts. World Energy Consumption 1980–2004 (from U.S. International Energy Annual 2004. Washington DC, US Energy Information Administration)*

	1980	2004
Oil	4.3	5.5
Gas	1.8	3.4
Coal	2.3	3.9
Hydroelectric	0.6	0.9
Nuclear	0.3	0.9
Wind, Solar, Bio Fuel	0.1	0.2

See also: ATMOSPHERIC SCIENCES, ENERGETICS, RADIATION BALANCE OF THE EARTH

Further reading

Crowell, B., 2007. *Conservation laws*. B. Crowell Publisher, New York.

ENERGY FLUX

An energy flux is the rate of heat flow across a unit area (such as Joules $m^{-2} s^{-1}$) and is composed of radiation, latent, sensible, and subsurface or soil heat fluxes. Radiative flux represents the amount of energy moving in the form of photons at a certain distance from the source per steradian/second. Latent heat flux is heat conveyed to and from the Earth's surface to the atmosphere that is associated primarily with evapotranspiration of water at the surface and from plants and subsequent condensation of water vapour in the atmosphere. When energy is added to water, it will change states or phases. The phase change of a liquid to a gas is called evaporation. At the molecular level, water is comprised of clusters of water molecules (H_2O), which are bound together by the bonding between hydrogen atoms. The heat added during evaporation breaks the bonds between the clusters, creating individual molecules that escape the surface as a gas. The heat used in the phase change from a liquid to a gas is called the 'latent heat of vaporization'. It is called 'latent' because the heat is stored in the water molecules and is later released during the condensation process. With evaporation, positive latent heat flux or transfer to the atmosphere occurs. Evaporation cools a surface due to energy that is removed from the water as molecules escape the surface. This process causes the surface temperature to decrease. Condensation is defined as a change from gas to liquid. During this phase change, the latent heat that accompanied evaporation is released from the water molecules and is transferred into the surrounding air. During this transfer process, latent heat is converted to sensible heat, causing an increase in the air's temperature.

When radiation is absorbed by the Earth, it raises the temperature of the surface. Conversely, when the surface is water, rather than used to entirely heat the water, a large portion of that energy is consumed in evaporation. As a result, with equal inputs of energy to land and water surfaces, land will heat up faster than water. Sensible heat flux is the time rate of flow for energy to be transferred from the warm Earth's surface to the cooler air above during the daytime, as long as there is a significant temperature difference between the two. Heat is also transferred into the air by conduction, as air molecules collide with those of the surface. As the air warms, air moves vertically upwards by convection. Because air is such a poor conductor of heat, convection is the most efficient way of causing a sensible heat flux into the air. The transfer of heat raises the air's temperature, and ends up cooling the surface. If the air is warmer than the surface, heat is transferred from the air to the surface, causing a negative sensible heat transfer. If heat is transferred out of the air, the air cools and the surface warms.

When the sun goes down, there is no input of solar radiation and strong surface inversions develop. At this time, the ground cools due to longwave emission and the air directly above the surface is warmer. Therefore, sensible heat is transported to the cool surface from above. Without available water, no transfer of latent energy occurs. Soil heat flux refers to the conduction of heat within the subsurface medium. A temperature gradient must exist between the surface and the subsurface for a heat flux to happen. Heat is transferred downwards when the surface is warmer than the subsurface (known as positive ground heat flux). If the subsurface is warmer than the surface, then heat is transferred upwards (known as negative ground heat flux). There is a considerable diurnal lag to soil temperatures with depth below the surface, as the conduction process is slower than the atmospheric processes.

See also: ENERGETICS, HYDROLOGICAL CYCLE, RADIATION BALANCE OF THE EARTH

Further reading

Bridgman, H.A. and Oliver J.E., 2006. *The global climate system, patterns, processes, and teleconnections.* CUP, Cambridge.

Hartmann, D.L., 1994. *Global physical climatology.* Academic Press, San Diego.

ENERGY AND MASS BALANCE

Energy and mass balance in climatology relates to the understanding of flows of energy and mass (e.g., water) from the Earth's surface to and from the atmosphere (and from space and to space) at scales ranging from global to local. Over a year, incoming energy in both the Earth and its atmosphere equals the outgoing energy. For the entire Earth–atmosphere system, the amount of radiation entering the system must equal to the amount leaving, or the system would continually heat or cool (see Table 38). Considering just the atmosphere, there is overall radiative cooling. The atmosphere is kept from a net cooling through the addition of latent and sensible heating from the Earth. The atmosphere has a warming effect on Earth's surface, which is commonly called 'greenhouse effect'. If there were no atmosphere around the Earth, the globally averaged surface temperature would be 255 °K or −18 °C (−0.4 °F).

Instead, the average surface temperature is 288 °K or 15 °C (59°F). The atmosphere acts as a greenhouse because of gases that selectively allow solar radiation to pass through but absorb and then re-emit terrestrial radiation. Collectively, these gases are called 'greenhouse gases' and include water vapour, ozone, carbon dioxide, molecular oxygen, methane, and nitrous oxide. These gases are only significant in certain wavelengths in which they absorb. As an example, ozone absorbs shortwave ultraviolet radiation, whereas water vapour absorbs infrared radiation. Most of the Sun's radiation that passes through the atmosphere to hit the Earth is in the visible part of the electromagnetic spectrum.

Table 38   *Water storages and flows*

Atmosphere storage	0.013×10^{15} m³
Precipitation to land	99×10^{12} m³/year
Precipitation to oceans	324×10^{12} m³/year
Land storage	33.6×10^{15} m³
Runoff/ground water	37×10^{12} m³ /year
Ocean Storage	1350×10^{15} m³
Evaporation/transpiration land to atmosphere	62×10^{12} m³/year
Evaporation ocean to atmosphere	361×10^{12} m³/year

From http://ww2010.atmos.uiuc.edu/(Gh)/guides/mtr/hyd/bdgt.rxml

Table 39   *Energy fluxes of the atmosphere and Earth (in watts per square meter) radiation budget*

Incoming solar radiation	342
Absorbed solar at Earth's surface	168
Reflected solar from surface	30
Absorbed solar by atmosphere	67
Reflected solar by clouds, aerosol and air	77
Total solar reflected to space	107
Surface emitted Infrared radiation	390
Absorbed infrared radiation at surface	324
Absorbed infrared in atmosphere from surface	350
Infrared emitted from atmosphere to space	165
Infrared from surface unimpeded to space	40
Infrared emitted by clouds to space	30
Total infrared emitted to space	235
Other energy sources	
Evapotranspiration from surface to atmosphere	78
Convection from surface to atmosphere	24

Mass balance – also known as a material balance – is an accounting of material entering and leaving a system. Fundamental to the balance is the conservation of mass principle, which states that matter can not be created nor destroyed. The term budget or balance is determined by identifying major inputs and outputs to the Earth–atmospheric system. Climatologists and hydrologists have resolved the quantities involved in the flows of energy and mass and these are summarized at the global scale in Tables 38 and 39 – cubic meters for water and energy in watts per metre squared. Local scale energy and mass budgets can be established for Earth phenomena, such as lakes, ice sheets, glaciers, cities, land cover units and biological systems.

See also: ENERGY FLUX, HYDROLOGICAL CYCLE

Further reading

Oke, T.R. 1997. *Boundary layer climates*. Methuen, London, 2nd edn.

Mather. J.R., 1978. *The climatic water budget in environmental analysis*. Lexington Books, Lexington, MA.

Kiehl, J.T. and Trenberth K.E., 1997. Earth's annual global mean energy budget, *Bulletin of the American Meteorological Society* 78, 197–208.

http://www.gewex.org/5thGEWEXConf_K.Trenberth.pdf

http://www.gewex.org/GEWEX-WMO_Bulletin.pdff

ENVIRONMENTAL CHANGE

Environmental change includes those changes that occur naturally in response to forcing factors operating over a range of timescales, those changes consequent upon human activity including those recent changes often described as global climate change. The period since the end of the major ice age known as the Pleistocene has seen the greatest alteration of the Earth's surface by humans. This has been most obvious on the land surfaces but scientific measurements have made clear that the oceans have also been changed, for example, by the addition of pollutants and by the inroads on whale and fish populations. The findings of palaeoecology, historical ecology and environmental history have given great depth to this story of the last 10,000 years, making possible some general findings as well as much local and regional detail.

In order to periodize that long stretch of time, it is convenient to think of human cultures as having been dominantly of the hunter-gather type until about 8000 BC, with replacement of these people by pre-industrial agriculture until c. 1750 AD. The 'industrial revolution' then penetrates to most parts of the planet (including the atmosphere) and is still present except that many of the human activities with environmental linkages accelerate after about 1950. Hunter-gatherers were once thought to be 'children of nature' and to have virtually no ecological effects. This view has been greatly challenged, especially by the realization that many such people had control of fire at the landscape scale and that many ecosystems could be burned by people possessing intimate knowledge of them: Australia, North America and sub-Saharan Africa are good examples.

The role of agriculture in altering both ecosystems and genotypes is obvious in areas such as the great irrigated areas of the world, both those of antiquity (e.g., Mesopotamia, Nile and Indus) but research shows that less intensive systems such as shifting agriculture and pastoralism also exert lasting influence upon ecosystems even if the human populations move on. Once coal- and oil-powered machinery becomes available, change is certain: the colonial impress upon many ecosystems (especially but not exclusively in the tropics) has been enormous, especially where a plantation economy was introduced, for example, for sugar in the West Indies or rubber in Malaya. Steam power meant that new technologies and new ideas were disseminated by the colonists to even very remote places. The process of population growth meant that even in an environment-conscious age such as the post-1950 period, the impacts have been enormous: the development of the tropics has accelerated and the emissions of has undergone an exponential increase. The new result is that few if any places on Earth can be said to be totally 'natural': at the very least there is likely to be some chemical which is widely disseminated through the oceans (as when DDT turned up in the body fat of penguins), or there is fallout from the atmosphere, as when lead from motor fuels accumulated in the Greenland ice-sheet or radioactive fallout from weapons resting contaminated the tundras of the high Arctic. Once the oceans and the atmosphere are involved then any adverse impacts are likely to be shared by everybody: the nature of the currents of the one and the air mass movements of the other ensure that effects are spread widely across the globe.

See also: CULTURES, DEFORESTATION, GREENHOUSE EFFECT, PALAEOECOLOGY, POPULATION GROWTH

Further reading

Simmons, I.G., 1989 *Changing the face of the earth*. Blackwell, Oxford, 2nd edn.

Simmons, I.G., 2008 *Global environmental history 10,000 BC to AD 2000*. Edinburgh University Press, Edinburgh.

http://www.sage.wisc.edu/pages/datamodels.html

EROSION RATES

Erosion takes several forms and processes, but generally modifies landforms. The river engineer or fluvial geomorphologist may be interested in rates of bank erosion and change of position of a river channel. People living along the coast are concerned with rates of coastal erosion. That is, how fast are the beach or sea cliff eroding and what is the importance to a particular community or location. Erosion rates from a geological perspective refer to the rates of denudation or general lowering of the landscape over time. Finally, in respect of agriculture, we are most concerned with rates of soil erosion. Rates may be reported as kg per hectare per year (soil erosion), mm per km^2 per 1000 yrs. (denudation), or mm to m per year (river bank erosion, coastal erosion).

Rates of erosion (denudation) from processes such as slope wash, landslides, and fluvial processes that lower the surface of the land or removal of soil are generally relatively low compared to rates for erosion of a riverbank or coastal erosion. General denudation of the land may be a fraction of a millimetre per year, which equates to a fraction of a metre per thousand years. These rates are generally much less than rates of uplift, which produce mountain ranges. However, where rates of uplift are high and produce steep valley-side slopes, rates of erosion are also higher than in regions with low rates of uplift.

There are several ways of trying to estimate rates of erosion, whether it is for soil erosion or general denudation of the land. One way is to attempt to make direct observation and measurements by placing erosion pins and measuring erosion over time. Over longer periods of time, rates of denudation or erosion may be measured by estimating the accumulation of the eroded particles found in lakes or ocean basins or other sites of deposition. When you know the volume of the material deposited and the time over which deposition occurred, you can back calculate the rate of erosion. For relatively small areas, it has been found that sediment may be measured (by repeated surveys) as it accumulates in human-made reservoirs and these measurements may be used to calculate rates of erosion as shown in Figure 45.

Measurement of rates of bank erosion of rivers is extremely variable from low rates of a few millimetres or centimetres per year to tens or even hundreds of metres per year for large rivers. These rates may be measured by examination of sequential aerial photographs over a period of years for large rivers. Where more detailed information is needed, new methods utilizing other forms of remote sensing may detect changes of the order of only a few centimetres.

Coastal erosion also is extremely variable from a few centimetres to a few metres per year. Aerial photographs are generally not effective in measuring retreat of coastal areas and coastal erosion unless the rates are high. On the other hand, new methods of remote sensing using light detection mechanisms (known as LIDAR) have proven effective to measure and evaluate rates of coastal erosion.

From an environmental perspective, it is important to be able to recognize and measure rates of erosion. In order to make estimates concerning future viability of soils or how large of

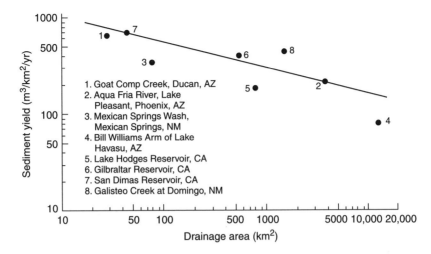

Figure 45 *Sediment yield for the southwestern United States based on rates of sediment accumulation in reservoirs. (Data from R.I. Strand. 1972. 'Present and Prospective Technology for Predicting Sediment Yield and Sources.' Proceedings of the Sediment-Yield Workshop. U.S. Department of Agriculture Publication ARS-5-40, 1975, p. 131)*

setbacks are required from river bluffs or coastlines for construction of buildings and other structures, rates of erosion must be known. Fortunately, our tools in measuring change at the surface of the Earth are being rapidly improved.

See also: EARTH SURFACE PROCESSES, ENVIRONMENTAL CHANGE

Further reading

Goudies, A.S., 1995. *The changing Earth. Rates of geomorphological processes.* Blackwell, Oxford.

Milliman, J.D. and Meade, R.H., 1983. World-wide delivery of river sediment to the oceans. *Journal of Geology* 9, 1–22.

Schumm, S.A., 1963. Disparity between present rates of denudation and orogeny. *US Geoiogical Survey Professional Paper 454-H*

Walling, D.E., 2006. Human impact on land–ocean sediment transfer by the worlds rivers. *Geomorphology* 79, 192–216.

FLOODS

A flood is inundation of land that is not usually submerged under water, and includes river floods and coastal floods. A river flood occurs when water exceeds the capacity of the river channel and flows out on to the flood plain for one of three main reasons:

1   Intense precipitation over all or part of a drainage basin produces more surface runoff than the channels of the drainage basin can discharge so that flooding occurs along part of the channel system. Intense precipitation triggered a river flood at Boscastle, Cornwall, UK on 16 August 2004 which damaged many buildings. Alternatively, prolonged and often continuous precipitation lasting for several days gives saturated ground conditions, delivering large amounts of quick flow to channels which may flood.

2   Snow melt floods are an annual occurrence in some parts of the world. In Arctic and mountain areas covered by snow in the winter the spring thaw produces high flood discharges when the snow melts. Snow melt floods can also occur in other parts of the world if occasional deep snowfalls melt rapidly, often associated with rainfall, producing high flood discharges.

3   Sudden releases of large amounts of water either by natural causes or collapse of man-made structures including dams. Natural causes include *Jokulhlaups* in Iceland which are flood waves produced when water drains from an ice-dammed lake, and along Canadian rivers for example, when large dams of trees break causing a flood wave immediately downstream. Failure of earth dams released large floods at a number of locations in the 19th century and sometimes landslides into a reservoir above a dam can instigate a large flood wave over the dam or sometimes weakens the dam structure.

Since 1860 there are 48 examples of recorded dam failures that killed more than 10 people, and a landslide in 1963 into the reservoir above the Vaiont dam in Italy caused water to overtop the dam sending a flood wave downstream causing 2600 deaths and much property damage. Hence techniques of flood frequency analysis and flood forecasting have been developed. Methods of flood mitigation have been achieved by using engineering works to enlarge the channel or produce a flood prevention scheme that may include a diversion channel; by modifying the flood (for example dams upstream that hold flood waters and release them slowly), by modifying the losses (flood insurance, structural measures such as buildings on stilts, or zoning land so that the most valuable is furthest away), or bearing the loss.

Coastal floods are produced by high tides, especially in combination with river flood events; by storm surges that are abnormally high sea levels about the time of spring tides; or tsunamis which are large waves produced by submarine earthquakes, volcanic eruptions, or landsliding. Coastal floods can be mitigated using the same methods as for rivers but engineering works, especially building sea walls, are often necessary to protect low-lying coastal areas. Any scheme of mitigation has to be produced for a particular design flood level and 80% of New Orleans, Louisiana, USA was flooded due to failure of flood walls in August 2005 (Fig. 39) following hurricane Katrina, which caused more than $200 billion in losses. Global climate change may increase the vulnerability of coastal and riverine areas to the flood hazard.

See also: COASTAL MANAGEMENT, DRAINAGE BASINS, HAZARDS, HYDROLOGICAL CYCLE

Further reading

Anderson, M.G., Walling, D.E. and Bates, P.W. (eds), 1996. *Floodplain processes*. Wiley, Chichester.

Baker, V.R., Kochel, R.C. and Patton, P.C. (eds), 1988. *Flood geomorphology*. Wiley, Chichester and New York.

Ward, R.C., 1978. *Floods. A geographical perspective*. Macmillan, London.

GEOCHEMICAL CYCLES

Geochemical cycling is the study of chemical dynamics, interactions, and processes of the Earth's surface. An important component is biogeochemical cycling, chemical dynamics between living organisms and the environment. Many important chemical changes take place as rocks and soils are formed and changed through volcanic, weathering, and sedimentation processes. Diagenesis is the weathering of rocks and the formation of soils and sediments. The rock cycle is one of the important components of geochemical cycling. Basic rock types include igneous, sedimentary, and metamorphic and many geochemical changes take place as material cycles between these different rock types. The weathering and breakdown of rocks contributes to global silicon and carbon cycles. Soil formation, soil erosion and transport, and sediment deposition are important processes of geochemical cycling. The geochemical cycling of organic matter provides insights into the production, dynamics, and decomposition of organic carbon.

Trace metal geochemistry concerns itself with the dynamics of metals that occur at very low concentrations in the environment. These trace metals, depending on concentration, can be either toxic agents or required nutrients. Trace metals include such elements as silver, copper, lead, zinc, cadmium, nickel, mercury and arsenic. Much material is added to the biosphere as the sea floor forms at mid-ocean ridges where volcanic activity produces material that then spreads out from the ridges. The geochemical dynamics of hydrothermal vents that occur around the ridges has provided an understanding of the chemical dynamics associated with vent communities (see oceanography) and sulphur cycling. Biological activity strongly affects the geochemical cycling of many elements including carbon, nitrogen, hydrogen, phosphorus, sulphur, oxygen and many minor elements such as heavy metals. Radioactive materials are also important in geochemical cycles. Lead 210 (^{210}Pb), for example, forms naturally as a result of interactions in the atmosphere and settles to Earth. Caesium 137 (^{137}Cs) is an atomic bomb product that is incorporated into the soil. Both these elements are used to determine the age of sediments.

See also: CARBON CYCLE, CYCLES, NITROGEN CYCLE

Further reading

Botkin, D.B. and Keller. E.A., 2007. *Environmental science – Earth as a living planet.* Wiley, New York.

Mitsch, W.J. and Gosselink, J.G., 2000. *Wetlands.* Wiley, New York. 3rd edn.

Nriagu, J. (ed), 1976. *Environmental biogeochemistry.* Ann Arbor Science. Ann Arbor, Michigan.

HYDROLOGICAL CYCLE

The term hydrological or water cycle expresses the way in which water in the hydrosphere (the water body of the Earth in liquid, solid or gaseous state and occurring in fresh and saline forms) moves between stores in the oceans (c. 97%), in ground water (c. 0.7%), in ice sheets and glaciers (c. 2%), with smaller amounts in the atmosphere, on land – in lakes and seas and rivers, and in the soil. Solar energy drives the hydrological cycle, providing the lift against gravity necessary as water vapour is transferred by evaporation to form atmospheric moisture and then precipitation. The energy required to drive the global hydrological cycle is equivalent to the output of 40 million major (1000 MW) power stations, and about one third of the solar energy reaching Earth is used evaporating water with about 400,000 km³ evaporated each year: the entire contents of the oceans would take about 1 million years to pass through the water cycle. Water (H_2O) that is constantly moving between the stores (Fig. 46) changes in phase from gaseous to liquid to solid in different parts of the cycle. Such changes or fluxes are extremely important and complex but some such as precipitation occur for a relatively small proportion of the time. Thus there are few places in the UK where rain or snow falls for more than 15% of the time. Over the land surface it is difficult to differentiate evaporation, which is a physical process, from transpiration which is part of a plant's active metabolism so that the two are grouped together as evapotranspiration. Although the time that water stays in the stores (residence time) of the atmosphere, the soil or the land surface is relatively short, it

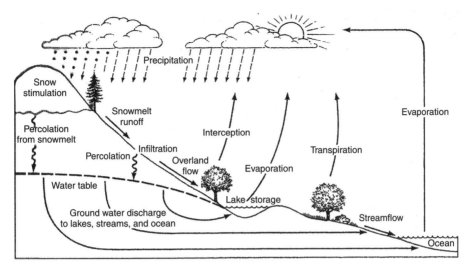

Figure 46 *The hydrological cycle. (From Dunne and Leopold, p. 5, with permission)*

can be several thousand years in the oceans, ground water or ice sheets, and in some arid areas the ground water (e.g., Algeria or Libya) is as much as 20,000–30,000 years old.

The water volumes in the different stores have changed in the past and some 10,000 years ago during the last ice advances there was more than twice as much ice over the Earth's surface (approx 72×10^6 km^3 then compared with 33×10^6 km^3 today). It has been proposed that laws guiding comprehensive water development and management need to recognize the integrity of the hydrological cycle because if the interconnections of the water cycle are arbitrarily divided up by the legal system then it is difficult to achieve a comprehensive, holistic approach.

See also: ENERGETICS, HYDROLOGY, HYDROSPHERE

Further reading

Davie, T., 2002. *Fundamentals of hydrology*. Routledge Publishing, London.

Dunne, T. and Leopold, L.B., *Water in environmental planning*. W.H. Freeman, San Francisco.

Jones, J.A.A., 1997. *Global hydrology: processes, resources and environmental management*. Longman, Harlow.

Newson, M.D., 1997. *Land, water and development: sustainable management of river basin systems*. Routledge, London, 2nd edn.

Walling, D.E., 1977. Rainfall, runoff and erosion of the land: A global view. In Gregory, K.J. (ed), *Energetics of physical environment*. Wiley, Chichester, 89–117.

LAKES

Lakes are lentic waterbodies, that is they are standing waterbodies. This is in contrast to lotic waterbodies, which are characterized by running water. Lakes are large enough so that wind-induced mixing plays a major role in mixing. Ponds, by contrast, are small enough that temperature-induced mixing dominates. Lakes form largely as a result of geological processes. The following is a general characterization of lake origins, but often several processes may interact to form lakes.

Glacial lakes are formed by the action of glaciers by several different mechanisms. Lakes can form in depressions on the surface of glaciers or even at the bottom of glaciers as subglacial lakes. Lakes can also form when blocks of glacial ice melt or when permafrost melts and lakes enlarge due to summer winds. Lakes also form in depressions caused by glacial scour and behind moraines. Tectonic lakes form in depressions caused by crustal instability such as warping, faulting, buckling and folding. Landslide lakes form when landslides block stream channels. Volcanic lakes form in collapsed volcanic craters and depressions in or formed by lava flows. Solution lakes form in areas where the substrate is dissolved. This is most common in calcium carbonate rocks. Such sinkholes are called cenotes in Mexico. Aeolian lakes form in arid areas where wind blown materials block water courses and water accumulates. Lakes associated with rivers form in several different ways. Levee lakes are linear and form behind the elevated natural levees bordering rivers. Oxbow lakes are cutoff river meanders. Shoreline lakes form behind coastal barriers such as beaches.

Lakes formed by organisms can be divided into two classes; those formed by humans by dam construction and those formed by other organisms such as beavers. Many lakes are stratified for all or part of the year. Normally, this is due to temperature differences between the surface and deeper waters. The upper warmer layer is called the epilimnion and the lower, colder water is called the hypolimnion. The two layers are separated by a steep temperature gradient called the thermocline. Typically, many lakes stratify in the warmer months when heating of the upper layer is highest. The stratification is broken down in the winter in a process called overturn when there are stronger winds and less thermal stratification. Lakes are often divided into a number of zones. The shallow water area around the periphery of the lake is called the littoral zone, which often supports rooted aquatic vegetation. The upper water layer in the centre of lakes is called the pelagic zone. It is often also the photic zone where there is enough light to support photosynthesis by phytoplankton. The deeper part of the lake including the bottom is called the profundal zone. Lakes are also classified as to their trophic state. Extremely low nutrient, clear lakes are called oligotrophic. Moderate nutrient levels result in mesotrophic conditions. Many lakes are eutrophic, that is they are over-enriched with nutrients. Eutrophic lakes are characterized by algal blooms, lowered water transparency, low oxygen conditions, and fish kills. Fresh water lakes are most often phosphorus limited and phosphorus reduction is a management goal.

See also: HYDROLOGICAL CYCLE, HYDROSPHERE

Further reading

Audesirk, T., Audesirk, G. and Byers, B.E., 2002. *Biology – life on earth*. Prentice Hall, Upper Saddle River, NJ, 6th edn.

Botkin, D.B. and Keller. E.A., 2007. *Environmental science – earth as a living planet*. Wiley, New York.

Brewer, R., 1994. *The science of ecology*. WB. Saunders, Philadelphia.

MINERALS

Most of the rocks in the Earth's crust consist of combinations of just 15 of the nearly 3600 known minerals. These 'rock-forming' minerals consist, in turn, of the most abundant elements by weight and volume in the Earth's crust (Table 40). Silica (SiO_2)

Table 40   *Common rock-forming minerals*

Mineral	Composition	Comprises rocks
Plagioclase	$NaAlSi_3O_6 - CaAl_2Si_2O_6$	Basalt, andesite, granite, gabbro
K-feldspar	$KAlSi_3O_8$	Granite
Olivine	$(Mg,Fe)_2SiO_4$	Basalt, gabbro
Quartz	SiO_2	Granite, sandstone
Muscovite	$KAl_2(AlSi_3O_{10})(OH,F)_2$	Granite
Biotite	$K(Mg,Fe)_3(Al,Fe)Si_3O_{10}(OH,F)_2$	Granite
Amphibole (hornblende)	$Ca_2(Mg,Fe)_4Al(Si_7,Al)O_{22}(OH,F)_2$	Andesite
Pyroxene (augite)	$(CA,Na)(Mg,Fe,Al,Ti)(Si,Al)_2O_6$	Andesite, basalt, gabbro
Calcite	$CaCO_3$	Limestone, marble
Dolomite	$(Mg,Ca)CO_3$	Dolostone
Gypsum	$Ca(SO_4)$	Evaporite
Chlorite	$(Mg,Fe)_3(Si,Al)_4O_{10}\cdot(Mg,Fe)_3(OH)_6$	Schist
Clay minerals	Al, (Mg, Fe) hydrous silicates	Shale
Alumino-silicates	Al_2SiO_5	Schist
Garnet group	$(Mg,Fe,Ca)_3Al_2Si_3O_{12}$	Schist, gneiss

Table 41   *Minerals in common igneous rocks. Ferromagnesian minerals in italics. Non-essential minerals in parentheses*

Basalt and gabbro	Andesite and diorite	Rhyolite and granite
Plagioclase	Plagioclase	Plagioclase
		Orthoclase
		Quartz
Olivine		
(Pyroxene)	*Pyroxene*	
	Amphibole	*(Amphibole)*
	(Biotite)	*Biotite*
		(Muscovite)

is an essential component in most of these minerals. Some minerals have a fixed composition (quartz and calcite), whereas the proportions of elements may vary in others (amphibole, pyroxene).

Rock-forming minerals are rarely found in the big, beautiful crystals prized by collectors, and their black or white colours seldom render them very exciting. As a result, many of the minerals in Table 40 are not common household names. One of them, plagioclase, is present in almost all of the igneous rocks in the Earth's crust (Table 36), making it one of the most common minerals in the world.

Basalt, the commonest rock in the Earth's crust, consists of olivine and plagioclase (Table 41). Olivine, pyroxene, amphibole, and biotite are dark-coloured minerals, because they contain iron (Fe) and magnesium (Mg) that impart a dark colour to them. Quartz, plagioclase, and orthoclase are light-coloured minerals, generally grey, white, or clear glassy, because they lack iron and magnesium.

The salient properties of the main rock-forming minerals are given below:

Quartz – is the hardest of the common rock-forming minerals with a hardness of 7 on the Mohs scale. It crystallizes typically in six-sided crystals with a sharp-pointed pyramid at each end. It has a vitreous lustre and when pure is clear and colourless. Its colour ranges from colourless, through yellow, grey-brown to black, pink and violet. Pink (rose quartz) has trace amounts of rutile needles and/or the presence of titanium in the crystal lattice, whereas the purple colour of amethyst is due to a trace amount of iron. Quartz is readily distinguished from other white minerals by its hardness and conchoidal fracture.

Feldspar – is the name for a group of minerals rather than a single mineral. The feldspars are combinations of potassium, or of calcium and sodium with aluminium, silicon, and oxygen. The feldspars are the most abundant mineral in the lithosphere and contain two main types: plagioclase and orthoclase.

Plagioclase – is the calcium- or sodium-bearing group of feldspars that forms a solid solution series between end members that contain solely calcium or solely sodium, in addition to aluminium and silica. It has a hardness of 6, slightly less than quartz, and a vitreous lustre. Most varieties are white or pale grey; some may be as clear as glass and resemble quartz, which is harder on the Moh's hardness scale than feldspar.

Orthoclase – is the potassium-bearing feldspar also with a hardness of 6. It has a pearly or vitreous lustre and is a common light grey or pink constituent of granite. It usually has a well-developed 90° cleavage. Orthoclase is a common constituent of granite, whereas a related mineral with the same composition but different crystal form, sanidine, is common in felsic volcanic rocks. Microcline, also of the same composition but with a different crystal structure, is common in metamorphic rocks.

Mica – is a group of minerals that have a sheet-like arrangement of atoms that results in a good to excellent cleavage parallel to those weak internal planes. Because of this weakness, mica is not common in clastic sedimentary rocks, owing to the breakdown and battering sediments take during weathering, erosion, and transportation. Two main varieties of mica are muscovite and biotite.

Muscovite – is the clear white mica in most felsic igneous rocks and in some low-grade metamorphic rocks as shiny flakes that make the rocks sparkle.

Biotite – is the black mica that contains iron and magnesium, which give it its characteristic colour. It exists as shiny, jet-black flakes and 'books' in igneous and metamorphic rocks such as granite and schist.

Ferromagnesian minerals comprise a great number of dark minerals so prevalent in igneous rocks. Compositionally they contain iron and magnesium that impart the dark colours. Hornblende, pyroxene, and olivine are the commonest minerals in this group.

Hornblende – is the principal mineral in a complex isomorphous series of rather variable composition called amphibole. It is typically dark green or black with a strongly vitreous lustre. It is readily distinguished from a similar common black mineral, pyroxene, by its cleavage planes that intersect at angles of 56° and 124°. Hornblende is usually one of the chief components of igneous rocks with compositions intermediate between basalt and rhyolite, and between gabbro and granite.

Pyroxene – is the name for a diverse group of minerals with great differences in physical properties. It is a dark black mineral, similar to hornblende in its composition, vitreous lustre, and occurrence but lacking the hydroxyl ion in its chemical formula. Its crystals are nearly equidimensional, and its cleavage planes intersect at 93° and 87°. Pyroxene crystal cross-sections are nearly square, whereas hornblende cross-sections are nearly hexagonal.

Olivine – is an iron, magnesium silicate, commonly identified by its olive green colour, glassy lustre, and conchoidal fracture in such mafic igneous rocks as basalt and gabbro. It breaks down easily in hydrous environments and so is almost unheard of in sedimentary rocks.

Calcite and Dolomite – are carbonate minerals in contrast to the silicate minerals listed above. They are normally white or pale yellow; calcite may even be colourless or even (rarely) blue, owing to a trace amount of copper. The amount and nature of impurities may yield a spectrum of earthy colours from yellow, through orange and brown, to black. Calcite has a hardness of only 3; dolomite is 3.5 with a slightly higher specific gravity. Both minerals are typical in sedimentary rocks and can be differentiated by the fact that calcite fizzes more readily in hydrochloric acid than does dolomite. Calcite has nearly perfect cleavage in three directions, dolomite in only one direction. Both minerals typically comprise sedimentary limestone, although the origin of dolomitic limestone is still debated.

Gypsum – is a name applied to a mineral and to rocks consisting of this mineral alone. It is very soft with a hardness of 2, it is white or colourless, but if impurities are included, then it may be grey, red, yellow, brown, or blue. Massive gypsum with a pearly lustre is known as alabaster. Gypsum and its close chemical relative, anhydrite, comprise evaporitic sedimentary rocks closely associated with rock salt.

Rock salt – is a mineral, which, because of its solubility in water, is rarely present at the Earth's surface except in extremely arid regions. It is colourless or white when pure, but is orange or red because of iron inclusion, and may also be grey, yellow, or blue. Rock salt is extracted at depth from former lake beds, marine layers, and from enormous 'salt domes' that have risen buoyantly into the shallow subsurface from deeper salt layers that were deposited originally on the sea floor.

See also: GEOLOGY, ROCK TYPES

Further Reading

Deer, W.A., Howie, R.A. and Zussman, J., 1992. *The rock-forming minerals*. Longman, Harlow, 2nd edn.

***Natural Hazards* see** Hazards

NATURAL PRODUCTION MECHANISMS OF MARINE FISHERIES

Natural production mechanisms are defined as those physical variables that have a direct relation to the abundance of biotic resources and control their magnitude, variability and persistence. In the coastal zone an ecological understanding of its systems requires an integrated analysis of the physical environmental and biological processes. For instance, fishery biology is applied ecology and thus of transcendent importance for the rational prospecting, evaluation, exploitation, and administration of demersal resources, is the integral ecological knowledge of biotic resources and its link to the physical environment.

World-wide, the state of tropical coastal demersal fisheries lacks the generalized information that would permit a deeper understanding of the problems of ecological interpretation and evaluation with good methodological bases. These multispecies tropical fisheries are the consequence of ecological interactions in the coastal zone and their decrease, increase or stability is a consequence of the natural variability of physical and biological processes. Diversity, distribution, abundance and persistence of fishery resources in the coastal zone are modulated by diverse and complex physical variables. These, because of their ecological dynamic, are considered as 'natural production mechanisms'; the most apparent are: (1) physical-chemical conditions of water, i.e., nutrient concentration, salinity, turbidity, temperature, (2) geographical latitude, (30 bathymetry and sediment types, (4) climate and meteorology, (5) river discharge, (6) tidal range and variation in sea-level rise, (7) areas of coastal vegetation, i.e., swamps, marshes, wetlands, coastal lagoons and adjacent estuaries, and (8) interactive dynamics between estuaries and the sea.

The responses of the ecosystem as a whole, are modulated by those 'production mechanisms' strongly integrated, such as rain, river discharge, tidal cycles and sea-level pulses, areas of coastal vegetation, and subsequent responses of ecosystem in aquatic primary productivity and fish biomass. Management decisions must be taken considering the integrated seasonal pulsing of such as 'natural production mechanisms', towards a more comprehensive fish catch and environmental planning.

See also: ECOSYSTEM-BASED MANAGEMENT

Further reading

Day, J.W., Hopkinson, C.S. and Conner, W.H., 1982. An analysis of environmental factors regulating community metabolism and fisheries production in a Louisiana estuary. In Kennedy, V.S. (ed), *Estuarine comparisons*. Academic Press, New York, 12–136

Deegan, L.A., Day, J.W., Gosseling, J., Yáñez-Arancibia, A., Sánchez-Gil, P. and Soberon, G., 1986. Relationship among physical characteristics, vegetation distribution, and fisheries yield in Gulf of Mexico estuaries. In Wolfe, D.A. (ed), *Estuarine variability*. Academic Press, New York, 83–100.

Pauly, D., 1986. Problems of tropical inshore fisheries: Fishery research on tropical soft bottom communities and the evolution of its conceptual base. In Mann Borgesse, E. and Ginsburg, N. (eds), *Ocean Year Book 6*. University of Chicago Press, Chicago, 29–54.

Yáñez-Arancibia, A., 2006. Middle America, coastal ecology and geomorphology. In Schwartz, M. (ed), *The Encyclopedia of coastal sciences*. Springer, Dordrecht, 639–645.

Yáñez-Arancibia. A., Soberón-Chávez, G.and Sánchez-Gil, P., 1985. Ecology of control mechanisms of natural fish production in the coastal zone. In Yáñez-Arancibia, A. (ed), *Fish community ecology in estuaries and coastal lagoons: towards and ecosystem integration*. UNAM Press, Mexico DF, 571–594.

NITROGEN CYCLE

The nitrogen cycle is one of the most complicated material cycles in nature. This is because nitrogen can exist in many different forms, because it is involved in so many biogeochemical processes, and because it has been so affected by human activities. Nitrogen can exist in a valence state of +2 to −5. It can exist in various gaseous states (N_2 and N_2O, NH_3), several inorganic dissolved states (NO_3^-, NO_2^-, NH_4^+), and as a number of organic compounds (i.e., urea, amino acids, proteins, DNA). Nitrogen is also a component of various rocks and is released with weathering. Nitrogen gas (N_2) makes up about 80% of the atmosphere, but it cannot be directly used by living organisms. Nitrogen fixation is the process of converting N_2 to nitrate or ammonia which can then be used. Major processes in the global nitrogen cycle include nitrogen fixation (by both biological and industrial processes), uptake and release by organisms, denitrification (the conversion of nitrate back to N_2), soil erosion, runoff, and flux in rivers, and burial in marine sediments.

The atmosphere contains most of the nitrogen on the Earth's surface but there are also significant amounts in living organisms and in the soil. Because nitrogen is present in so many organic chemicals, it is one of the most important elements in living organisms. Human activity has altered the nitrogen cycle in a number of ways. Air pollution introduces nitrogen oxides into the atmosphere and these are washed back to Earth in rainfall, sometimes as acidic forms. The rate of nitrogen fixation has been greatly increased through the production of fertilizers. When excess fertilizer and excess nitrogen in rainfall get into water bodies, it causes nitrogen enrichment and eutrophication. This is a world-wide problem in both fresh waters and coastal waters.

See also: BIOGEOCHEMICAL CYCLE, CYCLES

Further reading

Botkin, D.B. and Keller, E.A., 2007. *Environmental science – Earth as a living planet*. Wiley, New York.

Day, J.W, Hall, C.A.S. Kemp, W.M. and Yáñez-Arancibia, A., 1989. *Estuarine ecology*. Wiley, New York.

Mitsch, W.J. and Gosselink, J.G., 2000. *Wetlands*. Wiley, New York, 3rd edn.

PLATE TECTONICS

Plate tectonics was defined by Dennis and Atwater (1974, p. 1031) as 'A theory of global tectonics in which the lithosphere is divided into a number of *plates* whose pattern of horizontal movement is that of torsionally rigid bodies that interact with one another at their boundaries, causing seismic and tectonic activity along those boundaries'. Plate tectonics is a unifying theory, formulated in the mid-1960s from hypotheses expounded as early as 1915, that explains how mountains, ocean basins, and volcanoes form, and how earthquakes happen as a consequence of continuous movement of lithospheric 'plates' upon a convecting mantle. The theory states that the lithosphere is broken into a nine major 'plates' and six lesser ones that overlie the mantle much as the shell of an egg overlies the egg white. The plates move independently relative to one another, propelled by convective flow of the underlying mantle. Rates of movement are mostly in the range of 1–14 cm/yr. Crustal deformation occurs chiefly along the edges of the plates where they interact with adjacent plates. The boundaries of the plates are clearly defined by the long, narrow zones of earthquakes and active volcanoes and are of three main types: transform, divergent, and convergent (Fig. 47).

Some of the plates grind horizontally past each other along what is called a 'transform' plate boundary. Large, frequent earthquakes characterize and demarcate transform plate boundaries. The San Andreas fault in California (Fig. 48) is the locus of such a boundary where the Pacific plate moves northward, relative to the North American plate along a 1-15 km-wide zone that is 1100 km long.

Some of the plates move away from each other over millions of years, and the boundary between the two plates is a 'divergent' plate boundary. There basalt from the mantle wells up

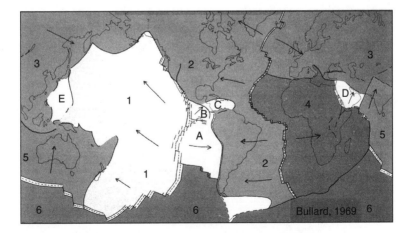

Figure 47 *Map of the major lithospheric plates. 1 = Pacific; 2 = North American and South American; 3 = Eurasian; 4 = African; 5 = Indo-Australian; 6 = Antarctic; A = Nazca; B = Cocos; C = Caribbean; D = Arabian; E = Philippine. Arrows indicate relative plate motions*

Figure 48 *Oblique aerial view of the San Andreas fault in the badlands of the Carrizo Plain, California. In this view the fault strikes from the lower left to upper right. Photo by Arthur G. Sylvester, Jan. 1982*

between the two plates, akin to the way blood wells up in a cut to 'heal' the 'wound'. The basaltic 'blood' creates mid-ocean ridges – great 'scars' on the Earth's surface such as the Mid Atlantic Ridge. Passive volcanism and very frequent, small earthquakes (M<6) typify the divergent boundaries.

Some of the plates move toward each other in 'convergent' plate boundary. Great mountain belts like the European Alps and mighty Himalaya in Asia buckle upward where continental Africa and India, respectively, collide into the belly of the Eurasian continental plate (Fig. 49). Associated volcanism is rare because the deformation involves only the upper part of the lithosphere.

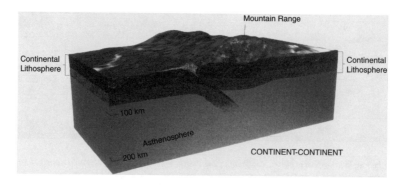

Figure 49 *Schematic block diagram of a mountain range formed at a continent-continent collisional plate boundary.*

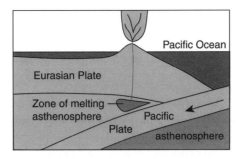

Figure 50 *Schematic diagram of subduction between the continental Eurasian and oceanic Pacific plates. The asthenospheric mantle melts with the aid of fluids derived from the subducted Pacific Plate. The magmatic melts rise through the east edge of the Eurasian plate to feed the volcanoes of the Japanese island arcs. (Source: Volcano World website: http://volcano.und.edu/vwdocs/volc_images/ north_asia/kuriles.html)*

Collision between dense oceanic plates and less dense continental plates also produces mountains in a process called subduction, whereby an oceanic plate slides beneath a continental plate (Fig. 50). The oceanic plate is sucked beneath the continental plate where its rocks are folded, metamorphosed, heated, partially melted, yielding magmas that intrude the upper part of the overriding continental plate (Fig. 50). Infrequent, very large earthquakes (M>7) and large explosive volcanoes characterize active convergent plate boundaries. One of the best known convergent plate boundaries is that along the length of the arcuate Japanese islands between the Pacific and Eurasian plates.

See also: EARTHQUAKES, GEOLOGY, TECTONICS

Further Reading

Dennis, J.G. and Atwater, T.M., 1974. *Terminology of Geodynamics*. AAPG (American Association of Petroleum Geologists) Memoir 7.

Kious, W.J. and Tilling, R.I., 1996. *This dynamic earth*. Reston VA, US Geological Survey; *also* http://pubs.usgs.gov/gip/dynamic/dynamic.html

Miller, R., 1983. *Continents in collision*. Time-Life Books, Alexandria, VA.

Moores, E.M. (ed), 1990. *Shaping the earth, tectonics of continents and oceans*. Freeman and Company, New York.

Redfern, R., 1983. *The making of a continent*. Times Books, New York.

Simkin, T. *et al.*, 2006. *This dynamic planet: world map of volcanoes, earthquakes, impact craters, and plate tectonics*. US Geological Survey Geologic Investigations Series Map I-2800, 1 two-sided sheet, scale 1:30,000,000.

Also http://www.minerals.si.edu/tdpmap

POWER AND ENERGY

The stages of historical change can be seen as marked by the nature of human access to energy. Hunter-gatherers have to garner energy from a wide area in the form of wild plants and animals, whereas agricultural people concentrate their energy sources in the form of fields of energy-rich crops like cereals and potatoes and herds of domesticated animals like cattle and sheep. With the discovery of how to use steam under pressure, the usefulness of fossil fuels (coal, oil and natural gas) became central to many economies and acted as a subsidy to the solar energy which had beforehand powered human societies. Knowledge of how to use nuclear fission to generate electricity added to the repertoire in the post-1950s period. Each stage since the invention of agriculture has had a period in which there has been an energy surplus above that needed to keep people alive and this has nearly always resulted in one class of people seizing control of the surplus and using it for such ends as their own lifestyle (luxury foods, impressive buildings, wonderful gardens, remarkable tombs) or in the cause of some perceived greater good such as territorial expansion via warfare or colonialism. The practice is alive today: most political and business leaders do not cycle to work. There is thus an interesting linkage between the ability of a society to produce energy, and politics. The use of energy often uses the vocabulary of 'power' ('power generation', 'power stations') and power in human societies is often concentrated in the hands of a few high-energy users, with their large cars, helicopters and private planes. This is mirrored in the energy consumption differences between e.g., a citizen of Europe or North America and of Nepal or Bangladesh.

See also: ENERGY, ENVIRONMENTAL CHANGE

Further reading

Nye, D.E., 1998. *Consuming power: a social history of American energies*. MIT Press, Cambridge MA.

Smil, V., 1994. *Energy in world history*. Westview Press, Boulder, CO.

PRECIPITATION

Precipitation is any form of water that falls from the sky to the ground. The form of water includes rain, snow, sleet, freezing rain, hail, and virga. A major component of the hydrologic cycle, precipitation is responsible for producing most of the fresh water on the planet. The standard way of measuring rainfall or snowfall is the standard rain gage, which can be found in 4-inch/100 mm plastic and 8-inch/200 mm metal varieties (Fig. 51). Condensation and coalescence play important roles in the water cycle. When relatively warm, moist air rises, precipitation begins to form. As air cools, water vapour begins to condense on condensation nuclei and forms clouds. After the water droplets grow large enough, one of two processes will likely occur to bring about precipitation. Coalescence takes place when water droplets join to create larger water droplets, or when water droplets freeze onto an ice crystal. Due to air resistance, water droplets in clouds generally remain stationary. When air turbulence takes place, water droplets collide and produce larger droplets. As these larger water droplets descend, coalescence continues, and drops become heavy enough to overcome air resistance; the drops then come down as rain.

Coalescence occurs most often in clouds above freezing. The Bergeron Process takes place when ice crystals acquire water molecules from nearby supercooled water droplets. As these ice crystals gain mass, they may begin to fall. As this happens, the ice crystals acquire more mass as coalescence occurs between the crystal and surrounding water droplets. Such a process is temperature-dependent, since supercooled water droplets are only present in a below-freezing cloud. Further, because of the great temperature differential between ground level and clouds, the ice crystals may melt and become rain as they fall.

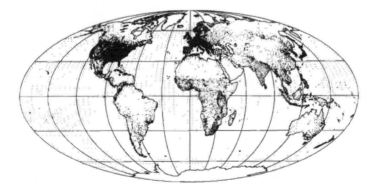

Figure 51 *Locations of 24 635 terrestrial stations and 2223 oceanic grid points at which 12 monthly mean precipitation avereges or estimates were compiled by Legates & Willmott (1990). An equal-area (Mollweide) projection is used so that stations densities among regions can be compared*

Precipitation is any product of the condensation of atmospheric water vapour that is deposited on the Earth's surface (Fig. 52). There are three categories. Types of liquid precipitation are rain or drizzle; freezing precipitation, freezing drizzle and freezing rain; types of frozen precipitation, snow, snow pellets, snow grains, ice pellets, hail, and ice crystals. Showery or convective precipitation occurs from convective clouds, including cumulonimbus or cumulus congestus. Convective precipitation is most prominent in the tropics. Hail and graupel are indications of convection. In mid-latitudes, convective precipitation is associated with cold fronts that usually come behind a front, squall lines, and warm fronts with significant moisture. Large-scale precipitation (stratiform) happens from slow (cm/s) ascent of air in synoptic systems, for instance, along cold fronts or before warm fronts. Caused by the rising air motion of large moist air across the mountain ridge, orographic precipitation usually takes place on the windward side of mountains. Such precipitation results in adiabatic cooling and condensation. In regions of the world where relatively consistent winds exist – such as trade winds – a wetter climate prevails on the windward side of a mountain than on the downwind or leeward side. Moisture is removed by orographic precipitation, leaving drier air on the descending and generally warming, leeward side where rain shadows can be seen. Likewise, the interior of larger mountain zones is usually quite dry– example is the Great Basin in the United States. Orographic precipitation generally happens on oceanic islands, for example, in the Hawaiian Islands. Often, rainfall on an island is on the windward side, while the leeward side tends to be quite dry and almost desert-like. This phenomenon results in substantial local gradients of average rainfall, with coastal areas receiving somewhere from 500 to 750 mm per year (20–30 inches), and interior uplands receiving over 2500 mm per year (100 inches). Leeward coastal

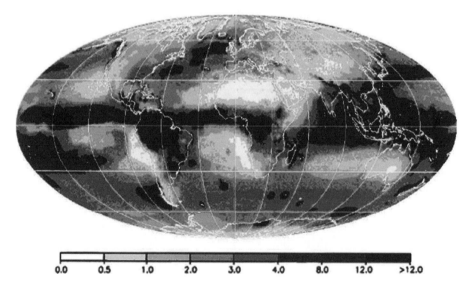

Figure 52 *Precipitation in mm per day. After Willmott, C.J. and D.R. Legates, Climate Research, Vol. 1, 1991, p. 181, given with permission of the journal through Inter-Research*

areas are especially dry, receiving 500 mm per year (20 inches) (example is Waikiki); whereas the tops of moderately high uplands are especially wet 12000mm per year (475 inches) (an example is Wai'ale'ale on Kaua'I).

See also: ATMOSPHERIC SCIENCES, CLIMATE CLASSIFICATION, HYDROLOGICAL CYCLE

Further reading

http://www.crh.noaa.gov/iwx/program_areas/coop/8inch.php

http://www.oardc.ohio-state.edu/williams.2157/Tipping_Bucket_Rain_Gage.htm

http://school.discovery.com/lessonplans/activities/weatherstation/itsrainingitspouring.html

http://cocorahs.org

http://www.globe.gov/fsl/welcome/welcomeobject.pl

http://www.nws.noaa.gov

PRODUCTIVITY

Productivity is a measure of the rate of biomass production in natural and agro-ecosystems. It is generally measured as the weight of a material produced per unit volume or area per unit of time. Most often, the material is carbon and the units of production are often g C per m^2 per year (or g C m^{-2} yr^{-1}). However many different units are used (i.e., Kg/ha/yr, bushels per acre per year, etc.). Productivity and biodiversity are probably the two most fundamental measures of the status or nature or essence of an ecosystem. Productivity is a measure of the potential of a system to produce organic carbon and biodiversity is a measure of the numbers and relative abundances of species (or biochemical pathways, habitats, etc.) of a system.

Productivity varies widely over the Earth. Tropical rain forests, coastal systems, and wetlands are among the most productive ecosystems. The productivity of these systems averages about 2000 g C m^{-2} yr^{-1}. Deserts, tundra, and central oceanic gyres are among the least productive with productivity of 125 g C m^{-2} yr^{-1} or less. A comparison of the productivity of major ecosystem types is presented in Table 42.

The most important factors that affect productivity include rainfall, nutrients, light, and temperature. Rainfall is the factor that is best correlated with broad patterns of terrestrial productivity. Areas with high rainfall have, on average, higher levels of productivity. Such areas

Table 42 *Average productivity of major world ecosystems (adapted from Whittaker and Likens 1975)*

Ecosystem type	Productivity (g C m^{-2} yr^{-1})
Rainforest	2000
Estuaries	2000
Wetlands	2000
Temperate Forests	600
Upwellings	500
Lakes	400
Prairies, Steppes	300
Continental Shelves	300
Deserts, Tundra	125
Open Ocean	125

include rainforests (both tropical and temperate), and temperate deciduous forests. For example, the eastern part of temperate North America and Western Europe have relatively high productivity due to abundant rainfall. The western part of temperate North America generally has low productivity due to lower rainfall. Deserts are among the least productive ecosystems because of their aridity. Temperature also affects productivity with higher latitudes being, other factors being equal, less productive than lower latitudes. High latitude tundra systems have low productivity. Nutrient concentrations are also correlated with productivity. For example, flood plain forests are more productive than adjacent upland forests without river flooding because of both water and nutrients associated with river flooding. Estuaries and coastal waters are highly productive because of fertilization by rivers entering these areas. Light has a greater effect on aquatic productivity than on terrestrial productivity. In grasslands and forests, there is abundant light, but understory forest species may be affected by shading. In water, light decreases exponentially with depth. So, aquatic primary productivity is strongly affected by light. One reason that estuaries and coastal waters are so productive is that a large proportion of the water column is in the lighted zone. Oceanic productivity is high near coasts, in upwelling zones (where deeper water flows to the surface bring high nutrient concentrations), and at high latitudes in the spring (when there is a combination of increasing light and high nutrients mixed up from deeper water by winter storms).

See also: BIOMASS, ECOSYSTEMS

Further reading

Audesirk, T. Audesirk, G. and Byers, B.E., 2002. *Biology – life on Earth*. Prentice Hall, Upper Saddle River, NJ. 6th edn.

Botkin, D.B. and Keller, E.A., 2007. *Environmental science – earth as a living planet*. Wiley, New York.

Brewer, R., 1994. *The Science of ecology*. W.B. Saunders, Philadelphia.

Campbell, N.A., Reece, J.B. and Mitchell, L.G., 1999. *Biology*. Addison Wesley Longman, Inc., Menlo Park, CA. 5th edn.

Day, J.W, Hall, C.A.S., Kemp, W.M. and Yáñez-Arancibia, A., 1989. *Estuarine ecology*. Wiley, New York.

Odum, E.P., 1971. *Fundamentals of ecology*. W.B. Saunders, Philadelphia. 3rd edn. (the most influential ecology text of the mid-20th century).

RADIATION BALANCE OF THE EARTH

The Radiation Balance refers to recognition of the fact that the Earth and its atmosphere for long time periods are in equilibrium neither drastically heating up or cooling down, although variability and trends may be observed and the system may be out of balance momentarily at the scale of geologic time. Scientists study the details of the components of radiation in short and longwaves of the electromagnetic spectrum to understand fluxes and their causes. Incoming solar radiation is shortwave, therefore the equation below is called the shortwave radiation balance SWB:

SWB = GR − RGR = DSR + DiSR − RGR
or depending on the albedo (back-reflection to space): = GR (1 − a)

GR = global radiation
DSR = direct solar radiation
DiSR = diffuse radiation
RGR = reflected portion of global radiation (*ca.* 4%)
a = albedo

In the infrared spectrum, the Earth's surface and atmosphere emit heat radiation. There is little overlapping between this and the solar spectrum. Since this is longwave radiation, this formula also is known as the longwave radiation balance (LWB):

LWB = ER = RS − AR

RS = radiation emitted from Earth's surface

AR = atmospheric radiation emitted back to Earth from atmosphere known as the greenhouse effect,

From those two equations for incoming and outgoing radiation, the total amount of energy now can be calculated (total radiation balance RB, net radiation) as:

RB = SWB + LWB.

Usually there is a gain from SWB, but because the Earth's surface is warmer overall than the atmosphere LWB is negative. To show the proportions in this balance equation by assigning 100 units to the solar radiation reaching our Earth's atmosphere at the top of the atmosphere (the solar constant of $1367 \, W/m^2$), the overall number of units of energy for RB at the Earth's surface is 29 units, with SWB = 45 units, and LWB = −16 units. The task is to precisely quantify various internal and external factors that influence the radiation balance. Internal factors are mechanisms affecting atmospheric composition (volcanism, biological activity, land use change, human activities). The external factor is solar radiation.

Over its lifetime, the Sun's average luminosity has increased by approximately 25% to date. External and internal factors are closely interconnected. For instance, increased solar radiation results in higher average temperatures and higher water vapour content of the atmosphere. Water vapour, a heat-trapping gas-absorbing infrared radiation emitted by the Earth's surface, can lead to either higher temperatures through radiation forces or lower temperatures from increased cloud formation and therefore increased albedo. The radiation balance can be specified for any scale of the environment from microclimates to global scales. In each instance, controlling factors of Earth–Sun geometry (orbital factors, distance to Sun, latitude, etc.), surface albedo, surface emissivity (emission relative to a black body emission amount according to surface temperature), atmospheric constituent fluxes and absorption must be understood to determine the overall net radiation at that surface (or the so-called radiation balance).

See also: ENERGY FLUX, ENERGY AND MASS BALANCE, RADIATION BALANCE OF THE EARTH

Further reading

Burt, J.E., 2007. *Understanding weather and climate.* Prentice Hall, Upper Saddle River NJ, 4th edn.

Thompson, R.D., 1998. *Atmospheric processes and systems.* Routledge, London and New York.

RIVER SYSTEMS

The system of rivers and stream channels in a drainage basin that make up the drainage network. The drainage network is a fundamental element of terrestrial landscapes because it acts as the artery along which water and sediment is transported and hence is critical to environmental processes. In semi-arid areas channels may be ephemeral and seldom contain flowing water. Elements in the drainage network of a river system have been denoted by systems of stream ordering in which the outermost unbranched tributaries are designated order 1 and in the most usually used system two first order streams combine to produce a second order, two second order streams will produce a third order and so on. Such ordering systems (Fig. 53) do not have any hydrological significance because a second order stream may receive numerous first order streams but will not increase in order until the confluence with a second order stream. Because the river system provides an interface between hydrology, ecology, geomorphology and other environmental sciences concepts have been developed including the river continuum concept (recognizing that structural and functional characteristics of stream communities are adapted to conform to the most probable state or mean state of the physical system) to express the typical way in which a river system

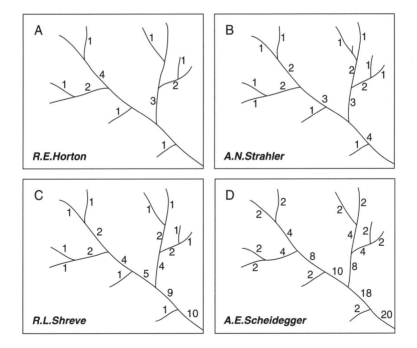

Figure 53 *Ordering systems. (From Gregory and Walling, 1973)*

changes downstream. River systems have also been classified according to the shape of drainage patterns including dendritic, trellised and annular patterns.

Studies of channel capacity, channel planform, and drainage density of the channel network have each involved attempts to establish equilibrium relations between indices of process and measurements of fluvial system form. Channel capacity, as the channel cross section area, has been related to flows which occur on average every one or two years, so that capacity or width of the channel can be employed to estimate likely flood discharge at a point along a river channel at ungauged sites. Additional research has focused upon river channel patterns, upon the controls on single thread and multithread patterns and what determines the thresholds between them. The density of the channel network varies with drainage basin characteristics including rock type, topography, vegetation and land use and the highest densities will generally produce greater flood discharges because water flow in a channel is more rapid than flow over slopes or through the soil.

As river systems evolve changes occur and the flood plain is controlled by the interaction between recent hydrological and sediment history together with the characteristics of the local area. Investigations of palaeohydrology and of river metamorphosis have employed contemporary relationships between channel form and process as a basis for interpreting river system changes resulting from a range of causes including dam and reservoir construction, land use change including urbanization and channelization. Such human-induced channel changes are extensive and superimposed upon the impact of shifts in sequences of climate which in some parts of the world such as Australia involved the alternation of periods of drought-dominated and flood-dominated regimes. Study of palaeofloods prior to gauged river flows have been important not only in enhancing flood frequency analysis, but also practically in the design or retrofitting of dams or other flood plain structures. Progress is now being made towards enhanced conceptual models of river systems with more integrated investigations embracing a range of spatial scales facilitated by enhanced remote sensing techniques, and GPS which enhances data capture and the speed of analysis.

See also: DRAINAGE BASINS, FLUVIAL ENVIRONMENTS, HYDROLOGY, INTEGRATED BASIN MANAGEMENT

Further reading

Coulthard, T.J., Kirkby, M.J. and Macklin, M.G., 1999. Modelling the impacts of Holocene environmental changes in an upland river catchment, using a cellular automaton approach. In Brown, A.G. and Quine, T.A. (eds), *Fluvial processes and environmental change*. Wiley, Chichester, 31–46.

Downs, P.W. and Gregory, K.J., 2004. *River channel management*. Arnold, London.

Gregory, K.J. and Walling, D.E., 1973. *Drainage basin form and process*. Arnold, London.

Harper, D.M. and Ferguson, A.J.D. (eds), 1995. *The ecological basis for river management*. Wiley, Chichester.

Walling, D.E., 2006. Human impact on land-ocean sediment transfers by the world's rivers. *Geomorphology 79*, 192–216.

ROCK TYPES

IGNEOUS ROCKS

Volcanic rocks are the attention grabbers of the geologic world, coming into being in the most dramatic ways: as basaltic lava flows oozing across sugar cane fields in Hawaii or flowing down hillslopes into the sea, or as explosive eruptions that produce tuffaceous pyroclastic rocks that blanket entire areas in the way Mount Vesuvius destroyed Pompeii in AD 79. The most abundant volcanic rock is basalt that forms the floors of the oceans, which cover 79% of the Earth's surface (Table 43). Basalt is derived by partial melting of upper mantle rocks and erupts relatively passively from the mid-ocean ridges like blood oozing from a cut.

The island arcs of the Earth, such as those around the northern and western Pacific Ocean (Aleutian, Kurile, and Japanese islands) consist of andesite that erupts from centralized stratovolcanoes like Mount Fuji in Japan and the major volcanoes of the Andes in South America, from which andesite takes its name. The really explosive, dangerous volcanoes, typified by Krakatoa and Santorini, are those composed primarily of rhyolite, which is the most viscous of the lavas because of its high content of silica and dissolved gases, the most abundant of which is H_2O. Volcanic rocks cool relatively rapidly so that they are characterized by small crystals, which typically range from less than 1 mm to as large as 10 mm, in contrast to those in plutonic rocks that range from 1 to 100 mm.

Table 43 *Common volcanic rocks*

Name	Minerals	SiO %	H_2O content	Viscosity	Explosivity
Basalt	Olivine, pyroxene, plagioclase	Low	Low	Low	Low
Andesite	Pyroxene, amphibole, plagioclase	Medium	Medium	Medium	Medium
Rhyolite	Biotite, plagioclase, orthoclase, quartz	High	High	High	High

Table 44   *Granitic rocks*

Rock name	Proportions of constituent white minerals
Granite	Quartz 40%, K-feldspar 30%, plagioclase 30%
Quartz monzonite	Quartz 10%, K-feldspar 45%, plagioclase 45%
Granodiorite	Quartz 35%, K-feldspar 35%, plagioclase 50%
Diorite	Quartz 5%, K-feldspar 5%, plagioclase 90%
Tonalite	Quartz 35%, K-feldspar 5%, plagioclase 60%

Plutonic rocks   cool slowly deep in the Earth's crust. The most common plutonic rock in the Earth's crust is granite; its several varieties are named and classified on the basis of the kind and proportions on their constituent minerals (Table 44).

Granitic rocks form by partial melting of the Earth's crust in subduction zones at depths of 15 to 60 km. The melts rise slowly up into the crust where they cool and crystallize at depths of 2–15 km over periods of tens of thousands to perhaps hundreds of thousands of years. Plutonic rocks typically form the hard core of some of the Earth's most scenic mountain ranges, including the Sierra Nevada of California, the Torre del Pain in Chile, and much of Scandinavia. Another type of plutonic rock commonly found with deeply formed granitic rocks is pegmatite, which forms masses and dikes of very coarsely crystalline rocks consisting typically of quartz, K-feldspar, and mica. Some crystals are huge – as large as a loaf of bread. Because plutonic rocks once formed at depth are now exposed at the surface, we may question how, and how quickly did the rocks thrust upward? Moreover, where did all the once-overlying rock – as much as 2–15 km – go? Geologists also argue about how granitic magmas form in the first place, how they rise upward into the crust, and how they make room for themselves. Granite bodies are very large, even many tens or hundreds of cubic kilometers, so to make room for them is a real achievement.

Table 45   *Types of sedimentary rocks*

Rock type	Description
Clastic sedimentary rock	A consolidated sedimentary rock composed principally of broken rock fragments derived from pre-existing rocks (of any origin) and transported as separate particles to their places of deposition by purely mechanical agents such as water, wind, ice, and gravity. AGI Glossary, 1987, with modifications
Evaporite	A consolidated sedimentary rock composed primarily of material formed directly by precipitation from solution or colloidal suspension (as by evaporation or by the deposition of insoluble precipitates), such as rock salt, chert, or gypsum. AGI Glossary, 1987, with modifications
Organic sedimentary rock	A consolidated sedimentary rock, such as limestone, diatomite, and coal, composed of organic remains.

SEDIMENTARY ROCKS

Sediment is defined as loose, solid rock fragments derived from weathering of pre-existing rocks (of any origin). Such sediment includes mud, silt, sand, gravel, and boulders. Sedimentary rocks are defined as consolidated agglomerates of loose sediment. Such rocks include mudstone, siltstone, sandstone, conglomerate, and breccia. Geologists divide sedimentary rocks into three classes, depending on whether the rock is an agglomeration of rock fragments (clastic sedimentary rock), whether it formed from precipitation of solids from water (chemically precipitated sedimentary rock, or more conveniently, evaporite), or if it formed from the remains of organisms, both plant and animal (organic sedimentary rock) (Table 46).

Clastic Sedimentary Rocks are classified primarily on the basis of their particle size (Table 42), but also constituent mineral grains, and grain shape. Adjectives are commonly used to distinguish some clastic sedimentary rocks, such as 'fine-grained' sandstone or 'coarse' arkose.

Compositional differences also enter into the naming of a sedimentary rock. Among the many kinds of sandstone are two main varieties: arkose and greywacke. Arkose is a coarse to very coarse-grained sandstone made predominantly of quartz and feldspar grains that are moderately angular, and its porosity may be high. Greywacke is commonly darker than arkose, and even though it contains quartz and feldspar, it has a much higher content of rock fragments, generally of the darker varieties of igneous and metamorphic rocks. The mineral grains in greywacke are also quite angular and unweathered, but unlike arkoses, these sand-sized particles are set in a clay or silty matrix, which makes the porosity lower than arkose. Grain or clast shape may be important in the naming of sedimentary rocks. If the clasts are large and rounded, the rock is a conglomerate, but if they are angular, then the rock is termed a breccia (Fig. 54).

Sediments are turned into rock during compaction when porosity is reduced, pore water is expelled, and the grains are cemented. The cement comes from precipitation of soluble

Table 46 *Classification of clastic sedimentary rocks based on particle size*

Sediment		Grain size (in mm.)	Rock
GRAVEL	Boulder	>256	CONGLOMERATE
	Cobble	64 – 256	
	Pebble	4 – 64	
	Granule	2 – 4	
SAND	Very coarse sand	1 – 2	SANDSTONE
	Coarse sand	$\frac{1}{2}$ – 1	
	Medium sand	$\frac{1}{4}$ – $\frac{1}{2}$	
	Fine sand	$\frac{1}{8}$ – $\frac{1}{4}$	
	Very fine sand	$\frac{1}{16}$ – $\frac{1}{8}$	
MUD	Silt particle	$\frac{1}{256}$ – $\frac{1}{16}$	SHALE or
	Clay	$<\frac{1}{256}$	MUDSTONE

Figure 54   *Naming of clastic sedimentary rocks based on particle shape. Left – conglomerate has rounded pebbles; red pocket knife for scale. Right – breccia has angular pebbles; blue pencil for scale. Photos by Arthur G. Sylvester, 1964, 1967*

substances in the pore water, such as calcium carbonate ($CaCO_3$), silica (SiO_2), or iron oxide (Fe_2O_3).

Evaporites are rocks that result mainly from evaporation of water that contained dissolved solids, and they are named according to their composition. The most significant and common of these rocks are rock salt and gypsum, both of which are largely products of the evaporation of sea water. Less common but of great economic significance are borates and potash. Some limestone strata may have been precipitated from sea water. More nebulous is the origin of siliceous sedimentary rock, especially chert, which may form from direct chemical precipitation of SiO_2 on the sea floor, or else the silica replaced the original host rock, much like the replacement of woody fibres with silica to make petrified wood.

The most abundant organic sedimentary rock is limestone, which is made of the tiny skeletons and shell fragments of animals such as coral and sea snails rather than quartz or feldspar grains. Some limestones have been built by the organisms themselves, such as lime-secreting algae or coral, that have built the great coral reefs of the tropics, including the Great Barrier Reef, which stretches for 1800 km along the coast of eastern Australia.

Limestone grades imperceptibly into dolomite when increasing amounts of magnesium enter into its composition, and it may also be deposited directly on the sea floor, although its origin is by no means agreed upon by all sedimentologists. Diatomite is a finely laminated, light-coloured shale that contains the microscopic remains of diatoms, which are ornate, single-celled plants that proliferate by the uncounted millions in the surface waters of the colder seas of the world. Diatomite is used for swimming pool filters and cat litter. Roughly comparable in origin is chalk formed from the calcitic remains of minute, free-floating, single-celled animals called foraminifera. Coal forms from the partly decomposed remains of land plants.

METAMORPHIC ROCKS

Metamorphic rocks are defined as any rock derived from pre-existing rocks by mineralogical, chemical, and or structural changes, essentially in the solid state, in response to marked change in temperature, pressure, shearing stress, and chemical environment, generally at depth in the Earth's crust. Metamorphic rocks are 'transformed' rocks – transformed in the solid state by recrystallization under elevated temperature and pressure. They are distinguished and classified on the basis of their mineralogy, texture, and fabric. These three characteristics give clues on the conditions of temperature and pressure that caused the metamorphism.

Two major classes of metamorphic rocks are recognized: those formed by contact metamorphism and those formed by regional metamorphism. Contact metamorphism occurs when hot magma intrudes relatively cold, pre-existing rocks and causes them to recrystallize. Thus, sandstone recrystallizes to quartzite, limestone recrystallizes to marble, and shale recrystallizes to slate or schist. Areas of such metamorphism are generally small and limited to the region around the intruded magma. Regional metamorphism occurs on the scale of entire mountain ranges when their cores are subjected to the high temperatures, pressures, and deformations associated with processes that form mountains. Wholesale recrystallization and growth of new minerals occurs at these depths, giving rise to large areas of schist and gneiss. At these depths, aqueous fluids may add or remove chemical components so that the composition of the pre-existing rocks is completely changed. The rocks may become so hot under high pressure that they become ductile and flow. Although no one has ever visited those depths in the crust to observe the processes, the fabrics of such metamorphic rocks contain evidence that flow was exceedingly viscous.

See also: MINERALS

Further reading

Bates, R.L. and Jackson, J.A. (eds), 1987. *Glossary of geology*, 3rd edn, American Geological Institute, Alexandria, Virginia, 788 pp.

Pitcher, W.S., 1993. *The nature and origin of granite*. Blackie Academic and Professional, New York.

SOIL MANAGEMENT

The realization that soils will not for ever maintain their desirable properties when used by human societies must be very ancient. Irrigated soils in the great river valleys of the Old World like the Euphrates needed protection against salinification as early as the 4th millennium BC and episodes of severe soil loss such as the Dust Bowl of the USA in the 1930s have all reinforced the ideas that careful management is likely to be needed if a valuable resource is to be protected: what may have taken millennia to form can be lost in a day. The management systems adopted vary with the type of soil concerned and with the aim of the management, so that very broad generalizations are difficult to make. There are for example 'natural' soils not subject to any intensive human uses that are valuable in watershed management because they (along with the vegetation and the subsoil) retain water and release it slowly to the runoff and hence are a form of flood control. Riparian zones of such soils may also slow down the release of excess soil nitrogen from chemical fertilizers into watercourses and thus prevent eutrophication. Modified soils are a very common category since they above all underlie the agricultural productivity upon which most humans depend for food. They are also present under plantation forestry, sports pitches and many golf courses. Their role in the spectrum of human use of the environment is unchallenged. There is a third category, of artificial soils. These have been created on substrates which would not occur in nature, such as mine wastes or the solid wastes from coal-fired power stations. They often contain levels of chemical elements which render them toxic to many plants and would poison organisms in watercourses. Gardens may have soils which have been so intensively modified by their owners that they could no longer be placed in a standard soil classification: clayey soils may be made open-draining by adding sand and the would-be rhododendron grower may have to pour in tonnes of peat to render soils acid enough for an imported species.

Management purposes are therefore unlikely to be uniform. They might aim to be some or all of the following. The prevention of erosion is paramount. Any soil which is unvegetated at any time is likely to lose some mineral and humic matter to runoff but ploughed slopes on steep slopes in areas with intensive rainfall events are very likely to lose material to sheet erosion and gulleying, the remediation of which is costly and slow. If there is one more important aim than the others, this is probably it. Second in order is the prevention of salinification in areas of irrigated soils. Many projects designed to improve food and fibre output in the tropics, for example, have eventually failed because the soil and water management was not careful enough to prevent the rise of mineral matter in the soil profile. Some adjustment of crop is possible (barley is more tolerant of salts than wheat) and GM crops may in time render the problem obsolete but salinified soils present a picture of wasted effort and capital as well as lower incomes. Acidification, on the other hand, is less widespread and its amelioration is usually in the hands of regional air management rather than local water control. It is also essential to keep up soil structure, usually by ensuring that there is adequate input of organic matter into the soils. This allows the crumb structure of the soils to be maintained and is a necessary (if not sufficient) guard against rapid erosion. It is also part of the maintenance of soil water levels which are essential to manipulate the drainage of soils, which is usually critical to crop production. Lastly, soils can be seen as reservoirs of carbon compounds which if released in gaseous form (e.g., as carbon dioxide or methane) would add to the greenhouse gases at the root of global warming. Equally they have the capacity to absorb carbon dioxide and help ameliorate the enhanced greenhouse effect.

Soils are thus hybrid entities with a base in natural processes but subject to varying degrees of human intervention and control, to the point that like vegetation types they represent a spectrum from those virtually unaffected by humans to those entirely created by human actions. Given their role in many economic activities, and not least in the production of the food supply needed to underpin population growth, their importance is hard to underestimate.

See also: GREENHOUSE EFFECT, VEGETATION AND ITS CLASSIFICATION, POPULATION GROWTH

Further reading

FAO, 1998. *World reference base for soil resources*. Food and Agriculture Organization of the United Nations, Rome.

Hillel, D., Hatfield, J.L., Powlson, D.S., Rosenweig, C., Scow, K.M., Singer M.J. and Sparks, D.L. (eds), 2004. *Encyclopedia of soils in the environment*, Four Volume Set, Volumes 1–4.

Horn, R. *et al.*, (eds), 2006. *Soil management for sustainability*. Catena Verlag, Reiskirchen.

SOLAR RADIATION

Solar radiation is radiant energy emitted by the Sun as electromagnetic energy. Generally about one-half of the radiation is in the visible short-wave part of the electromagnetic spectrum. The other half is mostly in the near-infrared part, with some in the ultraviolet part of the spectrum. Solar radiation is measured with a pyranometer or pyrheliometer. The average energy density of solar radiation just above the Earth's atmosphere is about $1367\,W/m^2$ – the solar constant (Fig. 55). The Earth receives a total amount of radiation determined by its cross-section, but as the planet rotates, this energy is distributed across the entire surface area. Therefore the average incoming solar radiation (or insolation) is one quarter of the solar constant or ~$342\,W/m^2$. The amount received at the surface depends on the state of the atmosphere and the latitude. The amount of radiation intercepted by a planetary body varies as the square of the distance between the star and the planet.

The Earth's orbit and obliquity change with time, sometimes achieving a nearly perfect circle, while other times stretching out to an eccentricity of 5%. The total insolation remains almost constant but the seasonal and latitudinal distribution and intensity of solar radiation received at the Earth's surface also varies. For example, at latitudes of 65 degrees the change in solar energy in summer and winter can vary by more than 25% as a result of the Earth's orbital variation. Because changes in winter and summer tend to offset, the change in the annual average insolation at any given location is near zero, but the redistribution of energy between

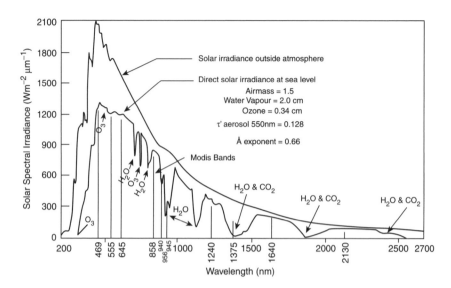

Figure 55 *The Solar Electromagnetic Spectrum showing amount of energy by various wavelengths from the sun. Note the difference between outside the atmosphere and at sea level – illustrating absorption by the atmosphere. Details of absorption by spectral band are shown.*

summer and winter does strongly affect the intensity of seasonal cycles. Two main components are received at the Earth's surface – direct beam and diffuse radiation (Fig. 56). The former comes in at an angle depending on solar position in the sky, the latter is scattered light in the atmosphere which comes in multi-directional forms. The Golden Colorado solar clear sky pattern illustrates the relative proportion of direct beam at a perpendicular angle during the day measured by an instrument facing directly at the Sun at all times. Global is the diffuse and Sun's

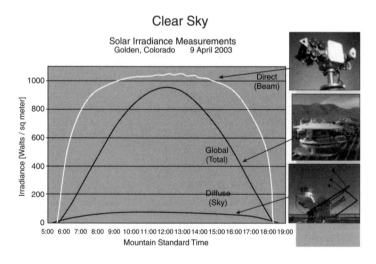

Figure 56 *From the National Renewable Energy Lab, U.S. Department of Energy. Direct beam is the energy received on a sensor always facing the sun at a 90 degree angle; global is the total of earth surface direct beam plus sky diffuse radiation; diffuse is the scattered radiation coming to the surface from the whole sky hemisphere.*

angular direct beam. Note that a surface faced perpendicular to the Sun's ray records more energy – a principle used in solar collector technology.

See also: ENERGY, ENERGY FLUX, RADIATION BALANCE

Further reading

Burton, J. and Taylor, K., 1997. *The nature and science of sunlight*. Gareth Stevens Publisher, Strongsville, OH.

Dogniaux, R. (ed), 1994. *Prediction of solar radiation in areas with a specific microclimate*. Commission of the European Communities, Kluwer Academic Publishers, Dordrecht.

Melnikova, I.N. and Vasilyev, A.V., 2004. *Short-wave solar radiation in the earth's atmosphere: calculation, observation, interpretation*. Springer Verlag, Amsterdam.

TECTONICS

In the AGI Glossary tectonics is defined as:

> A branch of geology dealing with the broad architecture of the outer part of the Earth, that is, the regional assembling of structural or deformational features, a study of their mutual relations, origin, and historical evolution. It is closely related to structural geology, with which the distinctions are blurred, but tectonics generally deals with the larger structures

Tectonic style is defined as 'the total character of a group of related structures that distinguishes them from other groups of structures in the same way that the style of a building or art object distinguishes it from others of different periods or influences'; graben as 'an elongate, relatively depressed crustal unit or block that may or may not be geomorphologically expressed as a rift valley'; and metamorphic core complex as 'a generally domal isolated uplift of anomalously deformed metamorphic and plutonic rocks overlain by a tectonically detached and distended unmetamorphosed cover of rocks'.

The term 'tectonics' derives from the Greek word, *tecton*, which means carpenter or builder. In geology, 'tectonics' is applied broadly to the building of Earth structures at the global, or macroscopic, scale. At the global scale, 'plate tectonics' refers to the constructional processes that form families of structures by interactions among the great plates of the Earth's crust. Three tectonic styles are recognized according to the type of plate interactions, and each has its own extensive vocabulary, literature, and examples.

Extensional tectonics is associated with divergent plate boundaries, that is, where plates drift apart from one another, stretching the crust, to produce such structures as rift valleys, grabens, tilted half-grabens, metamorphic core complexes, and listric normal faults. Collisional tectonics refers to convergent plate boundaries, where the crust is shortened, produces subduction zones, forearc and backarc basins, overthrust fault and fold belt mountains like the Alps, and magmatic arcs. Transform tectonics occurs between two, horizontally slipping plates produces strike-slip faults like the San Andreas, strike-slip basins, en echelon fold belts, and elongate uplifts.

Each of the global structures may be subdivided into specialized fields of study that cross the seemingly rigorous classifications imposed by plate tectonics. All basins formed by any type of plate tectonic mechanism, therefore, may be studied collectively as 'basin tectonics', or even narrowly as, for example, the tectonics of strike-slip basins, or of rift basins, or of forearc basins. The emplacement of granite, including how it makes room for itself during upward rise and intrusion into the crust, falls under the topic of 'granite tectonics'. 'Volcanotectonics' refers to all the processes involved in construction of a volcano or a chain of volcanoes. 'Seismotectonics' refers to all tectonic processes involved in producing earthquakes. 'Active tectonics' and 'neotectonics' pertains to crustal deformational processes in the last 2 million years.

See also: EARTHQUAKES, PLATE TECTONICS

Further reading

Bates, R.L. and Jackson, J.A. (eds), 1987. *Glossary of geology*, 3rd edn. American Geological Institute, Alexandria, Virginia, 788 pp.

Alterman, I.B., McMullen, R.B., Cluff, L.S. and Slemmons, D.B. (eds), 1994. *Seismotectonics of the central California coast ranges*. Geological Society of America Special Paper 292.

Keller, G.R. and Cather, S.M., 1994. *Basins of the Rio Grande rift: structure, stratigraphy, and tectonic setting*. Geological Society of America Special Paper 291.

National Research Council. 1986. *Active tectonics*. National Academy Press, Washington, DC.

Van der Pluijm, B.A. and Marshak, S., 2004. *Earth structure: an introduction to structural geology and tectonics*. Norton, New York, 2nd edn.

VOLCANOES AND VOLCANISM

'Volcano' is defined in the AGI Glossary as 'a vent in the surface of the Earth through which magma and associated gases and ash erupt; also, the form or structure, usually conical, that is produced by the ejected material' and tephra as 'a general term for accumulations of all hot fragments ejected from a volcano, including ash, cinders, blocks, and bombs – also called 'pyroclastic material'.

At every minute of every day somewhere on the Earth, both on land and beneath the sea, a volcano is actively erupting gas, tephra, and lava. It is a completely normal happening because volcanism is one of the ways the Earth continually loses heat and gas. Volcanism occurs primarily in three plate tectonic settings, and each type of volcano reflects the magma-producing processes that predominate in those settings. Rift zone eruptions occur along actively divergent plate boundaries, such as in the rift valleys of east Africa and all the mid-ocean ridges. When plates diverge, the underlying mantle partially melts at depth and thus supplies magma directly to the surface via extension cracks in the crust. The nature of rift volcanism is dramatically displayed where the Mid Atlantic Ridge rises above the sea in Iceland. There, typically low-silica, highly fluid basaltic magma, driven upward by magma pressure and gas, breaks out from ground cracks up to 10 km long in a spectacular curtain of fire that feeds very fluid, incandescent lava flows. Over a few days or weeks, the lava fountains coalesce into a few central vents around which tephra builds up into a cone, and from beneath which centralized lava flows emanate. The end result is a long valley peppered with linear chains of tephra cones amid a flood of black, hardened lava flows.

Hot spot volcanoes are located usually in the centres of tectonic plates, and their volcanic activity may last millions or tens of millions of years. As with rift volcanoes, hot spot volcanoes derive highly fluid, low-silica basaltic magma directly from the upper mantle. It is driven to the surface by magma pressure and gas. Hot spot eruptions are voluminous and almost continuous over geologic time, resulting in great, massive shield-like volcanoes, the best examples of which are Mauna Loa and Mauna Kea on the island of Hawaii. Located in the centre of the Pacific plate, they are the highest mountains in the world, rising 10,000 m from the ocean floor to peaks that are 3000 m above sea level. Their sub-sea extent is enormous. Other hot-spot volcanoes include the volcanic center at Yellowstone National Park in the North American continental plate. Its eruptions are more explosive, complex, and intermittent than those in Hawaii, because the supply of basaltic magma from the asthenosphere partially melts the lower crust, producing very viscous rhyolitic magmas capable of producing catastrophic eruptions which, in the past, have spread ash over more than half of the North American continent.

Subduction zone volcanoes are the large, classic, cone-shaped volcanoes we typically think of when we say 'volcano'. Well-known examples include Vesuvius in Italy, Fujiyama in Japan, Mount Rainier and Mount Shasta in the USA, and the many towering volcanoes in Central and South America. Their lavas include the entire range of volcanic rocks, from basalt through andesite to rhyolite, although they comprise predominantly intermediate, moderately viscous

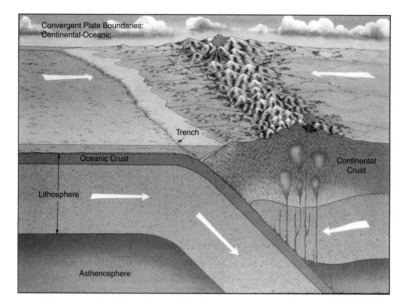

Figure 57 *Convergence between oceanic and continental plates causes partial melting of the lithosphere beneath the edge of the continent and above the down-going plate. The melts may rise as pods of magma high enough into the continental crust that they erupt from volcanoes along an arc-shaped chain. (source of diagram unknown)*

volcanic rocks, such as andesite and dacite. Their magmas are partial melts of the lower crust above the down-going, subducted plate, and they pierce the edge of the overriding plate as an arc-shaped chain of big, single volcanic centres (Fig. 57). Examples of such chains include the Japanese and Aleutian magmatic arcs.

Just how a volcano erupts depends on the composition of the magma and the gas contained in that magma. Mafic magmas like basalt erupt relatively passively, because they contain relatively little gas, and their low viscosity permits the relatively free escape of gas. In contrast, felsic magmas like rhyolite are gas rich, and their higher viscosity prevents easy escape of the gas, so that gas pressure builds as the magma rises to the surface and when the gas escapes, it usually escapes explosively and catastrophically, blasting the rock into ash that may spread in thick, choking blankets over vast areas.

See also: TECTONICS, ROCK TYPES

Further reading

Fisher, R.V., Heiken, G. and Hulen, J.B., 1997. *Volcanoes, crucibles of change.* Princeton University Press, Princeton, NJ.

Fisher, R.V. and Schmincke, H-U., 1984. *Pyroclastic rocks*. Springer-Verlag, New York.

Daniels, G.G. (ed), 1982. *Volcano*. Time-Life Books, Alexandria VA.

http://www.volcanoes.com/

WEATHERING

I s defined in the AGI Glossary as:

> The destructive process or group of processes by which earthy and rock materials on exposure to atmospheric agents at or near the Earth's surface are changed in color, texture, composition, firmness, or form, with little or no transport of the loosened or altered material; specifically. the physical disintegration and chemical decomposition of rock that produce an in-situ mantle of waste and prepare sediments for transportation.

Geologist F.Y. Loewinson-Lessing in 1936 wrote that 'Rocks, like everything else, are subject to change, and so also are our views about them'. Destructive changes in rocks are caused by moisture, interaction with organic acids beneath the ground, and the ceaseless activities of plants and animals. This process of rock alteration, weathering, is a very common and important process in geology.

Mechanical and chemical weathering (Table 47) may be difficult to separate in nature, because they commonly work together, although one or the other process may predominate in specific environments. In mechanical weathering, the particle size of rock changes from large to small as a result of energy exerted by physical forces, such as when a rock falls and breaks into smaller fragments, when tree roots or ice expand in a crack and break the rock even more, and when boulders smash together on the bed of a stream during a flood. Ants, worms, and burrowing animals work up weathered rock material and break it down even further.

In chemical weathering, the original rock is changed by addition or removal of rock material, usually by water (Fig. 58). Particle size is a principal factor in chemical weathering, because the greater the ratio of surface area to volume of a particle, the more vulnerable it is to

Table 47   *Types of weathering*

Mechanical weathering: rocks disintegrate by change in particle size, always from large to small, e.g., boulders break into cobbles.

Chemical weathering: rocks decompose by change in composition, e.g., feldspar decomposes to clay minerals.

Figure 58   *Chemical weathering of granite. Ground water and organic acids attack feldspars in granite, converting them to clay. When the granite is uplifted and exposed to the atmosphere, it breaks down into rock fragments, sand, silt, and clay ready to be washed away. Sunglasses give scale. Alabama Hills, California. Photo by Arthur G. Sylvester, July 1988*

chemical attack. For example, powdered sugar will dissolve faster in a cup of coffee than a sugar cube of equivalent volume. The rate and efficacy of chemical weathering is also determined by rock composition – certain minerals, for example, are more susceptible to chemical attack than others. Moisture accompanied by warmth speeds up chemical weathering as do plants and animals, which produce oxygen (O_2), carbon dioxide (CO_2), and organic acids that attack rocks.

Weathering provides sediments that go on to form sedimentary rocks. Chemical weathering of granite, for example, alters feldspar to clay, freeing the common and very stable minerals – quartz, clay, and iron oxides (Fig. 58). These products of weathering are then moved, usually moved by water and the influence of gravity – less commonly by wind and glacier ice, to a site of deposition. The water also carries soluble salts that may precipitate into other minerals like calcite and halite at the deposition site. Soil is one of the most valuable products of weathering. It is one of our planet's most precious resources that requires geologic time to make, but which has been flagrantly wasted by mankind over historic time.

See also: EARTH SURFACE PROCESSES, EROSION RATES

Further reading

Bates, R.L. and Jackson, J.A. (eds), 1987. Glossary of geology, 3rd edn. American Geological Institute, Alexandria, Virginia, 788 pp.

Boyer, R.E., 1971. *Field guide to rock weathering*. American Geological Institute, Fredericksburg, VA.

Robinson, D.A. and Williams, R.B.G. (eds), 1994. *Rock weathering and landform evolution*. British Geomorphological Research Group Symposia Series, Wiley, Chichester.

WETLANDS

Most of us are familiar with wetlands in some shape or form. The village pond, the trout stream, the river basin, or the local coastal lagoon or estuary, for example, are just some of the types of wetland that are widespread throughout temperate regions. But further south, in tropical and subtropical regions, there are muddy tidal flats, expansive floodplains and misty swamplands: three very different environments with very different plants and animals, but these are wetlands too. There are more than 50 definitions of wetlands in use throughout the world. Among these the broadest, and therefore that which is used most widely on an international scale, is provided by the *Ramsar Convention on Wetlands of International Importance, Especially as Waterfowl Habitat*. Ramsar is an Iranian city lying on the shore of the Caspian Sea, and it was here that the Wetland Convention was adopted in 1971. Designated to provide international protection to the widest possible group of wetlands ecosystems, the Ramsar Convention defined wetlands as '*areas of marsh, fen, peatland or water, whether natural or artificial, permanent or temporary, with water that is static or flowing, fresh brackish or salt, including areas of marine water the depth of which at low tide does not exceed six meters –20 feet'*.

Most representative wetlands around the world are: estuaries, mangroves and tidal flats: floodplains and deltas; freshwater marshes; lakes; peatlands; and forested wetlands. For 6000 years, river valleys and their associated floodplains have served as centres of human population, with many boasting sophisticated urban cultures. Their fertile soils brought in huge harvests upon which the peoples of the regions could depend. Today, the wetlands that nurtured the great civilizations of Mesopotamia and Egypt, and the Niger, Indus and Mekong valley, continue to be essential to the health, welfare and safety of millions of people who live by them. All wetlands are made up of a mixture of soils, water, plants, and animals. The biological interactions between these elements allow wetlands to perform certain functions including providing wildlife, fisheries and forest resources. The combination of these functions and products, together with the value placed upon biological diversity and the cultural values of certain wetlands, make these ecosystems invaluable to people all over the world.

Despite the importance of the range of resources and services which wetlands provide, humans have tended to take these for granted. As a result, the maintenance of natural wetlands has received low priority in most countries. But even as apathy and ignorance continue to permit conversion of wetlands, people are becoming increasingly aware of the loss of the services wetlands once provided free of charge, such as: groundwater and flood control, shore and storm protection, sediment and nutrient retention and export, plant and animal resources, energy resources, biological diversity, among others. Mitsch and Gosselink (2000) pointed out that any wetlands definition, for legal purposes, administrative concern, or ecological approach, must recognize the following: (1) wetlands are distinguished by the presence of water, either at the surface or within the root zone; (2) wetlands often have unique soil conditions that differ from adjacent uplands; and (3) wetlands support vegetation adapted to the wet conditions (hydrophytes) and, conversely, are characterized by an absence of flooding-intolerant vegetation. So the three-component basis of a wetland definition are: hydrology, physicochemical environment, and biota.

Based on the Weinstein and Kreeger up date monograph, there is no doubt that outwelling occurs from a number of wetlands to the adjacent ocean, where salt marshes are extensive and extremely productive, and tidal amplitudes large. Export pulses of organic matter and nutrients from marshes to the sea do not necessarily occur with every tidal cycle but may be intermittent associated with rain storms and high spring tides. The extent of outwelling is related to the level of productivity and extent of marsh cover within the estuary, the tidal amplitude and the geomorphology of the estuarine landscape.

See also: ESTUARINE ENVIRONMENTS

Further reading

Day, J.W., Gunn, J.D., Folan, W.J., Yáñez-Arancibia, A. and Horton, B.P., 2007. Emergence of complex societies after sea-level stabilized. *EOS, Transaction American Geophysical Union*, 88 (15): 170–171.

Diegues, A.C. (ed), 1994. *An inventory of Brazilian wetlands*. Gland, Switzerland, IUCN.

Dugan, P. (ed), 1993. *Wetlands in danger*. Mitchell Beazley/IUCN. London.

Dugan, P. (ed), 2005. *Guide to wetlands*. Phillips, London.

Hamilton, P. and MacDonald, K.B. (eds), 1980. *Estuarine and wetland processes*. Plenum Press Inc., New York, 653 pp.

Maltby, E., Dugan, P.J. and Lefeuvre, J.C. (eds), 1992. Conservation and development: the sustainable use of wetland resources. *Proceedings of the Third International Wetlands Conference, Rennes, France, 19–23 September 1988.* Gland Switzerland, IUCN.

Mitsch, W.J. and Gosselink, J.G., 2000. *Wetlands,* Wiley, New York, 3rd edn.

Pomeroy, L.R. and Wiegert, R.G. (eds), 1981. *The ecology of salt marsh.* Springer Verlag, New York, 271 pp.

Reid, G.K. and Wood, R.D., 1976. *Ecology of inland waters and estuaries.* D. Van Nostrand Co., New York, 485 pp.

Weinstein, M. P. and Kreeger, D.A., (eds), 2000. *Concepts and controversies in tidal marsh ecology.* Kluwer Academic Publishers, Dordrecht, The Netherlands, 875 pp.

PART V

SCALES AND TECHNIQUES

INTRODUCTION

One of the main ways in which science seeks to discover what is in the world and how it works is to find sub-units and see how they work, before trying to reassemble the whole. Thus an animal may be investigated via its organs (skeleton, skin, blood circulation and the like), the organs via their tissues (muscle, nerves, connecting tissue), the tissues via their cells, and so on down to the fundamental elements of matter. Apart from the intellectual issues in studying parts (a reductive approach) rather than wholes (holism), there is the severely practical matter of being able to manage to understand relatively simple systems as against the fearful complexity of wholes like a large ecosystem, or the whole of Earth's atmosphere. So, for example, geological time (like historical time) is in reality continuous but is best studied via a series of periods which seem to have something in common. Demarcating these periods is often subject to disagreement among scientists but they all acknowledge that some sub-division along the scale of time (i.e., a temporal scale) is needed. In the most recent period of geological history, the Holocene (invented Greek for 'all recent time'), the influence of humans becomes ever stronger and so the naming becomes more difficult as different disciplines other than geology become involved and suggest labels such as 'Anthropocene'. The idea of scale can apply also to human endeavour, so that it is possible to consider the environmental attitudes and policies of individual people, and then all the way to the work of the United Nations, which aims to give a total global perspective, though in fact it tends to be broken down by regions or topics.

One trouble with this approach turns out to be that the essential nature of units classified according to smaller and smaller sub-units is lost, since their qualities depend to a greater or lesser extent on their relationship with other entities, that is to say on the dynamics of their interactions, especially in the transfer of energy and matter between them. So at right angles as it were to the approach just discussed, is the idea of a functional study in which relatively large wholes are demarcated and then the ways in they interact with each other are investigated. Thus a large piece of African savanna may be studied in terms of how much solar energy is fixed as plant material in a year, how much of that is eaten by large herbivores and how much is dead before it becomes ant food, how much of the herbivores' tissue becomes lion, how much hyena and how much vulture. The energy incident upon Earth's atmosphere can be seen as a series of flows through large units, such as the land, ocean and ice surfaces, the tops of clouds. The flows comprise both the immediate reflection, for instance, and the absorption by soils, water and plants. This type of partition study is obviously critical to the flow of elements like nitrogen and carbon around the planet. It has its difficulties since the measurement of the dynamics of large units (how many wildebeest per 100 square km, how much carbon in the top metre of the Weddel Sea?) can be hard to calculate.

This type of question relates to a basic technique of the natural sciences when dealing with complex matters, which is to construct a model of the relationships between the components. In this context, a model is a representation in words, mathematical formulae or computer digits (or possibly all of these) of a system in the real world. It makes assumptions and sets up a framework that describes the logical and where possible quantitative relations between them. It can of course make false or incomplete assumptions as a step towards eventual improvement of the model. The idea is that by changing the relationships (sometimes called the parameters) of the model, it will get closer and closer to the observed reality. In the present context, global climatic models are probably the most important set of models currently being used, though many ecological models which predict the effects of human activity upon ecosystems are also vital. The value of models of this kind is that they can be used to suggest a spread of predicted outcomes, depending on the assumptions that are made and the values that are fed into the model. Public reception of models sometimes lacks an appreciation of the limitations of data and knowledge which underlie them and are scornful when predictions turn out differently: that is in the nature of models and needs to be understood as such.

Lastly, there are models of immense range, sometimes called meta-models. These attempt to characterize whole areas of approach to the world. The notions of aesthetics, for instance, takes the human senses as their starting point and views the universe in those terms: 'isn't that star beautiful?' rather than 'that is cMa in the Bayer system and its magnitude is x, and its solar mass is y'. Similarly, when from my study window I see a kingfisher's wing glinting blue in the

(a)   (b)

(c)   (d)

Figure 59   *Founding fathers: (a) Case Linnaeus, (b) Charles Darwin, (c) William Smith, (d)William Morris Davis*

sunlight (not a common occurrence), I do not immediately think of its Linnean binominal or which species was its Tertiary ancestor. But the environmental sciences depend, as we have said before, on the meta-model that there is order in the universe. This order is founded on the laws of physics in the first instance (especially those laws that deal with gravity and with thermodynamics) and in their absence there would be a kind of random chaos in which science as we know it would have no meaning. From the very start of western reflection on the nature of the universe, the Greek word κόσμος (cosmos) has meant order, even though we might smile at its use as the root of our word 'cosmetic'.

Individuals, including scientists have been extremely influential in the development of environmental sciences. More recently than people like Carl Linnaeus, Charles Darwin, William Smith and William Morris Davis (Fig. 59) individuals such as Al Gore have been influential – in relation to a film *An Inconvenient Truth*, which brings home Gore's persuasive argument that we can no longer afford to view global warming as a political issue – rather, it is the biggest moral challenges facing our global civilization.

AESTHETICS

This term refers to ways in which humans appreciate the beautiful in the world through their senses. Although aesthetics most frequently focuses on works of art, and is studied as a branch of philosophy, especially the philosophy of art, it is included in environmental science in relation to aesthetic quality of landscapes and of human environments which can be perceived to be of high or low aesthetic quality, but which may vary according to culture. Environmental aesthetics covers the interaction between an individual and the environment in relation to beauty: it includes the aesthetics of nature and also the appreciation of types of environment including urban ones. Environmental aesthetics is considered in a number of disciplines including landscape architecture, forestry, geography, psychology and philosophy. This range of disciplines produces diverse approaches including humanistic which focuses on understanding individual experience of environment; behavioural in psychology which takes the individual perceiving the landscape as the subject of study; and professional as used by architects and planners which employs principles including form, colour, texture horizontal and vertical planes and scale to analyze landscape aesthetics.

Concern with aesthetics was boosted by legislation in several countries such as the Natural Environmental Policy Act (NEPA) in the USA in 1969 which recognized the right of citizens to aesthetically pleasing surroundings, requiring agencies to take account of aesthetic considerations in management and planning decisions. A branch of land evaluation has therefore been concerned with aesthetic appreciation of aspects of environment and has been proposed either using aggregate schemes, for example to evaluate scenery as a resource as devised by

Figure 60   *Coarse wood debris in a small river channel in the New Forest, UK. River channels that are aesthetically highly valued are often perceived to be those unencumbered with wood debris, a view which may influence how channels are managed and restored, although natural channels in forested areas will normally contain wood debris. (Photograph K.J. Gregory)*

Linton in 1968, or has been based on perception of individuals by obtaining results from samples of people about their reactions to, or evaluations of, the way in which they perceive aesthetic characteristics of environment (Fig. 60). Environmental design is now a subject of interest to several disciplines, including ecological and other forms of restoration, so that studies of environmental aesthetics can contribute to understanding of what is 'natural' and so what should be the object of restoration.

See also: CULTURES, ENVIRONMENTAL PERCEPTION, ECOLOGICAL ENGINEERING, LAND EVALUATION, RESTORATION OF ECOSYSTEMS

Further reading

Porteous, J.D., 1996. *Environmental aesthetics*. Routledge, London.

Tuan, Yi Fu., 1993. *Passing strange and wonderful: aesthetics, nature, culture*. Island Press, Washington, DC.

BIOLOGICAL OXYGEN DEMAND (BOD)

Biological oxygen demand (BOD) is a measure of the oxygen required to completely oxidize organic matter in a water body. The concept of BOD comes from studies in the first half of the 20th century of sewage treatment and the impact of inadequately treated sewage on oxygen in streams. The amount of organic matter is the critical parameter, but when these concepts were developed, organic matter was difficult to measure and oxygen was easy, and BOD became the standard measure in sewage treatment. The units of BOD are in mg/l of dissolved oxygen required to fully oxidize organic matter. BOD is still widely used today. Before the development of modern sewage treatment plants, untreated sewage was routinely discharged into streams. Downstream of the plant, oxygen fell as organic matter in the sewage was decomposed (measured as BOD). As the organic matter dropped to low levels, oxygen levels recovered due to oxygen diffusing back into the water from the atmosphere. This is called the oxygen sag curve and was first described by Streeter and Phelps in the 1920s. A predicable succession of biological organisms developed downstream of the sewage discharge with pollution tolerant species near the discharge, such as Tubificide worms, and clean water species as oxygen levels recovered. With the development of modern conventional treatment systems, BOD was routinely used to describe the amount of organic matter in both raw and treated sewage. Typically raw sewage has a BOD in the range of 200–300 mg/l. Treated sewage has a BOD ranging from 30 to 5, depending on the level of treatment. When more natural systems of treatment are used as part of the treatment system, such as wetlands, higher levels of BOD (~30 mg/l) can be tolerated and still achieve good water quality. This is because nature is doing part of the work of treatment.

See also: POLLUTION OF AIR AND WATER, WASTE MANAGEMENT

Further reading

Kangas, P.C., 2004. *Ecological engineering – principles and practice.* Lewis Publishers, Boca Raton, FL.

Metcalf and Eddy, Inc., revised by G. Tchobanoglous. 2003. *Wastewater engineering, treatment, disposal and reuse.*, McGraw-Hill, Boston MA and London, 4th edn.

Mitsch, W.J. and Jørgensen, S.E., 2003. *Ecological engineering and ecosystem restoration.* Wiley, New York.

CLIMATE MODELLING

Climate modelling uses mathematics and principles of physics to produce numerical expressions and solutions for energy, mass, and momentum processes and characterizations for the atmosphere, oceans, land surface, and the cryosphere (snow/ice sheets/glaciers). These solutions are formed for global and regional scales to depict climate processes and typically solve the full equations for fluid motion on a spherical Earth. Models range from treating the Earth as a point to detailed Earth–atmospheric three-dimensional representation (see GENERAL CIRCULATION MODELS). A point model of Earth's radiation budget, for example, may involve using the well-known Stefan–Boltzmann law to calculate the radiative temperature, T, of Earth, by assuming S (the solar constant – the incoming solar radiation per unit area – about 1367 W m^{-2}), α (the Earth's average albedo, approximately 0.37 to 0.39), r is the radius of the Earth, and σ is the Stefan–Boltzmann constant – approximately 5.67×10^{-8} J K^{-4} m^{-2} s^{-1} (J = Joule, K = degrees Kelvin, m = meter, s = second). This modelling relies on the principle of conservation of energy between incoming solar radiation and energy emission from the Earth by this simple equation:

$$(1-\alpha)S\pi r^2 = 4\pi r^2 \sigma T^4$$

S shines to the cross-sectional area of Earth represented by πr^2; the spherical area of the entire Earth, $4\pi r^2$, emits at T – note in the equation both sides contain πr^2 and thus cancel out.

This yields a value of 246 – 248 °K or about -27 to –25 °C — for the Earth's average radiative temperature T. This is approximately 35 °K colder than the average observed surface temperature of 282 °K. This is because the above model represents the radiative temperature of the Earth with no blanketing atmosphere. The difference between this cold radiative temperature and the warmer surface temperature is the existence of the atmosphere's natural greenhouse warming effect. A radiative-convective model (RCM) simplifies the atmosphere by two energy transfer processes – up and down radiative transfer through atmospheric layers as well as vertical transport of heat by convection (especially important in the lower troposphere). The RCMs have advantages over the simple model above – they can resolve the radiative exchanges that impact the surface temperature, and the effects of varying greenhouse gas concentrations on the surface temperature. A zero-dimensional model may be expanded to estimate energy transported horizontally in the atmosphere. This kind of model typically is zonally averaged. The model also has the advantage of allowing a dependence of albedo on temperature – the poles can be allowed to be frozen and the equator warm. Depending on the nature of questions asked and the associated time scales, there are conceptual, more inductive models; and, general circulation models operating at the highest spatial and temporal resolution currently possible.

See also: ATMOSPHERIC SCIENCES, ENERGY FLUX, ENERGY AND MASS BALANCE

Further reading

IPPC Climate Change, 2001. *Working Group I: The Scientific Basis 8. Model Evaluation*. IPCC Secretariat, Geneva.

Hanse, *et al.* at http://pubs.giss.nasa.gov/abstracts/submitted/Hansen_etal_2.html

Intergovernmental Panel on Climate Change, WMO/UNEP at http://www.ipcc.ch/

Rennó, N.O., Emanuel, K.A. and Stone. P.H., 1994. Radiative–convective model with an explicit hydrologic cycle. 1. Formulation and sensitivity to model parameters. *Journal of Geophysical Research*, 99, No. D7, 14429–14442.

COST-BENEFIT ANALYSIS (CBA)

Most human actions involve an intuitive calculation, often unconscious, of the costs and benefits of a particular activity. Shall I kill this small animal, with a lot of noise, or shall I wait in case a bigger meal comes along, albeit without any certainty? With the coming of industrialization and the chance to commit large amounts of private or public finance, and the very real likelihood of causing environmental damage, a method of systematizing the likely cost of a project and the return from it was needed. The result was CBA. The process involves monetary value of initial and ongoing expenses versus expected return. Constructing credible measures of the costs and benefits of specific actions is often very difficult. In practice, analysts try to estimate costs and benefits either by using survey methods or by drawing inferences from market behaviour. because of the timescale over which a project is expected to last, all the costs must be put on a single time-horizon. Cost-benefit analysis attempts to put all relevant costs and benefits on a common temporal footing. A discount rate is chosen, which is then used to compute all relevant future costs and benefits in present-value terms. This is all much more difficult where the environment is concerned. Putting a monetary value on the return of a commercial crop is possible, but on a much-loved view, or the amenity value of a stand of trees, is very difficult and subject to many controversies. The flow of future benefits is also difficult to deal with; in an era of rapid technological change any development may be rendered obsolete well before its expected expiry date in the CBA. In general, CBA is best used as one of a number of tools in a decision-making process but not allowed to determine the outcome by itself. Like computers, it is susceptible to the GIGO principle.

See also: CULTURES, ENVIRONMENTAL VALUES, ENVIRONMENTAL LAW

Further reading

Nas, T.F., 1996. *Cost-Benefit analysis: theory and application*. Sage, Thousand Oaks, CA.

Pearce, D., 1998. Cost-benefit analysis and environmental policy. *Oxford Review of Economic Policy* 14, 84–100.

Turner, R.K., *et al.*, 2003. Valuing nature: lessons learned and future research directions. *Ecological Economics* 46, 493–510.

DATING TECHNIQUES, RADIOMETRIC AND ISOTOPIC DATING

A number of techniques have been used to date past events. Many of these use either radioactive or stable isotopes. But other approaches have also been used. Tree rings are an example of this.

Many elements have several stable and radioactive isotopes. These isotopes can be used to provide dates for a number of different processes from days to millions of years. Radioactive isotopes have been used to measure rates of sedimentation, water movement, and photosynthesis. Stable isotopes have been used to gain an understanding of climate change, food web dynamics, and the fate of terrestrial organic matter in the marine environment. Radioisotopes are atoms that spontaneously decay at a fixed rate; this is termed radioactive decay. Radioactive isotopes decay at an exponential rate and this can be used to determine the half-life of different isotopes. The half-life can vary dramatically. For example, ^{232}Th has a half-life of 14 billion years (nearly the age of the universe!) while that of ^{131}I is about 0.02 years. By knowing this and the relative proportions of different daughter products, different processes can be dated. There are a group of primordial radionuclides that are left over from the beginning of the universe and that are very long-lived. Included in this group are nuclides of U, Pa, Th, Ac, Ra, Rn, Fr, Po, Bi, Pb, and Tl. Although these isotopes are long-lived, the daughter products in the decay chains most often have shorter half-lives and are sometimes used for shorter-term data. Some of the daughter products are transported to the sea floor where they become associated with sediment particles and can be used to date sedimentation rates. The relative activities of ^{40}K and ^{40}Ar has been used to determine the time since the solidification of rocks. The K–Ar ratio has been used to determine the age of the Earth and to date magnetic field reversals. The latter has been used to support the theory of plate tectonics and to measure rates of sea-floor spreading, as well as ash layers in the distant past.

Cosmogenic radionuclides are formed in the atmosphere by the collision of atmospheric gases with low-energy cosmic rays. For example, an incoming cosmic ray proton shatters the nucleus of an atmospheric atom producing a neutron. This neutron enters the nucleus of a ^{14}N atom and knocks out a proton producing a ^{14}C atom. The uptake of ^{14}CO$_2$ by plants marks organic matter that then can be used to date the organic matter for up to several thousand years. Other cosmogenic radionuclides that have proved useful for dating purposes are ^{3}H, ^{7}Be, ^{10}Be, ^{26}Al, and ^{32}Si.

There are a number of artificial radionuclides that have been introduced into the environment as a result of human activities. Fallout from nuclear weapons testing and leakage from nuclear power plants are the two major sources of artificial radionuclides. ^{90}Sr and ^{137}Cs are the two most abundant longer-lived isotopes. ^{90}Sr can be taken up by human and deposited into bone tissue where it can be a health threat. ^{137}Cs becomes associated with sediment particles and be used to measure sedimentation rates on the order of decades. Bomb testing peaked around 1963, and the highest concentration of ^{137}Cs in estuarine soil core occurs at 1963, giving an estimate of the rates of soil accumulation. Tritium (^{3}H) and bomb radiocarbon have been used to obtain information about rates of ocean circulation.

Stable isotopes are also useful in dating techniques as well as in other applications in environmental sciences. Stable isotopes are non-radioactive isotopes of elements. Most studies have focused on stable isotopes of carbon, hydrogen, oxygen, sulphur, and nitrogen. The lightest isotopes of these elements are most abundant. The relative abundances of the isotopes are normally reported as values relative to an accepted standard. Variations in the relative abundances of stable isotopes of the same element are caused by the preferential reaction or transport of the lighter isotope. This fractionation can occur in physiochemical and biological processes. The selective reaction of a lighter isotope causes the remaining isotopes in the pool to become depleted in the lighter, more reactive isotope. For example, rainfall in clouds becomes depleted in ^{18}O as air masses move towards the poles. This causes the water in polar ice caps to become depleted in ^{18}O and sea water to become enriched in ^{18}O. This is reflected in biogenic carbonate shells of marine organisms. As these shells fall to the ocean bottom, the ration of ^{18}O/^{16}O varies depending on the temperature of the overlying water. Thus, this ratio can be used to determine past history of ice volume and thus sea level. Oxygen isotopes can also help determine the temperature that rocks crystallized. The variation of ^{13}C and ^{12}C in plantonic cores can be used to determine the temperature of the oceans at different times in the past since the isotopic composition varies with temperature.

Tree rings can be used as a direct measure of age. Dendrochronology is the dating of past events (climate, fires) using tree rings. For example, trees generally have wide rings in wet years and narrow rings in dry years. Normally, many trees have one annual growth ring. But many tropical trees and some temperate trees (e.g., Bald cypress, *Taxodium distichum*) may have false rings. However, these difficulties can be overcome by cross-referencing. In addition, isotope analysis can be done on different trees to help establish age. The oldest living trees can be several thousand years old. Bristlecone pines (*Pinus longaeva*) can live for more than 4000 years and cross-dating of live and dead trees have been used to establish chronologies for almost 9000 years in the past. Tree rings have also been used to show that cypress trees increased their growth when they were fertilized with nutrient rich water.

See also: ISOTOPIC DATING, RADIOMETRIC DATING

Further reading

Faure, G., 1977. *Principles of isotope geology*. Wiley, New York.

Libes, S., 1992. *An Introduction to marine biogeochemistry*. Wiley, New York.

ECOSYSTEM APPROACH

The ecosystem approach is defined as a strategy for management of land, water and living resources that promotes conservation and sustainable use in an equitable way, and was adopted at the Second Conference of the Parties of the Convention on Biological Biodiversity (CBD) in 2003 as the primary framework for action under the Convention. For coastal zone ecosystems, an early paper dealing with the ecosystem approach of estuarine systems was published by Day and Yáñez-Arancibia in 1982.

What is distinctive about the ecosystem approach? It provides a framework for planning and decision-making that balances the objectives of the CBD. People are placed at the centre of biodiversity management. Capturing and optimizing the functional benefits of ecosystems is emphasized. The importance of biodiversity management beyond the limits of protected areas is emphasized, while protected areas are recognized as being vitally important for conservation. The flexibility of the approach with respect to scale and purpose makes it a versatile framework for biodiversity management. Transboundary biodiversity problems can be addressed using the ecosystem approach and regional political structures.

The objectives of management of land, water and living resources are a matter of societal choice. Ecosystem managers should consider the effects (actual or potential) of their activities on adjacent and other ecosystems. Recognizing potential gains from management, there is usually a need to understand and manage the ecosystem in an economic context. Management should be decentralized to the lowest appropriate level.

Any such ecosystem-management programme should: (a) reduce those market distortions that adversely affect biological diversity, (b) align incentives to promote biodiversity conservation and sustainable use, and (c) internalize cost and benefits in the given ecosystem to the extent feasible. The ecosystem approach should consider all forms of relevant information, including scientific and indigenous and local knowledge, innovation, and practices. The ecosystem approach should involve all relevant sectors of society and scientific disciplines.

See also: ECOSYSTEM, SYSTEMS: THEORY, APPROACH AND ANALYSIS

Further reading

Day, J.W. and Yáñez-Arancibia, A., 1982. Coastal lagoons and estuaries: ecosystem approach. *Ciencia Interamericana* OAE Washington, D.C 22 (1–2), 11–25.

De Fontaubert, A.C., Downes, D.R. and Agardy, T.S., 1996. *Biodiversity in the seas: implementing the convention on biological diversity in marine and coastal habitats.* IUCN, Gland and Cambridge.

IUCN, 1999. Global Biodiversity Forum, 13th Session. San Jose, Costa Rica, May 1999, Abstract of Proceedings The World Conservation Union, Gland.

Smith, R.D. and Maltby, E., 2003. *Using the ecosystem approach to implement the convention on biological diversity: key issues and case studies.* IUCN Gland, Switzerland and Cambridge.

ECOTECHNOLOGY

Ecotechnology is an environmental technology where the forces of nature are used to a great extent in achieving desired goals. For example, sewage can be treated using a highly engineered, advanced treatment plant or the ecotechnological approach of wetland assimilation. Ecotechnology is strongly related to the fields of restoration ecology and ecological engineering. Properly done, ecotechnology should use less energy and natural resources and have less of an impact on the environment. Both ecotechnology and ecological engineering involve creating and restoring sustainable ecosystems that have value to both humans and nature. But ecotechnology is broader than this. It is a technology that is greener; that has a lower environmental impact than conventional technology. Solar and wind energy are forms of ecotechnology as are green roofs and green buildings. There are many different examples of ecotechnology and below a few examples are given.

» Eco-friendly farming attempts to achieve food production without the negative impacts of industrial farming, such as soil erosion, water quality deterioration, and high energy inputs. This type of farming is often organic, uses organic fertilizer, efficient irrigation, multiple crops planted together, crop rotation, and crops with deep root systems.
» Assimilation wetlands can replace or augment conventional treatment methods. Not only do they result in better water quality, they also can restore and create wetlands. Assimilation wetlands are less costly and use less energy. For example, in conventional treatment, for every kilogram of carbon reduced as biological oxygen demand reduction, 2–3 kg of carbon are put into the atomosphere as CO_2 due to the energy use in the treatment process. Assimilation wetlands can sequester carbon by plant uptake and burial in sediments.

» Solid waste management can be dealt with using ecotechnology. Much solid waste is organic and non-toxic and can be composed and reused. In an energy scarce world with climate change, ecotechnology offers an approach that can help with these problems.

See also: ECOLOGICAL ENGINEERING, RESTORATION OF ECOSYSTEMS

Further reading

Botkin, D.B. and Keller, E.A., 2007. *Environmental science – Earth as a living planet*. Wiley, New York.

Brewer, R., 1994., *The science of ecology*. Saunders, Philadelpha.

Kangas, P.C., 2004. *Ecological engineering – principles and practice*. Lewis Publishers. Boca Raton, FL.

Mitsch, W.J. and Jørgensen, S.E., 2003. *Ecological engineering and ecosystem restoration*. Wiley, New York.

ENVIRONMENTAL MODELLING see Modelling

EXPERT SYSTEMS

An expert system is a structured programme that represents knowledge of one or more human experts. This class of programme was first developed by researchers in artificial intelligence during the 1960s and 1970s. An expert system contains a set of rules and may recommend a set of actions as outcomes. Expert systems make use of reasoning capabilities to reach conclusions. They aid, for example, in helping with tasks in the fields of human resources, accounting, medicine, process control, financial service, production, environmental science, meteorology, and climatology. In most applications, problem expertise is encoded in both programme and data structures. The expert system approach is one that emphasizes data structures.

The general architecture of an expert system involves two major components: (a) a problem-dependent set of data declaration called the 'knowledge base' or 'rule base', and (b) a problem-independent programme that is called the 'inference engine'. The knowledge base of an expert system consists of a large number of 'if, then' type of statements that are interrelated that are equivalent to a series of mental steps in a human reasoning process. The storage capacity of personal computers has increased to a point that it is now possible to run some types of simple expert systems on personal computers. An example in environmental area is a neural network approach to solving problems. A study was designed to relate tree growth to precipitation in Arizona.

'The neural network technique relies on developing generalized relationships between tree growth and cool-season precipitation after being trained in a self-organizing learning procedure. The neural network technique allows for the definition of complex relationships between input/output data including those that may be nonlinear. Neural network techniques rely on an iterative learning process that functions by repeatedly training the input data used in the model to minimize error. In a sense, a neural network functions like a human's brain in that it learns through repeated trials. The results of these repeated trials are stored as a system of weights that are applied to data used in the final model. The advantage of neural networks lies in their ability to learn and represent complex and often nonlinear relationships directly from the data being modeled' (from Knutti *et al.* 2003).

See also: PRECIPITATION

Further reading

Ignizio, J.P., 1991. *Introduction to expert systems: the development and implementation of rule-based expert systems.* McGraw-Hill, New York.

Joseph C., Giarratano, J.C. and Riley, G., 2005. *Expert systems, principles and programming.* Thompson Course Technology, Boston. 4th edn.

Jackson, P., 1998. *Introduction to expert systems.* Addison-Wesley, Harlow.

Knutti, R., Stocker, T.F., Roos, F. and Plattner, G.-K., 2003. Probabilistic climate change using neural networks. *Climate Dynamics,* 21, 257–272.

http://www.ispe.arizona.edu/climas/research/paleoclimate/methods.html

FIELD STUDIES AND SURVEYING IN GEOLOGY

Geologists go into the field to study the Earth because that is where the rocks are. The rocks contain the record of Earth history as well as mineral and energy resources geologists seek for eventual societal exploitation.

A geologic field study documents much useful information about the types of rocks in the study area, including their spatial and temporal relations to one another; the thickness and extent of strata; the kind, size, and extent of structural features in the rocks, such as folds and faults; the kinds of faults that cut the rocks, as well as which way and how much the rocks have been displaced by the fault; the locations of lava flows, relics of glaciation; and the locations of mineral concentrations.

These field data are typically recorded on a geologic map (Fig. 61). From that map, the geologist may construct a geologic cross-section, which is an interpretative representation of a vertical slice across the land (Fig. 62).

Frequently a geologist may gather data digitally with a laptop or handheld computer connected to a GPS unit for precise determination of location (Fig. 63).

Thus, the geologic map may be viewed at any stage of construction, and the laptop's software can simultaneously construct a cross-section, structure contour map, or any sort of interpretative portrayal of the map. Later the geologist may combine his/her data digitally with that of other geologists into a regional compilation that can be downloaded immediately by any remote client or user.

The specific information obtained in a geologic field study may be crucial for engineering applications, including the strength of rocks for building foundations, landslide-prone areas, and the presence of mineral, petroleum, and water resources. Routes for pipelines, electrical transmission lines, tunnels, dam sites, canals, and transportation are mapped and studied by

Figure 61 *Making a geologic map in the field. The geologist draws the locations of rock strata, faults, and many other geologic features on a topographic map, as in this case, or aerial photograph. Photo by Ivar Midkandal, 2004, with permission*

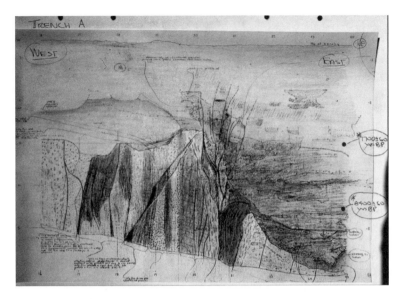

Figure 62   *Field sketch geologic cross-section. The sketch infers how the arrangement of rocks and faults would look like in a vertical slice across the land. Photo by Arthur G. Sylvester, Nov. 1984*

geologists to determine geologic conditions that may pose problems for construction. Geologists study glaciers, both ancient and modern, active and inactive volcanoes, earthquake faults, landslides, sinkholes, and other natural hazards. Geologists explore for mineral resources, both metallic and non-metallic, and for energy resources coal, oil, gas, oil shale, and uranium. Many governmental agencies require geologic studies as essential components of land use evaluations, such as Environmental Impact Reports.

Geophysicists conduct field studies to obtain information about what is beneath the Earth's surface. The most commonly used techniques include the use of reflection seismology and radar to image the depth, thickness, and orientation of subsurface layers to depths of nearly 10 km. Petroleum geologists obtain additional information from cuttings taken from drill holes and from sundry instruments lowered by wirelines into drill holes to measure various rock properties, including resistivity, magnetic susceptibility, and trace amounts of radioactive elements in the rocks.

Still other geologists and geophysicists measure ongoing slip of faults, rise and fall of mountains, and subsidence of the land as a result of ground water withdrawal. They use classic surveying fundamentals that a land surveyor uses but with high-precision GPS because of the need to compare measurements over time wherein the displacement rates may be only a few millimetres per year.

Once a specific land area is mapped geologically, is it necessary to map it again, given the fact that the rocks and structures will hardly change in a human's lifespan? The answer to that

Figure 63   *Geologist mapping with a hand-held microprocessor wirelessly connected to GPS receiver via a Bluetooth. The GPS antenna pokes out of his backpack. Photo by Gary Raines, January 2007, with permission*

question assumes the area was mapped well the first, or any, time. Usually a geologic study has a specific purpose and a specific kind tools and methods not necessarily used in another study. Therefore, several geologic and geophysical maps may exist of a specific land area; each of them may have shortcomings needing rectification or benefit from study by new techniques in the future.

See also: GLOBAL POSITIONING SYSTEMS

Further reading

Bevier, M.L., 2005. *Introduction to field geology*. McGraw-Hill Ryerson, New York.

Burger, H.R., Sheridan, A.F. and Jones, C.H., 2006. *Introduction to applied geophysics*. Norton, New York.

Compton, R.R., 1985. *Geology in the field*. Wiley, New York.

Spencer, E.W., 2000. *Geologic maps*. Waveland Press, Long Grove, IL. 2nd edn.

Geologic Maps and Mapping: http://ncgmp.usgs.gov/ncgmpgeomaps/

FOSSIL RECORD

From the fossils preserved in thousands of metres of sedimentary rocks worldwide, representing the last billion or so years of geologic time, we know that many creatures evolved from relatively simple forms to complex hierarchies of plants and animals. Some organisms have changed little over the passage of time, whereas others have become increasingly complex. In some cases, the older rock layers also contain the remains of organisms that lack living counterparts, revealing evidence of past extinctions.

The extinctions occurred when organisms became too inflexible or specialized, or were unable to adapt to a changing environment. The most spectacular extinction is that of the dinosaurs, which flourished during Mesozoic time – the Age of Dinosaurs – and then abruptly died off. Other geologically important, now extinct, groups that thrived in past geologic times include trilobites, which are curiously segmented marine arthropods distantly related to spiny crabs and were prevalent in Paleozoic time, and ammonites, which are complexly partitioned marine molluscs, distantly related to the squid and octopus and which prevailed in Mesozoic time.

The oldest fossils are nebulous traces of soft-bodied animals, probably like present-day jellyfish and worms, which lived about a billion years ago when animals were incapable of making preservable hard parts. More animal forms made their appearance over the next 500 million years, and by early Cambrian time several animals, notably trilobites, began to make hard parts out of silica. In mid-Cambrian time, a multitude of diverse forms of life suddenly appeared in the fossil record, probably just because they could make preservable hard parts. These included corals, which made hard parts out of calcite.

Aside from algae, which have existed for at least a billion years, the earliest plants appeared in the geologic record in Silurian time – about 425 my ago. They diversified and multiplied in the Devonian Period when fishes became prevalent. Reptiles and amphibians started climbing out of the sea onto the land in latest Devonian time – about 360 my years ago. The oldest mammals date from late Carboniferous time (315 my ago). Some of the reptiles evolved into dinosaurs in late Triassic time – about 215 my ago.

The Cenozoic Era is regarded as the 'Age of Mammals'. It is subdivided into just two periods, the Paleogene (65.5 to 23 my ago) when mammals became the dominant land creatures, and the Neogene (23 my ago to present), sometimes called the 'Age of Grasses'.

The boundaries between the Eras tend to be times of mass extinctions, such as the boundary between the Mesozoic and Cenozoic Eras when the dinosaurs abruptly became extinct and mammals became dominant. Some paleontologists estimate that as much as 90% of life on Earth became extinct at the time designated as the boundary between the Paleozoic and Mesozoic Eras. The causes of the extinctions are argued to include massive meteorite impacts, increased planetary volcanism, and catastrophic climate change induced by global changes in ocean temperatures and circulation.

One important challenge facing palaeontologists is to establish adequate correlations between the marine and non-marine rock sequences. Especially in Mesozoic and Cenozoic rocks, successions of reptiles and mammals permit correlation of non-marine sediments in the same way that classic marine faunas are used to correlate more widespread marine sediments.

See also: FOSSILS AND PALAEONTOLOGY, GEOLOGIC TIME, TIMESCALES, GEOLOGIC

Further reading

Simpson, G.G., 1983. *Fossils and the history of life*. Scientific American Books, New York.

Walker, C.D. and Ward, D., 2000. *Fossils*. Smithsonian Institution Handbooks, Washington, DC.

FOSSILS AND PALAEONTOLOGY

'Fossil' in the AGI Glossary in 1987 is defined as 'Any remains, trace, or imprint of a plant or animal that has been preserved in the Earth's crust since some past geologic or prehistoric time; loosely, any evidence of past life'.

Once regarded as 'sports of the Devil placed in rocks to confuse man', fossils were scientific mystery until Leonardo da Vinci correctly maintained that they are impressions or remains of once-living organisms preserved in stone. Fossils are more than just bones and shells; they include plants, even pollen, excrement, eggs, and cast-off shells of living animals. In many instances, an organism's teeth may be all that comprises parts hard enough to fossilize. Several processes are required to fossilize organic remains: reduction of the organism to its hard parts; burial, usually by water bearing sediment such as mud; and impregnation by water carrying dissolved minerals which, through cell-by-cell replacement, turn both bony and woody material to stone. So-called 'petrified wood' is an example of this process – cell by cell wood of a tree is replaced by silica until every bit of organic matter has become stone. Lastly, the rock encasing the fossil has to be uplifted and eroded to reveal the fossil.

Students of fossils, called palaeontologists, recognize several types, including megafossils, which are enough to be seen with the unaided eye; microfossils, which require a microscope to study; and trace fossils, which are nebulous imprints or impressions of soft-bodied animals (Fig. 64), and footprints, tracks, trails, and burrows made by walking, crawling, or slithering animals (Fig. 65). Megafossils include vertebrates (animals with backbones) like giant dinosaurs, horses, birds, turtles, and snakes, and invertebrates (animals without backbones) like clams, snails, and corals. Among the microfossils are single-celled forms like diatoms, dinoflagellates, and radiolarians. Diatoms are algae that produce a 'shell' of silica; dinoflagellates leave a tiny 'shell' or cyst from 20 to 150 thousandths of a millimetre long; radiolarians also leave a siliceous 'shell' between 100 and 200 thousandths of a millimetre. Soft-bodied animals may leave only delicate impressions of their form in sediment. Tracks, trails, and burrows give palaeontologists clues about how an organism's environment and the organism lived and moved around in it. Coprolites (fossil excrement) may indicate what the organism ate.

The coexistence of several kinds of fossils may provide clues about the plant and animal communities. Palaeontologists must be careful about jumping to conclusions, however, when various fossils are found in proximity to one another. In some instances, winds may have blown thin shells, or floods may have carried skeletons into disordered piles called a death assemblage, which can give a false view of the environment and life of the animals.

Fossils of all sizes are important for geologic and palaeontologic studies because they provide critical information about the age and environment of deposition of the rocks in which

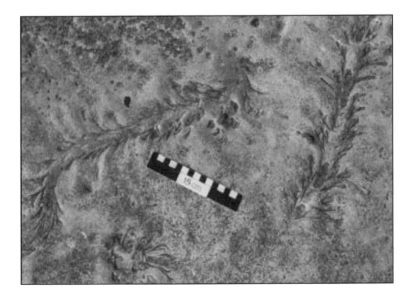

Figure 64 *Traces or tracks of an unknown animal or plant in mudstone. Pt. Lobos, California. Photograph by Arthur G. Sylvester, Nov. 1999*

Figure 65 *Swirly tracks and trails of an unknown critter in limestone bedding surface. Deep Spring Valley, California. Photograph by Arthur G. Sylvester, July 1969*

they are found. Marine microfossils are especially good for age determinations, because their life forms evolve rapidly and disperse so rapidly. The best environmental indicators are fossils whose life forms evolved hardly at all.

See also: GEOLOGY

Further reading

Gould, S.J., (ed), 1993. *The book of life: an illustrated history of the evolution of life on earth.* Norton, New York.

Prothero, D.R., 2004. *Bringing fossils to life: an introduction to palaeontology.* McGraw-Hill, Boston and London, 2nd edn.

Simpson, G.G., 1983. *Fossils and the history of life.* Scientific American Books, New York.

Walker, C.D. and Ward, D., 2000. *Fossils.* Smithsonian Institution Handbooks, Washington, DC.

For Vertebrate Fossils:

Maisey, J., 1996. *Discovering fossil fishes*. Westview Press, Boulder CO.

Norman, D., 1994. *Prehistoric life*. Boxtree Publishers, London.

For plants:

Gould, S.J., (ed), 1993. *The book of life: an illustrated history of the evolution of life on earth*. Norton, New York.

Prothero, D.R., 2004. *Bringing fossils to life: an introduction to palaeontology*. McGraw-Hill, Boston and London, 2nd edn.

Simpson, G.G., 1983. *Fossils and the history of life*. Scientific American Books, New York.

Walker, C.D. and Ward, D., 2000. *Fossils*. Smithsonian Institution Handbooks, Washington, DC.

Willis, K.J. and McElwain, J.C., 2002. *The evolution of plants*. Oxford University Press, Oxford.

GEOGRAPHICAL INFORMATION SYSTEMS (GIS)

This acronym stands for Geographic Information Systems and refers to the collection, analysis, storage and display of data which are spatially referenced to the surface of the Earth. In a simple form it was used by land use planners who overlaid a sheet of transparent film over a topographic map in order to relate e.g., housing projects to flood plains or earthquake zones. Now, however, it is entirely dependent on the digital computer. In its present form it is largely a creation of surges in development in the 1960s and the 1980s. Since the latter time, it has entered curricula in higher education in order to satisfy a large commercial demand for skilled operatives. Since GIS plots phenomena from the material world, it has to deal with (a) discrete objects such as a factory or a supermarket and (b) continuous 'fields' such as rainfall or elevation above sea level. GIS uses the notions of (a) raster data where each cell contains a single value; cells are arrayed in rows and columns and (b) vector data which use points, lines or polygons to locate data: this method is customarily used for continuously variable data such as rainfall or height. Both methods may be given 'depth' by tying characteristics such as colour to the cell or other shape being captured in the computer programme. Both, too, have their advantages and disadvantages in terms of the amount of storage space that is needed and the speed of retrieval of information and the flexibility with it which can be depicted for interpretation by the operator. Though excellent at finding patterns and apparent relationships between phenomena and processes, the human brain is also a key element in the final product which, as so often, takes the form of words.

See also: CARTOGRAPHY

Further reading

Duckham, M., Goodchild, M.F. and Worboys, M.F., (eds), 2003. *Foundations of geographic information science.* Taylor and Francis, New York.

Longley, P., Goodchild, M.F., Macguire, D.J. and Rhind, D.W., 2005. *Geographic information systems and science.* Wiley, Chichester, 2nd edn.

GLOBAL CIRCULATION MODELS (GCMs)

A general circulation model (GCM) or global climate model is designed to determine climate behaviour by integrating various fluid-dynamical, chemical, or biological equations that are either derived from physical laws empirically constructed. There are atmospheric GCMs (AGCMs) as well as ocean GCMs (OGCMs). The two models can be coupled to form an atmosphere–ocean coupled general circulation model (AOGCM). With additional components (e.g., a sea ice model or a land model), the AOGCM becomes the basis for a full climate model. The first AOGCMs were constructed in the late 1960s by Syokoru Manabe and Kirk Bryan at the Geophysical Fluid Dynamics Laboratory in Princeton, NJ. Since its first construction in the 1960s, over a dozen institutions around the world now use AOGCMs for climate predictions. A recent trend in GCMs is to extend them to become Earth system models to include sub-models for atmospheric chemistry or a carbon cycle model to better predict changes in carbon dioxide concentrations from changes in emissions. This approach provides feedback between the systems. Chemistry–climate models look at the possible effects of climate change on the recovery of the ozone hole. Climate prediction uncertainties depend on uncertainties in both models and the future course of industrial growth, development, and technology. Significant uncertainties and unknowns remain. Sophisticated models contain parameterizations for processes (such as convection) on scales too small to be resolved. More complete models may include representations of carbon and other cycles. A simple general circulation model (SGCM) which is a minimal GCM, consists of a dynamical core, the primitive equations, energy input into the model, and energy dissipation in the form of friction, so that atmospheric waves with the highest wave numbers are the ones most strongly attenuated. Such models are used to study atmospheric processes within a simplified framework; they

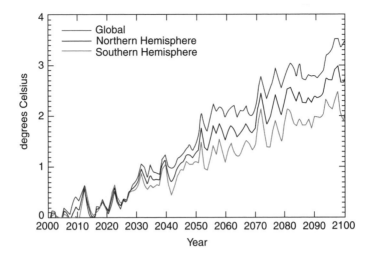

Figure 66   *Annual average surface air temperature change from HadCM3 IS92a. Source: Hadley Center for Climate predicition and research, the Met office*

are not necessarily suitable for future climate projections, but are the basis for detailed model predictions of future climate, such as those discussed by the IPCC. Typical AGCM resolutions are between 1 and 5 degrees in latitude or longitude: the Hadley Centre model (Fig. 66) HadAM3 uses 2.5 degrees in latitude and 3.75 in longitude, providing a grid of 73 by 96 points with 19 levels in the vertical. HadGEM1 uses a grid of 1.25 in latitude and longitude.

See also: ATMOSPHERIC SCIENCES, MODELLING, OCEANOGRAPHY, CLIMATE CHANGE, CLIMATOLOGY

Further reading

McGuffie, K. and Henderson-Sellers, A., 2005. *A climate modelling primer.* Wiley, Chichester, 2nd edn.

Washington, W.M. and Parkinson, C.L., 2005. *An introduction to three-dimensional climate modeling.* University Science Books, Suasalito CA, 2nd edn.

http://www.aip.org/history/climate/GCM.htm

http://www.ecmwf.int/products/forecasts/guide/The_general_circulation_model.html

http://www.iop.org/activity/policy/Publications/file_4147.pd

http://www.metoffice.gov.uk/research/hadleycentre/models/modeldata.html

GEOLOGIC TIME

The Dutch author and historian Hendrick Van Loon portrayed the staggering immensity of geologic time with the following allegory:

High up in the North in the land called Svithjod, there stands a rock. It is one hundred miles high and one hundred miles wide. Once every thousand years a little bird comes to this rock to sharpen its beak. When the rock has thus been worn away, then a single day of eternity will have gone by. (*The Story of Mankind*, 1922.)

Time is measured by the events that take place within it. For geologists – for whom geologic time is a particular possession not shared by other scientists – Earth history is recorded by the rocks and the events that have affected them. That history rests on the assumption that 'the present is the key to the past'. This statement is the principle of uniformitarianism, formulated by 'the father of geology', James Hutton in 1785, and published widely by the Scottish geologist, Charles Lyell, in 1830.

The uniformitarian principle rests on the circumstance that no known facts contradict it. It applies to all physical, chemical, and biological processes in the Earth. The opposite of uniformitarianism is catastrophism, and in reality geologic changes occur by means of many small catastrophic events, such as floods, landslides, and earthquakes. To medieval people, these changes were very impressive, particularly when the life expectancy was less than 50 years. If that big earthquake happens in *your* lifetime, then you may view it as a message from the deities, and the earthquake is added to your legends and sagas. Medieval catastrophists needed a short Earth history to match their literal interpretation of their Bible, so it was inconceivable to them that splashing rain, ocean surf, and intermittent earthquakes could have played a significant role in shaping the Earth's surface. Instead, they surmized that the large features of the Earth necessarily required tumultuous creation at a past time when the formative processes were violent cataclysms. With this background, it is understandable why it was a dramatic philosophical leap by James Hutton to propose that the Earth is several millions of years old, that it was formed and shaped by the very processes seen today at the same rates. Now it is recognized that the Earth is 4.6 billion years old, and that its history has been very long, complex, and wondrous.

Geologic time is divided into units similar to the way in which historic time is divided into eras and epochs, none of which are necessarily equivalent in length. Because the planet is so tectonically active in the way it builds, deforms, and reforms the crust, the more recent geologic events leave more complete and decipherable records than those grown increasingly fragmentary over a long lapse of time. Thus the later part of the geologic timescale has more sub-divisions of time periods than the earlier.

See also: DATING TECHNIQUES, ISOTOPIC DATING, TIMESCALES, UNIFORMITARIANISM

Further reading

Gould, S.J., 1989. *Wonderful Life: The Burgess Shale and the nature of history.* W.W. Norton and Company, New York.

Gradstein, F.M., Ogg, J.G. and Smith, A.G., 2004. *A geologic time scale.* Cambridge University Press, Cambridge.

GLACIATIONS

Glaciation refers to the processes involved with covering of parts of Earth with land-bound moving masses of ice known as glaciers. Throughout Earth history, there have been several major Ice Ages or glaciations that have been identified by study of the geologic record. The earliest of these glaciations occurred several billion years ago and the best-studied glaciation or Ice Age is the Pleistocene glaciation that has been ongoing for approximately the last one million years. The geologic record in the Northern Hemisphere suggests that during that time there have been about 14 major advances of glacial ice that spread outward and to the south, covering the northern parts of Europe and North America. The major cycle of glaciations is about 100,000 years, with minor cycles at 40,000 and 20,000 years. These cycles reflect global climate changes of 5 °C to 7 °C that correlate with changes in Earth's orbit around the Sun and thus the amount of solar energy received by Earth. However, these changes in solar energy are thought to be forcing mechanisms and by themselves are not strong enough to produce the changes in climate observed during glaciation.

With environmental science and glaciation, the conditions associated with glaciers and the results of glacial processes are of most concern. The most recent glaciation, sometimes called the Last Glacial Maximum, occurred about 20,000 years ago in North America and Europe when glaciers several kilometers thick caused sea level to drop by over 100 m. This was a harsh time for humans who had only recently evolved. Since the Last Glacial Maximum, there have been several minor glaciations including the Younger Dryas that occurred from approximately 11500–13000 B.P. The Younger Dryas was a time of rapid cooling, and the UK may have been as much as 5 °C colder than today. Following the end of the Younger Dryas, the planet moved into a much more equable period of climate, which favoured increasing human population and development of processes such as agriculture several thousand years later. Of particular importance to human society was the Medieval Warm Period that lasted several hundred years from about 11000 AD until approximately 1300 AD During that period Vikings colonized Iceland, Greenland, and North America. Sea temperature was probably several degrees warmer than now and this helped warm the land and allowed the Vikings to raise crops including corn and other grains. Unfortunately, the Medieval Warming Period was followed by a minor glaciation known as the Little Ice Age, which lasted from about 1400 AD to approximately 1850 AD Cooling of the Little Ice Age brought severe storms, wet periods, drought, heat and cold that led to crop failures and created treacherous conditions at sea that restricted trade and commerce. The difficult conditions caused the abandonment of Viking settlements in North America and parts of Greenland. During that same period, there were episodes of famine and the spread of the Black Plague. It is thought that the cold and wet conditions of Europe contributed to the severity of the plague, which reached Italy in 1348 AD and shortly after that, Spain, France, Scandinavia and Central Europe within a year or so. In England, one-quarter to one-third of the population died within a single decade. As a result, entire towns were abandoned and the production of food for the population that remained was severely jeopardized. The history of Viking settlements in the New World and the occurrence of the bubonic plague in Europe suggest that climate change and its accompanying effects can have dire consequences for humans. A change in global climate with return to a more glacial-like condition can significantly affect our global food supply and produce a catastrophe to human culture.

Today in environmental science, glaciation is much focused upon the perspective of melting glaciers. Recent warming has evidently resulted in retreat of many alpine glaciers in North America, South America, Europe and other areas. The melting of these glaciers, many of which formed during the Little Ice Age or perhaps the Younger Dryas, will bring water resource problems to those areas that depend upon glacial meltwater for human consumption and agriculture. In Iceland, glaciers are also melting at a fast pace and in some cases this is producing flood hazards as glacial meltwater periodically bursts from beneath the ice.

Study of glaciation and the glaciers themselves has greatly increased our understanding of the global climate system. Through the study of ice cores and small bubbles of past atmosphere, it has been possible to reconstruct the recent history of climate change. Warming can be fast, for instance, and is often followed by slower cooling. The warm periods are known as interglacials, in one of which all life lives today, whereas the glacial events are characterized by much cooler conditions than today. The survival of humans in the future is in part dependent upon the climate and how rapidly it may change. A climate change that would affect only a few per cent of the world's food supply could be catastrophic and so studying of natural changes during glacier periods is enabling a better understanding of Earth history and what it may mean humans in the future.

See also: ANTARCTIC ENVIRONMENT, GLACIAL GEOMORPHOLOGY, GLACIOLOGY

Further reading

Benn, D.I. and Evans, D.J.A., 1998. *Glaciers and glaciation*. Arnold, London.

Gillespie, A.R., Porter, S.C. and Atwater, B.F., 2004. *The Quaternary Period in the United States*. Developments in Quaternary Science 1. Elsevier, Amsterdam.

Lamb, H.H., 1977. *Climate: Present, past and future. Vol. 2, Climatic history and the future*. Barnes and Noble Books. New York.

Marsh, W.M. and Dozier, J., 1981. *Landscape*. Addison-Wesley, Reading, MA.

Menzies, J., 2000. *Modern and past glacial environments*. Heinemann-Butterworth, Oxford.

Pelto, M.S., 1996. Recent changes in glacier and alpine runoff in the North Cascades, Washington. *Hydrological Processes* 10, 1173–1180.

GLOBAL POSITIONING SYSTEMS (GPS)

The Global Positioning System (GPS) is a navigation system that uses a constellation of 24 satellites placed into orbit by the US Department of Defense.

First launched into orbit in 1978 by the US Department of Defense (USDoD) for its exclusive use, the system became fully available for international public use in 1994. The 'man-made stars' circle 20,000 km above the Earth at about 11,000 km per hour, twice a day in very precisely determined orbits, and they continuously transmit signal information back to Earth. GPS works anywhere in the world, in all weather conditions, 24 hours a day. The signals travel by line of sight, so that they pass through clouds, glass, and plastic, but they cannot go through most solid objects such as buildings and mountains, or underwater or underground.

To calculate a position on the Earth, a user needs a signal receiver that can determine how far it is from the satellites. Four satellite distances are required to calculate the user's 3D position: latitude, longitude, and altitude. The receiver determines how far it is from each satellite by comparing the time each signal was transmitted with the time it was received, then displays that position as coordinates or a point on an electronic map. Because precise timing is key to accurate distance measurements, the satellites are equipped with on-board atomic clocks. Even so, the default uncertainty is about 30 m with simple, over-the-counter receivers, but sub-centimetre precision is obtained with sophisticated receivers. Once the user's position has been determined, other information, such as speed, bearing, track, trip distance, distance to destination, sunrise and sunset time, can be calculated and displayed.

A GPS signal contains three essential bits of information to determine a position: a pseudo-random code, ephemeris data, and almanac data. The pseudorandom code is an identification code that specifies which satellite is transmitting information. Ephemeris data tell where each satellite should be at any moment throughout the day. Almanac data contain the current date and time, as well as important information about the health of the satellite.

The Earth's ionosphere and atmosphere may cause the GPS signals to be delayed, causing errors in position determination. Some of these errors can be partly eliminated with models built into the receiver that calculate an average amount of delay. Other error sources may come from the satellite clocks, timing inaccuracies in the receiver, and multipath reception, wherein the GPS signal reflects off large rock surfaces or buildings before reaching the receiver. At times a satellite's reported location may be inaccurate, although the USDoD constantly monitors each satellite's location in the network and can adjust the orbits if necessary. Other errors in the system may be magnified by less than optimal configurations of the satellites in the sky. The USDoD may also degrade or even turn off the signals to prevent adversaries from using them, but it has not done so since 2000.

See also: FIELD STUDIES AND SURVEYING, REMOTE SENSING

Further reading

Dana, P.H., 1999. *Global positioning system overview,* The Geographer's Craft Project. Department of Geography, The University of Texas at Austin, Austin, TX.

Hoffmann-Wellendorf, B., Lichtennegger, H. and Collins, J., 1994. *GPS: Theory and practice.* Springer Verlag, New York, 3rd edn.

http://www.colorado.edu/geography/gcraft/notes/gps/gps_f.html

HOLOCENE

The Holocene is the most recent interval of geologic time. The Holocene extends from the present back to approximately 11,000 calendar years BP (10,000 BC). The Holocene began following the retreat of the most recent Pleistocene glaciation and generally, it brought with it warming more constant climatic conditions. Minor variations include the Holocene climatic optimum in which the global temperature was a degree or two Celsius warmer than today. The climatic optimum ended approximately five to six thousand years ago when sea levels became more or less stabilized until the most recent more rapid rises due to global warming. Minor exceptions to generally rather consistent climate include the Medieval Warm Period from approximately 1100–1400 AD, which was followed by the Little Ice Age, which lasted until approximately 1850 AD.

The Holocene is the time during which human civilizations began and flourished. Earliest signs of civilization that include agriculture and division of labour probably began somewhere between five to seven thousand years ago during the mid-Holocene. The late Holocene, from present back to about four thousand years ago, is the time of tremendous development of human knowledge and technology now present. During the very latest Holocene, that is the last one hundred years, there has been a tremendous increase in information technology that drives civilization today.

See also: DATING TECHNIQUES, QUATERNARY, TIMESCALES

Further reading

Bell, M. and Walker, M.J.C., 2005. *Late quaternary environmental change: physical and human perspectives.* Longman, Harlow, 2nd edn.

Bull, W.B., 1991. *Geomorphic response to climate change.* Oxford University Press, NY.

Gillespie, A.R., Porter, S.C. and Atwater, B.F., 2004. *The Quaternary period in the United States.* Developments in Quaternary Science 1. Elsevier, Amsterdam

ICE CORES

These are the columns of ice, usually c. 10–12 cm in diameter and up to 3000 m long, which are extracted from Antarctica, Greenland and other ice masses for analysis. Each core records annual layers of ice and represents a major way in which the study of the Quaternary has been revolutionized since the 1960s. Characteristics of the layers from the surface downwards provides information about climate changes over periods of up to 800,000 years. This constituted a significant advance: not only is the record composed of annual layers of snow which became compacted into ice, (although deeper into the ice the layers thin and annual layers become indistinguishable), but also the record is uninterrupted. Analysis of ice cores provides various types of information.

- » Measured annual ice increments can provide data about changes in snow accumulation and ablation.
- » Analysis of the isotopic composition of the water comprising the ice provides estimates of global ice volumes and temperatures and is analogous to methods used for ocean cores.

Water contains naturally occurring isotopes (of different weights) and when the air temperature falls the heavier water molecules condense faster than the normal water molecules; the relative concentrations of the heavier isotopes based on the oxygen isotope ratio (O_{18}/O_{16}) can be used to indicate temperature at the time when the ice was deposited. Over short timescales the change in temperature from summer to winter produces a very clear oscillation in the O_{18}/O_{16} ratio which can be used to determine the age of the core at different depths. Deuterium, a heavy isotope of hydrogen, when present in a sample of ice in relatively large amounts signifies a higher temperature when the ice was originally formed.

» Analysis of the dissolved and particulate matter in the ice from atmospheric fallout including traces of volcanic ash recording volcanic eruptions, dust from aeolian activity, and sea salt.

» Analysis of physical characteristics of the ice: air bubbles provide samples of the atmosphere at the time they were formed, giving indications of former concentrations of greenhouse gases such as carbon dioxide, nitrous oxide and methane. Current concentrations of CO_2 are higher than at any time in the last 440,000 years.

The European Project for Ice Coring in Antarctica (EPICA) at Dome Concordia and Kohnen Station (1996–2005), supported by the European Commission and by 10 national contributions, aimed to obtain full documentation of the climatic and atmospheric record archived in Antarctic ice, by drilling and analyzing two ice cores, and to compare the results with those from Greenland. Information going back more than 700,000 years has furnished information about eight glacial cycles.

See also: ANTARCTIC ENVIRONMENT, ARCTIC ENVIRONMENTS

Further reading

Alverson, K., Oldfield, F. and Bradley, R., (eds), 2000. *Past global changes and their significance for the future.* Quaternary Science Reviews Series. Elsevier Science, Amsterdam.

Bradley, R.,1999. *Paleoclimatology: reconstructing climates of the Quaternary.* Academic Press, San Diego. CA, 2nd edn.

EPICA community members, 2004. Eight glacial cycles from an Antarctic ice core. *Nature* 429, 623–628.

Websites relating to international programmes:

GNIP Global network for isotopes in precipitation: http://isohis.iaea.org?GNIP.asp

NOAA: http://www.ngdc.noaa.gov/paleo/

PAGES Past Global Changes: http://www.pages-igbp.org

ISOTOPIC DATING

Stable isotopes are also useful in dating techniques as well as in other applications in environmental sciences. Stable isotopes are non-radioactive isotopes of elements. Most studies have focused on stable isotopes of carbon, hydrogen, oxygen, sulphur, and nitrogen. The lightest isotopes of these elements are most abundant. The relative abundances of the isotopes are normally reported as 'del values' relative to an accepted standard. This refers to the difference or delta between the sample and the standard. Variations in the relative abundances of stable isotopes of the same element are caused by the preferential reaction or transport of the lighter isotope. This fractionation can occur in physiochemical and biological processes. The selective reaction of a lighter isotope causes the remaining isotopes in the pool to become depleted in the lighter, more reactive isotope. For example, rainfall in clouds becomes depleted in ^{18}O as air masses move towards the poles. This causes the water in polar ice caps to become depleted in ^{18}O and sea water to become enriched in ^{18}O. This is reflected in biogenic carbonate shells of marine organisms. As these shells fall to the ocean bottom, the ration of $^{18}O/^{16}O$ varies depending on the temperature of the overlying water. Thus, this ratio can be used to determine past history of ice volume and thus sea level. Oxygen isotopes can also help determine the temperature that rocks crystallized. The variation of ^{13}C and ^{12}C in plantonic cores can be used to determine the temperature of the oceans at different times in the past since the isotopic composition varies with temperature.

See also: DATING TECHNIQUES, RADIOMETRIC DATING

Further reading

Faure, G., 1977. *Principles of isotope geology*. Wiley, New York.

Libes, S., 1992. *An introduction to marine biogeochemistry*. Wiley, New York.

LAND EVALUATION

The estimation of the potential of land for specific kinds of use, including productive uses such as arable farming, livestock production and forestry together with uses that provide services or other benefits such as water catchment areas, recreation, tourism and wildlife conservation. Land evaluation identifies variations in landscape quality as a basis for developing strategies for landscape conservation, management and enhancement; it can therefore involve comparison of the requirements of land use with the resource potential offered by the environment. Whereas landscape ecology is more general, landscape evaluation has the advantage that the approach is use-specific, but it cannot be utilized for other purposes and can be very labour-intensive.

Landscape evaluation has been undertaken by:

» Development from systematic data sources, so that national surveys of rock types, of superficial deposits, and of soils have provided the bases for landscape evaluation applications. Thus in the case of soils national surveys have often been utilized to produce land capability surveys including US Soil Conservation Service Systems from which capability systems can be derived. It is possible to proceed from maps of soil bodies, to soil quality maps and soil limitation maps, and subsequently to land classification in terms of soil crop response, present use, use capabilities, and recommended use. Developments by the FAO (Food and Agriculture Organization of the United Nations) in assessment of land performance for specified purposes based on soil data have occurred since the 1970s, and a journal entitled *Soil Survey and Land Evaluation* was inaugurated in 1981. Soil survey can provide a major component for the derivation of systems of land capability, well illustrated for agricultural uses by Bridges and Davidson in 1982, and can be developed to indicate yield potential. However translating evaluation into economic terms is very dependent upon prices at a particular time.

» Developments from land systems or landscape ecology surveys by adapting the results to be specific for a particular use. In Russia landscape geochemistry was developed and extended to forecast environmental impact including the extension of soil geography to accommodate the impact of technology. Land capability surveys were also developed by Ian McHarg to indicate the value of specific areas for particular kinds of land use in the course of evaluation of land suitability for urban extension and development.

» Other approaches to landscape evaluation, greatly facilitated by application of GIS, have developed for specific purposes such as the value of scenery as a resource. Three phases of landscape evaluation were recognized by Unwin in 1975 as:

1. measurement; to gain an inventory of what actually exists in the landscape;
2. value; to obtain value judgements or preferences in the visual landscape;
3. evaluation; to assess the quality of the objective visual landscape in terms of individual or societal preferences for different landscape types.

The original synthetic method used weights attributed to landscape components such as relief, to give spatial patterns of scenic resources, later complemented by methods based upon preference techniques which utilized individual perceptions of landscape quality, which can be obtained by individual responses to photographs of landscapes.

The types of methods available are listed in Table 48 showing the importance of GIS to refine and amplify the results.

Table 48 *Examples of aspects of environment that have been the subject of landscape evaluation*

Environmental characteristic	Way in which evaluation has been undertaken
Geology/rock type	Suitability for residential development and for use of septic systems
Soil type	Necessary drainage treatment for arable land use
	Land use capability classification
Relief	Slope categories in relation to agricultural implements
Vegetation type	Map primary productivity, possibly based on evapotranspiration and soil capability classes
Climate	'Clo index' indicates human comfort expressed in terms of clothing required to maintain thermal equilibrium
	For agriculture accumulated day degrees in excess of a threshold value; potential water deficit; accumulated frost as degree-days below freezing
Integrated evaluations	For cultivation of vines; based upon soil type, possible sunshine, danger of late frost, general aptitude of land.

See also: GIS, LANDSCAPE ARCHITECTURE, LANDSCAPE ECOLOGY

Further reading

Bridges, E.M. and Davidson, D.A., (eds.), 1982. Agricultural uses of soil survey data. In Bridges, E.M. and Davidson, D.A., (eds.) *Principles and applications of soil geography*. Longman, London, 171–215.

Dent, D. and Young, A., 1981. *Soils and land use planning*. Allen and Unwin, London.

Glazovskaya, N.A., 1977. Current problems in the theory and practice of landscape geochemistry. *Soviet Geography* 18, 363–373.

McHarg, I.L., 1969. *Design with nature*. Natural History Press, New York.

McHarg, I.L., 1992. *Design with nature*. Wiley, Chichester.

Unwin, K.I., (1975) The relationship of observer and landscape in landscape evaluation. *Transactions of the Institute of British Geographers* 66, 130–133.

Vink, A.P.A., 1983. *Landscape ecology and land use*. Translated from Dutch and edited by D.A. Davidson. Harlow, Longman.

LANDSCAPE ECOLOGY

Landscape ecology includes those methods of describing physical environment in ways which indicate how the environment may be utilized for agriculture, forestry, residential or other purposes. In this sense it differs from land evaluation which relates use to a specific purpose. Therefore, according to Vink in 1983, a major task of landscape ecology is to describe and characterize landscape according to relationships between the biosphere and the anthroposphere, and in more detail it has been defined by Kupfer in 1995 as:

> ... the study of how spatial scale and heterogeneity affect ecological processes. Landscape ecological principles are drawn from a diverse array of disciplines and fields, including physical and human geography, biology, geology, forestry, wildlife management, landscape architecture and planning ... The central focus of landscape ecology is the interrelationship between landscape structure – the spatial patterning of ecosystems across space – and landscape functioning – the interactions or flows of energy, matter and species within and among component ecosystems.

An early development of landscape ecology was description of terrain characteristics of the Flanders battlefield in World War 1 (1914–1918) and there were many attempts in Russia to characterize the major physiographic zones of the USSR and later to use a historico-genetic approach. The term was first used by Troll, to connote the interaction between geography (landscape) and biology (ecology) with applications to land development, regional planning and urban planning, it has subsequently been used by various environmental science disciplines to interpret landscape as supporting interrelated natural and cultural systems. Landscape ecology has now evolved to become a distinctive discipline of particular relevance to spatial ecology, biogeography, with landscape ecological theories integrating existing principles from applied biogeography and population biology, and the journal *Landscape Ecology* which was created in 1987, aims to draw together expertise from biological, geophysical, and social sciences to explore the formation, dynamics and consequences of spatial heterogeneity in natural and human-dominated landscapes.

It is possible to undertake landscape ecology in several ways: systematic, quantitative and integrated. Systematic approaches depend upon expressing environmental characteristics in a way which is relevant to environmental use and management such as the conversion of the contours of a topographical map into a slope map which is directly pertinent to land use. Quantitative methods of expressing the character of environment were greatly aided by remote sensing and by GIS and land resource information systems can be developed from soil survey records. Integrated approaches require the synthesis of characteristics of physical environment, topography, climate, soil and land use for example together as in the land systems approach where a land system is an area or groups of areas with recurring patterns of topography, soils and vegetation and having a relatively uniform climate. Whereas such approaches originally had to be undertaken manually, and were successfully used by Commonwealth Scientific Industrial Research Organization (CSIRO) in Australia, they can now be achieved quantitatively and GIS has provided an important way of achieving quantitative landscape ecology surveys.

In addition to thinking about ecosystems as physical objects, Haines Young in 2000 suggested that it is possible to visualize them in terms of attributes with value for people as natural assets or 'natural capital' to combine the scientific and cultural traditions of landscape ecology in managing landscapes. Characterization of environmental management in terms of the 'natural capital' paradigm provides an understanding of how the physical and biological processes associated with landscapes have value in an economic and cultural context, so that it is the study of natural capital from a dynamic, evolutionary and landscape perspective. It is not a steady state that is sought but rather a sustainable trajectory for ecosystems and landscapes so that equilibrium models rarely apply with no single sustainable state but a whole set of landscapes that are more or less sustainable.

See also: LAND EVALUATION

Further reading

Davidson, D.A. and Jones, G.E., 1986. A land resources information system (LRIS) for land use planning. *Applied Geography* 6, 255–265.

Farina, A., 1998. *Principles and methods in landscape ecology*. Elsevier, Dordrecht.

Haines-Young, R., 2000. Sustainable development and sustainable landscapes: defining a new paradigm for landscape ecology. *Fennia* 178, 7–14.

Klopatek, J.M. and Gardner, R.H., (eds), 1999. *Landscape ecological analysis: issues and applications*. Springer, New York.

Kupfer, J.A., 1995. Landscape ecology and biogeography. *Progress in Physical Geography* 19, 18–34.

Landscape Ecology. Journal published since 1987 by Kluwer Publishers.

Turner, M.G., Gardner, R.H. and O'Neill, R.V., 2001. *Landscape ecology in theory and practice*. Springer, New York.

Vink, A.P.A., 1983. (Trans and ed by Davidson, D.A.) *Landscape ecology and land use*. Longman, London.

MODELLING

A model is any abstraction or simplification of a system. A model of an ecosystem is simpler than a real ecosystem just as a model of a car is simpler than a real car. Models are widely used in many aspects of society from regulating industrial production processes to predicting future global climate change. Modelling is done to aid the conceptualization and measurement of complex systems and to predict the consequences of an action that would be expensive, difficult, or impossible to do in the real world. This entry specifically deals with ecological and environmental models. Two general types of models exist; analytic and simulation models. The analytic approach refers to a mathematical set of procedures for finding exact solutions to differential and other equations. Analytical models can be used relatively simple sets of equations using algebraic or differential equations. But they are not useful in dealing with complex sets of equations that describe the functioning of ecosystems. A simulation model does not give an exact solution to an equation. But simulation models can solve many equations nearly simultaneously, and it is possible to include all manner of non-linear equations. A distinction can be made between mechanistic and descriptive equations. A mechanistic equation describes the exact relationship between two variables. For example, one could model photosynthesis with equations showing the absorption of photons of light and a series of exact chemical reactions. On the other hand, a general relationship between sunlight reaching the surface of the Earth and plant growth could also be used. Most ecological and environmental models use a combination of both of these types of equations. Models can be used for understanding, assessing and optimizing our understanding of natural systems. Models can be used to test the validity of field measurements and our assumptions based on this data. Models can be used to predict or assess various actions. For example, models have been used to help understand the impacts of toxic materials in the environment. Models can also be used to optimize environmental decision-making.

The development or construction of a model involves a series of steps. Most of the time, but not always, the following steps are involved in modelling. Often a conceptual model is developed. Sometimes this is just in the head of the modeller. But mostly it involves a more formalized process. Essentially, the conceptual models says, 'This is how I think my system is'. Conceptual models may be *a priori* (based on pure logic) or empirical (based on experience) or a mixture of both. Development of a conceptual model can be thought of as a kind of thought experiment as with a food chain where little fish are eaten by big fish and the factors that might affect this. Very often, a conceptual model is followed by a diagrammatic model where symbols are used to formally illustrate the conceptual model. A series of boxes (or circles or some other symbol) is used to show the major components or state variables in the model, and arrows can be used to show how the boxes are connected. Forcing functions are inputs to the model that are outside the system boundaries of the particular system being modelled. The Sun, for example, is outside of ecological models because the Sun's energy comes from outside the Earth. If one were modelling our solar system or our galaxy, however, the Sun would probably be included as a state variable. In the example of the food chain, there would be a box for small fish and big fish with an arrow showing a flow from the small fish box to the big fish box. The degree to which different entities (i.e., different species of small fish) are included as part of the same state variable is an express of the degree of aggregation of the model. From the diagrammatic model, equations can be developed. In the food chain example, if it is thought that the feeding (F) by big fish is affected by the concentration of small fish (S) and big fish (B) and temperature (T), a general equation can be written:

$$F = f(SBT)$$

where feeding is a function of the concentrations or densities of both groups of fish and of temperature.

The degree of aggregation of a model is based on the questions asked by the model and by the nature of the system. For example, in a trophic model, it is common to use trophic levels such as plants, herbivores, lower carnivores, and higher carnivores, but groups of species or individual species can also be used. In building mathematical models, there are various relationships between state variables and forcing functions. A variable is independent if it is not affected by model interactions; light from the Sun for example. A dependent variable is affected by model interactions. For example, the population of little fish is affected by big fish because of predation in the model. The model is initialized using information or data about the system. For our example, this includes values for forcing functions such as sunlight, the size of populations of big and small fish, and the rate of feeding of big fish on little fish. Based on these values and relationships, mathematical equations are developed which describe the various interactions. Various mathematical relationships that are commonly used to describe functional relationships include linear, exponential, logistic, and Michaelis-Menton (these are descrbed in the references cited below). Once a model is calibrated using data about the forcing functions, state variables, and variable relationships, it is most often run on a computer. The model is then validated using an independent set of data. Models have been widely used in all aspects of ecology and environmental science. Some of the most exciting models today are those being used to predict future conditions associated with global climate change.

See also: CLIMATE MODELLING, ECOSYSTEM, SYSTEMS: THEORY, APPROACH AND ANALYSIS

Further reading

Haefner, J.W., 1996. *Modeling ecological systems – principles and applications.* Chapman and Hall, New York.

Hall, C.A.S. and Day, J.W., 1977. *Ecosystem modeling in theory and practice.* Wiley, New York.

PLEISTOCENE

The Pleistocene epoch is the period of geologic time from approximately 1.8 million years ago to 10,000 years ago. The Pleistocene is most noted for the tremendous shifts in climate as a result of periodic glacial (cold) and interglacial (warm) conditions. Of primary interest to environmental science is the latest Pleistocene, which is from 20,000 years before present to the beginning of the Holocene at 10,000 years ago. The latest Pleistocene is of interest because that was a time when the great melting of the Pleistocene ice sheets occurred and sea level rose rapidly. Near the end of the Pleistocene, about 11,000 years before present, a minor cooler period known as the Younger Dryas occurred and this was followed by rapid warming into the Medieval Warm Period of the Holocene.

Our study of the Pleistocene shows that the continents were in roughly the same position that they are today and although there was a lot of climatic variability, most is known about the Pleistocene because it is a recent part of geologic time. Most of the landscape of today is a product of erosion and deposition along with tectonic processes that occurred during the Pleistocene.

See also: QUATERNARY, TIMESCALES, GLACIATION

Further reading

Bull, W.B., 1991. *Geomorphic response to climate change.* Oxford University Press, New York.

Frenzel, B., *et al.* (eds), 1992. *Atlas of palaeoclimates and palaeoenvironments of the northern hemisphere: late Pleistocene, Holocene.* Fischer, Stuttgart and New York.

Gillespie, A.R., Porter, S.C. and Atwater, B.F., 2004. *The Quaternary Period in the United States.* Developments in Quaternary Science 1. Elsevier, Amsterdam.

Jones, R.L. and Keen, D.H., 1993. *Pleistocene environments in the British Isles.* Chapman and Hall, London.

Elias, S., (ed), 2007. *Encyclopedia of Quaternary science.* 4 vols. Elsevier, Amsterdam.

Munn, T., (ed), 2001. *Encyclopedia of global environmental change.* 5 vols. Wiley, Chichester.

QUATERNARY

The Quaternary is that part of geologic time known as the Quaternary Period, which extends from approximately 1.8 million years ago to present. The subdivisions of the Quaternary are the Pleistocene and Holocene epochs. The Quaternary, as with the Pleistocene, is characterized by large shifts in climate. Not only did global temperatures vary by several degrees Celsius, but also there were numerous glaciations and interglacial times. During the Quaternary, some areas became cooler and drier while others became warmer and wetter and these periods in time were variable across the landscape. For example, the Quaternary in Northwestern Europe is characterized by long periods of cold climates lasting several hundred thousand years interspersed with warmer conditions that lasted from approximately ten to twenty thousand years. The vast majority of time in the last three-quarters of a million years has been cold compared with conditions of today.

When global warming is considered, most of today's concerns are about the melting of sea-ice and glaciers along with rise in sea level and more intense storms such as hurricanes. Study of the Quaternary shows that climate changes may occur rapidly and that understanding the potential effects of human interference with that system is only just beginning. Study of the Quaternary record from ice cores, as well as sediments in lakes, bogs, and the ocean, is providing important information on past environmental and climatic conditions on Earth. This will inform a better understanding of what human societies may be facing in the future as conditions change.

See also: PLEISTOCENE, HOLOCENE

Further reading

Bull, W.B., 1991. *Geomorphic response to climate change*. Oxford University Press, New York.

Ehlers, J. and Gibbard, P.L., (eds), 2004. *Quaternary glaciations: extent and chronology*. 3 vols. Elsevier, Amsterdam.

Elias, S., (ed), 2007. *Encyclopedia of Quaternary science*. 4 vols. Elsevier, Amsterdam.

Gillespie, A.R., Porter, S.C. and Atwater, B.F., 2004. *The Quaternary period in the United States*. Developments in Quaternary Science 1. Elsevier, Amsterdam.

Munn, T., (ed), 2001. *Encyclopedia of global environmental change*. 5 vols. Wiley, Chichester.

Quaternary Science Reviews – journal published since 1984.

RADIOMETRIC DATING

Radioactive isotopes have been used to measure rates of sedimentation, water movement, and photosynthesis. The half-life of isotopes can vary dramatically. For example, ^{232}Th has a half-life of 14 billion years (nearly the age of the universe!) while that of 131I is about 0.02 years. By knowing this and the relative proportions of different daughter products, different processes can be dated. There is a group of primordial radionuclides, which are left over from the beginning of the universe and that are very long-lived. Included in this group are nuclides of U, Pa, Th, Ac, Ra, Rn, Fr, Po, Bi, Pb, and Tl. Although these isotopes are long-lived, the daughter products in the decay chains most often have shorter half-lives and are sometimes used for shorter-term dating. The relative activities of ^{40}K and ^{40}Ar have been used to determine the age of the Earth and to date magnetic field reversals. The latter has been used to support the theory of plate tectonics and to measure rates of sea-floor spreading.

Cosmogenic radionuclides are formed in the atmosphere by the collision of atmospheric gases with low-energy cosmic rays. For example, in this way a stable ^{14}N atom can be transformed into a radioactive ^{14}C atom. The uptake of ^{14}CO$_2$ by plants marks organic matter that then can be used to date the organic matter for up to several thousand years. Other cosmogenic radionuclides that have proved useful for dating purposes are ^{3}H, ^{7}Be, ^{10}Be, ^{26}Al, and ^{32}Si.

There are a number of artificial radionuclides that result mainly from fallout from nuclear weapons testing and leakage from nuclear power plants. ^{90}Sr and ^{137}Cs are the two most abundant longer-lived isotopes. ^{137}Cs becomes associated with sediment particles and be used to measure sedimentation rates on the order of decades. Tritium (^{3}H) and bomb radiocarbon have been used to obtain information about rates of ocean circulation.

See also: DATING TECHNIQUES, ISOTOPIC DATING

Further reading

Faure, G., 1977. *Principles of isotope geology*. Wiley, New York.

Libes, S., 1992. *An introduction to marine biogeochemistry*. Wiley, New York.

http://pubs.usgs.gov/gip/geotime/radiometric.html

REMOTE SENSING

The commonplace meaning of this term has become the collection of data from orbital satellites. This is an important application but not the only one, since aerial photography, radar, radiometry and sound reflectance are all equally eligible for this category of data collection, which senses electromagnetic radiation at some distance to provide information about the environment. The range of sensors now available have become invaluable adjuncts to the environmental sciences since they make it possible to collect data from inaccessible or dangerous areas. Further, their methodology often makes it possible to scan large areas at a time: the progress of ice-melt over Antarctica, for example, or a time-series estimation of deforestation in tropical forests such as those of the Amazon. Areas of political and military tension such as disputed borders can be 'patrolled' remotely without the risks generated by military hardware on the ground. Thus military use is widespread so that the latest and most accurate data are normally classified information until something more precise comes along.

Basically two kinds of remote sensing are employed. Passive sensors detect natural energy (radiation) that is emitted or reflected by the phenomena under observation. Film photography and infrared detection are major sources of this type, whereas active collection gives off energy in order to scan objects and areas. A sensor then detects and measures the radiation that is reflected. Precision can be very high: laser-based detection of variations in terrain (using an aircraft platform) can be detected to an accuracy of about 10 cm vertically and 50 cm horizontally. In many fields, remote sensing of data has been a revolution in the quantity of data collectable and only the parallel development of the digital computer for processing the information and putting it into comprehensible form such as maps and pictures has allowed to achieve its current status, with a considerable potential yet to be exploited. LiDAR (Light Detection And Ranging) is an important development with many applications including detection of faults, measurement of uplift, monitoring changes of glaciers and measurement of forest canopy heights, biomass and leaf areas. LiDAR can be used very effectively in conjunction with GIS. Examples of the recent use of remote sensing (Table 49) illustrate some of the potential available.

Table 49 *Applications of types of remote sensing*

Application	Method and sensor	Reference example
Effects of atmospheric pollution	Demonstrate extent of atmospheric pollution from metal smelters on vegetation	
Landslide assessment	Airborne laser altimetry (LIDAR) data used to produce high-resolution DEMs characterizing a large landslide complex.	McKean, J. and Roering, J., 2004. Objective landslide detection and surface morphology mapping using high-resolution airborne laser altimetry. *Geomorphology* 57, 331–351.

(Cont'd)

Application	Method and sensor	Reference example
Floodplain topography for flood prediction	LiDAR used to characterize topography of floodplains to enhance flood prediction by hydraulic models.	French, J.R., 2003. Airborne LiDAR in support of geomorphological and hydraulic modelling. *Earth Surface Processes and Landforms* 28, 321–335.
River pollution	Mineral sensing from hyperspectral data to map pollution from mine tailings discharged into rivers.	
Aufeis (overflow icings)	SAR interferometry used to assess growth rates of overflow icings	
Biochemical composition of plant canopies	Use of laboratory based near infrared spectrometry (NIRS) to estimate concentration of 12 foliar biochemicals from reflectance spectra of dried and ground slash pine needles	Curran, P.J., Dungan, J.L. and Peterson, D.L., 2001. Estimating ther foliar biochemical concentration of leaves with reflectance spectrometry. *Remote Sensing of Environment* 76, 349–359.
Regional- to global-scale land cover mapping.	Use of Envisat's Medium Resolution Imaging Spectrometer (MERIS) data for mapping eleven broad land cover classes in Wisconsin	Dash, J., Mathur, A., Foody, G.M., Curran, P.J., Chipman, J.W. and Lillesand, T.M., 2006. Land cover classification using multi-temporal MERIS vegetation Indices. *International Journal of Remote Sensing* 1–23.

See also: CARTOGRAPHY, GIS (GEOGRAPHICAL INFORMATION SYSTEMS)

Further reading

Lillesand, T.M., Kiefer, R.W. and Chipman, J.W., 2004. *Remote sensing and image interpretation.* Wiley, Hoboken, NJ. 5th edn.

http://www.fas.org/irp/imint/docs/rst/index.html – a tutorial

http://lidar.cr.usgs.gov

http://science.hq.nasa.gov/ — 'mission to planet earth' materials

ROCK CLASSIFICATION

The Earth's lithosphere is made up of three broad categories of rocks that are classified primarily on the basis of their origin, but they are identified on the basis of the kind, size, and shapes of the minerals in them. With inevitable exceptions, borderline cases, and overlaps, the rocks of Earth are placed in the following broad categories: Igneous, Sedimentary, and Metamorphic.

IGNEOUS ROCKS

Igneous rocks were once molten and have solidified either upon the Earth's surface as extrusive rocks called volcanic rocks – or deep beneath the Earth's surface as intrusive rocks called plutonic rocks. A third class is also distinguished: those molten rocks that have been injected within 1 km or so of the Earth's surface are called hypabyssal rocks and typically comprise narrow zones of intrusive rocks called dikes (Fig. 67).

Figure 67 *Black basaltic dike, 30 cm thick, intruded into white granite. Rock hammer for scale. Emigrant Pass, Death Valley. Photo by Arthur G. Sylvester, 2001*

The Main types of valcanic rocks are:

> »   Magma: molten rock beneath the Earth's surface.
> »   Lava: molten rock on the Earth's surface.
> »   Tephra: pyroclastic volcanic ejecta, including ash, cinders, blocks, and bombs.

Igneous rocks are either volcanic or plutonic, depending on their environment of cooling and crystallization. Compositionally, some volcanic rocks may be identical to some plutonic rock; the two are distinguished by crystal size. Volcanic rocks are glassy or finely-crystalline because they cool quickly – even in days or weeks – when erupted on the Earth's surface. By contrast, plutonic rocks have large, interlocking crystals because they cool 5–20 km deep in the Earth's crust at rates measureable in thousands or tens of thousands of years.

Volcanic rocks are comparatively familiar as the steaming, smoking volcano threatening to erupt over a town, or the incandescent lava flow streaming toward the sea seen in vivid video footage. Of the volcanic rock on the Earth's surface, basalt is the most abundant; andesite comprises the bulk of island arcs and volcanic mountain ranges. Rhyolite is less common but erupts from volcanoes with the capability of causing widespread catastrophic havoc by blasting tens or hundreds of cubic kilometres of tephra over half a continent.

Plutonic rocks, as exemplified by granite, the plutonic equivalent of rhyolite, are familiar because they are currently a popular rock type for kitchen countertops, and they comprise much of the world's spectacularly scenic mountains, such those in Yosemite National Park (Fig. 68). The pluton equivalent of andesite is granodiorite, which is a variety of granite. Gabbro is the plutonic equivalent of basalt. It is a comparatively uncommon rock on the Earth's surface but constitutes a large component of the crust, because it underlies the basalt of the ocean floors.

SEDIMENTARY ROCKS

Sedimentary rocks may be thought of as 'derived' or secondary rocks, because they are the breakdown products of any and all pre-existing rocks. Rocks break down as a result of *in situ* weathering, then they are eroded and transported to a site of deposition, buried, and made into rock. Common products of these processes are sandstone, conglomerate, and shale. Some sedimentary rocks are precipitates of soluble substances by evaporation of sea water, including rock salt, gypsum, nitrates, and some kinds of limestone. Still other sedimentary rocks form from the accumulated organic remains of plants and animals – examples of these include coal and limestone.

Like all other rocks, sedimentary rocks are aggregates of minerals and rock fragments, but unlike igneous and metamorphic rocks, their grains are not interlocking.

In general, sedimentary rocks are deposited on the surface of the Earth: on land, on the sea floor or lake bottoms where their formation can be observed. Most sedimentary rocks are layered (Fig. 69) but not all (glacial till, for example), and not all layered rocks are sedimentary

Figure 68 *Spectacular outcrops of granite conspire to make awe-inspiring scenery in Yosemite National Park, California. Photo by Arthur G. Sylvester, October 2005*

Figure 69 *Layered tuffaceous sedimentary rocks. Owens River Gorge, California. Photo by Arthur G. Sylvester, July 1969*

(banded metamorphic rocks, for example, Fig. 70). The layered nature of sedimentary rocks is a result of their having been deposited in a body of water, or having been transported by flowing water.

METAMORPHIC ROCKS

Metamorphic rocks are the most complex of the three rock types, because unlike sedimentary rocks and some igneous rocks whose processes of formation can be observed on the surface of the Earth, metamorphic rocks are products of heat, pressure, and chemical activity operating deep in the Earth over long periods of time. These factors cause the pre-existing rocks to recrystallize, either partially or completely, perhaps cause new minerals to form, and may cause the rock to attain a wholly new fabric or preferred orientation of minerals.

Metamorphic rocks are hard to identify and classify because they have no single mode of origin. They can be made from igneous, sedimentary, and even previously metamorphosed rocks under a multitude of combinations of temperature, pressure, and fluid activity. The characteristic that all metamorphic rocks have in common is their crystallinity. Like igneous rocks, their crystals may be interlocking, but unlike igneous rocks, some metamorphic rocks have a strongly layered fabric owing to segregation of minerals into bands that may cause them to

Figure 70   *Banded gneiss. Rock hammer for scale. Rognlien, Bergen, Norway. Photographed by Arthur G. Sylvester, 1974*

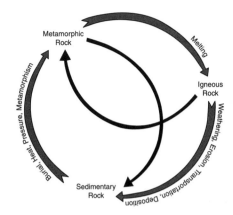

Figure 71 *Rock Cycle. Schematic depiction how each rock type can be derived from, and transformed into, any other rock type.*
Source: http://csmres.jmu.edu/geollab/fichter/Wilson/PTRC.html

resemble sedimentary rocks. The layering or banding is called foliation, which is characteristic of gneiss (Fig. 70). Slate and schist are common foliated metamorphic rocks wherein the foliation planes are closely spaced because of the planar alignment of micaceous minerals that the rock to split readily. Such rocks are said to have a cleavage.

These three families of rocks are distributed unevenly over the face of the Earth, most of which is basalt of the ocean floors. Only 5% of the Earth's crust consists of sedimentary rocks, whereas they cover about 75% of the total land surface like a discontinuous blanket spread thinly and unevenly over much more abundant crystalline rocks, which are the true foundations of continents.

The Rock Cycle chart (Fig. 71) is a convenient scheme to show the genetic relationships among the three rock categories. For example, a metamorphic rock can be melted to produce an igneous rock; the igneous rock may be changed into metamorphic rock. Similarly, any rock may be weathered, eroded, and its fragments transported, deposited, and made into sedimentary rock.

See also: GEOLOGY, MINERALS

Further reading

Pough, F.H., Scovil, J. and Peterson, R.T., 1998. *A field guide to rocks and minerals*. Peterson Field Guides, Houghton Mifflin, Boston.

Sorrell, C.A. and Sandstrom, G., 2001. *Rocks and minerals: a guide to field identification*. St. Martin's Press, New York.

SEA-LEVEL CHANGE

Eustatic sea-level rise (ESLR) is the rise in the level of the sea that is due to the increase in the volume of the world's oceans. It increased at a rate of 1–2 mm yr-1 over the last century and it is predicted to accelerate over the coming century by a factor of 3–4. ESLR is due to two main factors, an increase in the surface layers of the oceans due to thermal expansion and melting of land-based ice masses (glaciers and ice caps). There is a broad consensus in the scientific community (articulated by the Intergovernmental Panel on Climate Change in 2007) that the main cause of the acceleration of sea-level rise is due to human-induced global warming. Accelerated sea-level rise threatens both natural coastal ecosystems as well as human development in the coastal zone. Coastal wetlands are especially sensitive to increasing sea level because they exist within a relatively narrow elevation range, sometimes less than 10 cm. When coastal wetland vegetation is flooded for longer periods due to ESLR, it becomes stressed and ultimately dies. It has been shown that coastal wetlands in a number of locations in the world are disappearing due to accelerated sea-level rise. Low-lying coastal settlements are also vulnerable to ESLR and some low-relief island nations may completely disappear in the 21st century. Rising sea level combined with lower freshwater input will lead to increased saltwater intrusion and salinity stress. This especially threatens the extensive tidal freshwater wetlands. This combination of high RSLR, increased temperature, and lower freshwater input results in the north central Gulf having the highest vulnerability to climate change in the United States.

Relative sea-level rise (RSLR) is the total change in the level of the sea relative to the land surface in coastal areas due to a combination of ESLR and sinking of the land. A major cause of RSLR in coastal ecosystems is subsidence. Subsidence is common in deltas. Deltas naturally sink as the soft sediments deposited on them by rivers consolidate and compress under their own weight. Essentially, deltas and their associated wetlands are subsiding, often at a faster rate than they are being built up by addition of organic and inorganic soil formation. Where the sediment supply from rivers has been greatly reduced, the rate of build up is less than RSLR. Because of this, the wetlands experience more flooding. For example, while the current rate of sea-level rise is between 1–2 mm/yr, RSLR in the Mississippi delta is about 10 mm/yr. Other deltas such as the Nile, Rhone, and Ebro also have high rates of RSLR ranging from 2–5 mm/yr. Because of their high rate of RSLR, deltas serve as models for the effects of accelerated sea-level rise on coastal ecosystems in general and on approaches to manage coastal systems sustainably to survive accelerated sea-level rise.

See also: CLIMATE CHANGE, DELTAS, WETLANDS

Further reading

Day, J. and Templet, P., 1989. Consequences of sea-level rise: implications from the Mississippi Delta. *Coastal Management* 17, 241–257.

Day, J.W., Boesch, D.F., Clairain, E.J.H., Kemp, P., Laska, S., Mitsch, W., Orth, K., Mashriqui, H., Reed, D., Shabman, L., Simenstad, C., Streever, B., Twilley, R., Watson, C., Wells, J., and Whigham, D., 2007. Restoration of the Mississippi delta: lessons from hurricanes Katrina and Rita. *Science*. 1679–1684.

Gornitz, V., Lebedeff, S. and Hansen, J., 1982. Global sea-level trend in the past century. *Science* 215, 1611–1614.

Ibáñez, C., Prat, N. and Canicio, A., 1996. Changes in the hydrology and sediment transport produced by large dams on the lower Ebro river and its estuary. *Regulated Rivers* 12, 51–62.

IPCC (Intergovernmental Panel on Climate Change). 2007. *Climate change 2007: the science basis*. Contribution of Working Group 1 to the Fourth Assessment Report, Cambridge University Press, Cambridge.

Pont, D., Day, J.W., Hensel, P., *et al.*, 2002. Response scenarios for the deltaic plain of the Rhône in the face of an acceleration in the rate of sea-level rise, with a special attention for *Salicornia*-type environments. *Estuaries* 25, 337–358.

Stanley, D.J. and Warne, A., 1993. Nile delta: recent geological evolution and human impacts. *Science* 260, 628–634.

TIMESCALES, GEOLOGIC

Geologic time is determined in two senses: absolute and relative. In human terms, absolute time is akin to saying an event happened in the third week of June 2007, plus or minus 2–3 days. Because geologic time is so much vaster than human time, however, radiometric age dating is used to give the absolute age of a rock as, say 175 million years (my) plus or minus 15 my. Relative age, in human terms, is like saying something happened in the latter half of the Ming dynasty. A rock might be said to be Jurassic in age.

RELATIVE TIME

The standard geologic timescale The geologic timescale is subdivided into Eras, Periods, Epochs, and Stages, each corresponding to when specific sequences of sedimentary rocks were deposited. It is a theoretical representation of a stack of layered rocks that permits world-wide correlation of strata. It is based on real rocks that can be examined in their type localities. Because nowhere on Earth has sedimentation been continuous, the entire column has been pieced together from stratigraphic sequences based principally on correlations of fossils from place to place. For example, the rocks in the Jura Mountains of Switzerland were laid down in the Jurassic Period, the rocks at Wenlock, Shropshire, England were laid down in the Wenlockian Epoch of the Silurian Period, and the rocks in the Maastricht district of Belgium were laid down in the Maastrichtian Stage of the Upper Cretaceous Period.

The boundary problems between various sub-divisions of the timescale emphasize the fact that the standard column is an arbitrary man-made set of sub-divisions. As long as geologists agree to draw the boundaries at the same places according to the same criteria, then it is extremely useful to communicate ages of strata and times of events to one another. The geologic timescale is under constant study and refinement by the International Commission on Stratigraphy, especially to tie down the boundaries and events within the timescale by absolute dating methods. The latest version of the timescale may be accessed at http://www.stratigraphy.org.

The geomagnetic polarity timescale The Geomagnetic Polarity Timescale developed from worldwide studies of rocks that contain magnetic minerals whose (orientations are opposite to that of the current magnetic field. The studies concluded that the Earth's magnetic pole reverses aperiodically so that a compass needle would point to the south pole at times instead of the north pole as it does today. The reversals are recorded mainly in volcanic rocks by little needles of the mineral magnetite, which 'feel' the pull of the Earth's magnetic field and align accordingly at the time the lava solidifies.

The reversals happen over a few thousand years and as frequently as a few tens of thousands of years of each other, but they average about every 200,000 years over the past 100 million years. The 'palaeomagnetic' record proves reversals have happened many times over the past billion years. The last reversal was about 750,000–780,000 years ago. How and why reversals occur, when the next one will occur, and consequences for the Earth and life are fascinating topics of current research.

The irregularity of the reversals over time provides a unique pattern against which marine geologists match palaeomagnetic measurements to determine the age of the basaltic ocean floors of the world (Fig. 72).

Absolute timescales Rocks can be dated in an absolute sense by 'rock clocks' that use the principals of radioactive decay of chemical isotopes, and start 'ticking' at the moment a molten rock cools and solidifies. Some elements, such as uranium, thorium, potassium, rubidium, and carbon decay into daughter products whose abundance in a mineral or rock can be measured with a mass spectrometer. For example, naturally occurring potassium ^{40}K (parent) decays to stable ^{40}Ar (daughter). Because potassium's decay rate has been determined experimentally and therefore is well known, the ratio of the parent isotope (^{40}K) to the daughter isotope (^{40}Ar) tells how long that particular decay process has been going on in a rock or mineral. The method assumes that only the parent existed when the rock solidified from a magma or was at its peak of metamorphism, and that neither addition nor subtraction of ^{40}K or ^{40}Ar has occurred, or else the measured ratios would be phony.

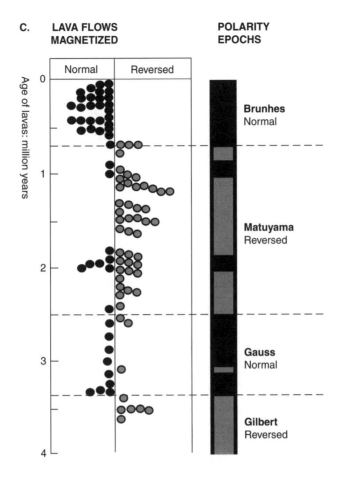

C.   **LAVA FLOWS MAGNETIZED**   **POLARITY EPOCHS**

Figure 72 *Geomagnetic polarity epochs during the last 4 million years as determined by the alignment of magnetic minerals in lava flows. Source unknown*

Each element possesses a specific decay rate, so elements with slow decay rates are suitable for dating old rocks, whereas those with fast decay rates are more suited to dating young rocks. The well-known dating technique called carbon 14 is useful for only the past 70,000 years because its decay rate is so rapid. Various isotopes of uranium and thorium, on the other hand, decay so slowly they are useful for dating rocks as old as a thousand million years.

Radioactive decay of certain chemical elements provides the basis for absolute time determinations, especially in igneous rocks and some metamorphic rocks. Absolute ages of sedimentary rocks cannot be determined, because their constituent mineral grains formed long before they were transported, deposited, and formed into rock (see section on radiometric dating). They can be given a relative age, however, by the fossils they contain, by the known ages of rocks above and below them, or by intrusive rocks that cut them.

See also: GEOLOGICAL TIMESCALES, GEOLOGY

Further Reading

Faure, G., 1986. *Principles of isotope geology*. Wiley, New York, 2nd edn.

Gradstein, F.M., Ogg, J.G. and Smith, A.G., (eds), 2005. *A geologic time scale 2004*. Cambridge University Press, New York and Cambridge.

International Commission on Stratigraphy, 2006. International Stratigraphic Chart, available at http://www.stratigraphy.org/

Levin, H.L., 2006. *The earth through time*. Wiley, Hoboken, NJ, 8th edn.

UN PROGRAMMES

Founded in the aftermath of World War II, the United Nations is best known for its debates in the Security Council and General Assembly. In an environmental context, its work is addressed mainly through the role of specialized agencies, though from time to time the General Assembly will mandate an especially important gathering, such as the 1992 'Earth Summit' conference in Rio de Janeiro. The leading agency is the United Nations Environmental Programme (UNEP) which collects data, provides information and encourages action across the whole field of environmental processes in atmospheric and terrestrial ecosystems as affected by humans and also in conservation areas. Another key agency is the World Meteorological Organization (WMO) which deals among other things with the standardization of data, their collection and dissemination. Jointly with UNEP, it founded in 1988 the Intergovernmental Panel on Climatic Change (IPCC), which is the most authoritative source on current climatic changes and their relationship to human activities. A key agency since 1945 in the developing nations has been the Food and Agriculture Organization (FAO), which attempts to make sure that the latest scientific findings can be used in poorer countries. These agencies have now made sustainable development central to their missions. Less obviously relevant but also involved in the field are the agencies concerned with health and with population, since the number and distribution of people (and indeed migrations caused by conflict) have environmental linkages. The UN's Educational, Scientific and Cultural Organization (UNESCO) is concerned with the Earth as a human habitat and also sponsors the list of World Heritage Sites, which may be either purely human-made or nearly-natural, or a hybrid of both. The political work of the UN is often criticized for its inadequacy but the agencies are in general regarded as beneficial though not immune to dominant paradigms in science and economics.

See also: SUSTAINABLE DEVELOPMENT, CLIMATE CHANGE

Further reading

http://www.fao.org/ – Food and Agriculture Organization.

http://portal.unesco.org/en/ev.php – UNESCO

http://www.unsystem.org/ – a list of the many UN organizations.

VEGETATION CLASSIFICATION

The word 'vegetation' is not a precise term. It is more inclusive than 'flora', which is confined to the taxonomic varieties found at a particular place, but does not take in the data about climate and soils which are found in the definitions of biome. The term is spatially undefined and can legitimately be applied to e.g., the whole of the tropical forests, a patch of relict woodland in an agricultural landscape, a field of soy beans or even a lawn. It is no surprise that there is no one system of classification of vegetation types. In Europe, for instance, the dominant species tends to give its name to the vegetation type, without any mention of climate: oak woodland/beech woodland/birch woodland might be found in any topographical guide to the types. Classifications which recognize the role of domestic escapes or of weed species are also common in lands without great areas of wild terrain such as the British Isles. In North America, a classification which is truly hierarchical (as with soil classification) is being adopted in official circles: this goes down from climatic zone, through plant habit and growth form (tree, shrub, herb etc.) to the dominant species. Both types of classification can be used of 'natural' vegetation as well as of wild or semi-wild areas with distinct human influence.

The architecture of vegetation is another key to its description and classification. Intuitive classes such as forest, grassland, cropland, desert or even waste land are used and these depend very often on plant habit, which is the form of the mature plant (e.g., herb, low shrub, cushion cactus) and its foliage (e.g., deciduous tree, evergreen shrub). All descriptions and classifications have to acknowledge that they may be snapshots in time of a dynamic system, as was acknowledged by the synoptic accounts at world scale in the 1980s. Vegetation changes through two main influences. The first is endogenous change, when a biotic community creates the conditions for its own replacement with a later flora and fauna. Thus on bare rock after glaciation, bacteria, mosses and lichens gradually create enough loose rock and humic material for flowering plants, which in turn provide a habitat for shrubs and then trees. This is called succession. It contrasts with exogenous change which is forced upon an ecosystem by external events.

The blanketing of a forest by volcanic ash is one example, or the drowning of a reclaimed area by the breakdown of a sea-wall. Many human influences exert exogenous pressures on plant communities and indeed may provide a bare surface from which endogenous succession may start: mining waste heaps provide an example. Vegetation mapping has now moved strongly in the direction of land cover mapping which includes human-dominated systems and which can be used in for example inventories of sequestered carbon.

See also: CLIMATE ZONES, ECOSYSTEM, SOIL SCIENCE, VEGETATION TYPES

Further reading

Archbold, O.W., 1995. *Ecology of world vegetation*. Chapman and Hall, London.

Collinson, A.S., 1988. *Introduction to world vegetation*. Unwin Hyman, London, 2nd edn.

Eyre, S.R., 1982. *Vegetation and soils: a world picture*. Arnold, London, 2nd edn.

Küchler, A.W., (ed), 1965–1970. *International bibliography of vegetation maps*. University of Kansas, Lawrence, KS.

Olson, D.M., *et al.*, 2001. Terrestrial ecosystems of the world: a new map of life on earth. *Bioscience* 55, 993–938.

PART VI

ENVIRONMENTAL ISSUES

Introduction

The entry titles in this section of the Companion probably carry the most familiar words and concepts – for most of us at any rate – of any in the book. Topics like climate change, deforestation, recycling and risk are the currency of all the media and opinion surveys sometimes declare that we are fed up with them and wish they would go away. However they keep intruding, whether at a local level in the annoyance caused by having to sort garbage before it is collected or at a world scale by the apparent effects of increasing carbon dioxide and other gases in the atmosphere. What this shows is that the environmental sciences are inextricably tied in with the social lives of humans and the attitudes (both deliberate and unconscious) which we take towards our non-human surroundings.

There has probably never been a human society which had no set of perceptions of its surroundings. Hunter-gatherers lived in hourly contact with weather, soils and their food sources; pre-industrial agriculturalists were attuned to the weather, the condition of the soil, water management and the lives of pests. The contacts of western societies after the 19th century have been rather different, though, since we have used the power granted by the harnessing of fossil fuels to buffer us against nature. Not that we have ceased to use the resources of the natural world, simply that most of us, now living in cities, have only limited contact with nature and that tends to be at times of stress and disaster: the results of hurricane Katrina (Fig. 42) are an obvious example, as are those of the 2004 tsunami around the Indian Ocean. This distancing is largely conveyed to us via technology: the more technology-rich a society has been, the less it felt in contact with its environments; the Swiss architect and playwright Max Frisch (1911–1991) said in his novel *Homo Faber* of 1957 that 'technology was a way of organising the world so that we didn't have to experience it'. One result of the power of technology during the 19th and much of the 20th centuries was a widespread attitude that everything was possible. If there were enough machines and enough technological knowledge (underlaid by science) then any problems created by the natural world could be overcome. It was so pervasive a position that some commentators have labelled a 'myth' in the proper sense of that word, i.e., a generally accepted story about the true nature of the world put into a short and often graphic form. The industrial myth recalls that of Prometheus in the classical world. He stole fire 'and all the arts' from the gods for human use. But, readers, follow up the rest of the story.

Humans have a unique sense of the future. Science has added to this awareness by its practice of making predictive models but the notion has been around in the west for thousands of years in the form of teleology (that all life was directed towards a particular goal that would be achieved when the world ended) or since the Enlightenment of the 18th century in the belief that human society was perfectible once oppression and tyranny had been removed. One form in which this has been expressed is in the writing of utopias, that is places where life is as perfect as it could be; Aldous Huxley's novel *Island* (1962) has a strong environmental content and its is a foil to his far better-known dystopia, *Brave New World* of 1932. There is an explicit environmental orientation to Nigel Calder's futurist polemic *The Environment Game* (1967) when he puts all human economies under plexiglass domes and leaves the outside biomes to themselves – a sort of decoupling of economy and ecology. But an optimistic sense of the future is now somewhat of a minority view and has largely been replaced with a pervasive anxiety (Fig. 73).

This anxiety is carried on the backs of two major issues, both with direct environmental linkages, and a third which is less direct but may be argued to be equally important. The first is resource scarcity. Given rising demands for industrial energy as e.g., vehicle fuels and industrial power as well as domestic electricity, can enough energy be supplied to (a) keep up the standards to which the west is accustomed, and (b) feed energy to poor countries to enable them to follow the western path to better lives? Since oil has been so versatile a commodity in the western path, the prediction that we have passed 'peak oil' contributes to a pessimistic outlook. 'Alternative' energy sources may be renewable in a way that oil is certainly not, but sunlight is very diffuse by comparison. The second issue is that of the contamination of the world by wastes that are cast out after their use as resources. This is detectable on all spatial scales from styroform food packaging littering city streets to the greenhouse gases which are more or less evenly spread through the upper atmosphere, where they break down only very slowly.

Figure 73   *Environment is now the subject of great public awareness. The extent of public concern, in relation to environmental issues can be demonstrated by bodies such as Greenpeace shown in the demonstration above (photograph by Garry Austin with permission)*

'The environment' has thus been seen in the past as equally a cornucopia of resources and a bottomless sink for wastes. This is no more the case.

The next question is, what do we as humans do about these anxieties and what are the roles of science in any future programmes of work? There is no one simple and acceptable answer. To some, it has to be governmental regulation of the flows of energy and materials: that is, of consumption. To others, the free market will provide answers if only left to itself: once the price of materials and energy gets too high, substitutes will emerge from science and technology. Yet behind all these arguments there is the third consideration: the size of the human population that drives pretty well all these environmental worries. Science which illuminates the effects of rapid population growth and helps to slow it down peacefully could not be better employed.

The verb 'to issue' means to come out or emerge. The entries in this section are not the only ones, nor will they be the last, to issue from the environmental sciences. More than ever, and at a time when some still find it convenient to be sceptical about humanity's environmental impacts, we must continue to encourage informed debate to underpin better decision-making. The former US President Harry S. Truman (in office 1945–1953) had a sign on his desk 'The buck stops here' – nowadays the buck stops with each of us.

AGENDA 21

Agenda 21, the Rio Declaration on Environment and Development, and the Statement of principles for the Sustainable Management of Forests were adopted by more than 178 Governments at the United Nations Conference on Environment and Development (UNCED) held in Rio de Janerio, in June 1992. Agenda 21 is a comprehensive plan of action to be taken globally, nationally and locally by organizations of the United Nations (i.e., not only UNEP), governments, and significant groups in every area with human impacts on the environment. The Commission on Sustainable Development (CSD) was created in December 1992 to ensure effective follow-up of UNCED, to monitor and report on implementation of the agreements at the local, national, regional and international levels. The full implementation of Agenda 21, the Programme for Further Implementation of Agenda 21 and the Commitments to the Rio principles, were strongly reaffirmed at the World Summit on Sustainable Development (WSSD) held in Johannesburg in 2002. The core elements of the Declaration are focused on the use of the Earth for the benefit of humanity and the notion that this should be attained by sustainable development. To that end, all the UN Agencies have had to orient their programmes to encompass such ideas, involving the diminution of environmental impacts, the conservation of energy resources and where appropriate the slowing of population growth. Many local communities in developed countries have launched their own Agenda 21 programmes, often involving the re-use and recycling of wastes and the lowering of embedded energy content in their activities. Explicit mention of Agenda 21 is muted in the 2000 period onwards, either because it was too ambitious in its scope, because the concern about environment has overtaken it, or perhaps because its vocabulary has been widely accepted and some moves towards implementation have been made.

See also: CLIMATE CHANGE, GLOBAL CONVENTIONS AND TREATIES, POPULATION GROWTH, RECYCLING, SUSTAINABLE DEVELOPMENT, UN PROGRAMMES

Further reading

Lafferty, W.M. and Eckerberg, K., 1998. *From the earth summit to Local Agenda 21: working towards sustainable development*. Earthscan, London.

http://www.un.org/esa/sustdev/documents/agenda21/index.htm

CLIMATE CHANGE

C limate change refers to the variations of Earth's climates over various timescales relative to variability or average weather from decades to millions of years. These changes evolve due to variations from within the Earth–atmosphere system, from extraterrestrial processes, or from human activities, such as greenhouse gas emissions and land-use changes. The term 'climate change' is generally used to refer to ongoing climate changes in contemporary times, which are highlighted, for example, by the average rise in air temperature, commonly known as global warming. The extraterrestrial factors include variability of solar radiation and what can be called the 'Earth–Sun geometry' (how the Earth's orbital shape differs, and the tilt and wobbling of Earth's axis of rotation differs).

Within the Earth's environment, there are changes that can control climate. These changes involve the deep oceans and ocean surface temperatures, as well as terrestrial changes in the biosphere and periodic events such as volcanic eruptions. Time and space scales must be considered an analysis of climate change. These can range from millions of years for the entire Earth to regional and local short-term changes. Glaciers, for example, are recognized as sensitive indicators of climate change, advancing due to climate cooling such as Little Ice Age, and retreating during climate warming such as during the last century. Since the last century, however, glaciers have not been able to regenerate enough ice in winter months to account for the ice and snow lost in the summer. In fact, the most important climate processes of the last several million years have been the glacial and interglacial cycles of the ice ages. Orbital variations explain overall the ice age cycles. However, continental ice sheets and 130 m sea-level change played a role in climate response geographically. Glacial variations can influence climate without major effects of orbital changes. On the scale of decades, climate changes result from less-understood variability within the ocean/atmospheric systems. Many climate conditions, notably El Niño/Southern Oscillation (ENSO), the Pacific Decadal Oscillation (PDO), the North Atlantic Oscillation (NAO), and the Arctic Oscillation (AO), have all been recognized as drivers of short-term phenomena within the climate system. On longer timescales, the thermohaline circulation plays a significant role in altering heat storage in the climate system, and as a result, impacts climate variability. Diverse aspects of the environment react at variable rates to a climate change and thus reinforce in the direction of that change or may act to reduce the impact of that change. These responses are feedbacks in the climate system discussed under **CLIMATE FORCING AND FEEDBACK**. Current studies indicate that radiative forcing by greenhouse gases is the primary cause of global warming (Table 50 after Miehl *et al.* 2004). Greenhouse gases have also been important in understanding Earth's climate history. Over the last 600 million years, carbon dioxide concentrations have varied from >5000 ppm to less than 200 ppm, due primarily to the impact of geological processes and biological changes. The Paleocene–Eocene thermal maximum, the Permian–Triassic extinction event, as well as the end of the Varangian snowball Earth event are examples of rapid changes in greenhouse gases in the Earth's atmosphere that have in turn led to global warming.

Since the 1950s, increased carbon dioxide levels have been related to be a major cause of global warming. Notations of solar activities are based on observing sunspots and beryllium isotopes. On the longest timescales, the Sun itself is getting brighter as it continues to evolve. Early in Earth's history, the Sun was too cold to bring about liquid water to the Earth's surface. This early history is known as the Faint young sun paradox time. On more modern time scales, there are also many forms of solar variation, including the 11-year solar cycle, and longer-term modulations. These variations have triggered the Little Ice Age and caused the beginning of warming observations in the 1900s. Milankovitch cycles are due to mutual interactions of the Earth, its moon, and other planets. These variations are considered the driving factors underlying the glacial and interglacial cycles of the present ice age. Anthropogenic factors or human

Table 50 *Global temperature change relative to 1900 (in degrees Celsius)*

	1940	1970	1994
Greenhouse gases	0.10	0.38	0.69
Sulphate emissions	−0.04	−0.19	−0.27
Solar forcing	0.18	0.10	0.21
Volcanic forcing	0.11	−0.04	−0.14
Ozone	−0.06	0.05	0.08

After data from Miehl *et al.* (2004).

activities have changed the environment and influenced climate. The biggest factor of present concern is increases in CO_2 levels due to emission from fossil fuel combustion, followed by aerosols which counter the warming effect through filtering out energy in the atmosphere. Other factors, for example, which impact climate are land use, ozone depletion, and deforestation. Starting with the Industrial Revolution in the 1850s, the human consumption of fossil fuels has elevated CO_2 levels from a concentration of ~280 ppm to more than 370 ppm to present day. These increases are projected to reach more than 560 ppm before 2100. Prior to widespread fossil fuel use, humans' largest impact on local climate resulted from land use. Urbanization and agriculture basically changed the environment. For example, the climate of Mediterranean countries was changed by widespread deforestation between 700 BC and 0 BC, with the result that the modern climate in the region is significantly hotter and drier and the species of trees in the ancient world being extinct. A number of important feedbacks exist in the climate system. It has been documented that orbital variations provide the timing for the growth and retreat of ice sheets. However, the ice sheets themselves reflect sunlight back into space and hence promote cooling and their own growth, known as the ice-albedo feedback. Falling sea levels and expanding ice decrease plant growth and in turn lead to declines in carbon dioxide and methane, which leads to further cooling. Rising temperatures caused by anthropogenic emissions of greenhouse gases lead to retreating snow lines, revealing darker ground underneath, and as a consequence absorbing more sunlight. It is unclear whether rising temperature promotes or inhibits vegetative growth. Depending on the direction of the change, carbon dioxide would be effective and in turn could impact the temperature of the Earth. Alterations in water vapour, clouds, and precipitation may occur with global warming which can further produce feedbacks in the carbon cycle in Earth's temperature.

See also: CLIMATE FORCING AND FEEDBACK

Further reading

Cowie, J. 2007. *Climate change biological and human aspects.* Cambridge University Press, Cambridge, UK. 487pp.

Emanuel, K.A., 2005. Increasing destructiveness of tropical cyclones over the past 30 years. *Nature* 436, 686–688.

Jones, C., 2001. What effects are we seeing now and what is still to come? Climate change: facts and impacts. In Miller, C., and Edwards, P. (eds), *Changing the atmosphere: expert knowledge and environmental governance, politics, science and the environment.* MIT Press, Cambridge MA.

Meehl, G.A., Washington, W.M., Ammann, C.A., Arblaster, J.M., Wigleym, T.M.L. and Tebaldi, C., (2004). Combinations of natural and anthropogenic forcings in twentieth-century climate'. *Journal of Climate,* 17, 3721–3727.

Ruddiman, W.F., 2003.The anthropogenic greenhouse era began thousands of years ago, *Climatic Change,* 61, 261–293.

Ruddiman, W.F., 2005, *Plows, plagues, and petroleum: how humans took control of climate.* Princeton University Press, Princeton NJ.

Ruddiman, W.F., Vavrus, S..J. and Kutzbach, J. E., 2005. A test of the overdue-glaciation hypothesis, *Quaternary Science Reviews,* 24, 1–10.

Climatic change futures: http://www.climatechangefutures.org/

IPCC Fourth Assessment Report published in 2008 by the Intergovernmental Panel on Climate Change. Can be seen at http://www.ipcc.ch/pub/online.htm

NAS: National Academy of Sciences: Understanding and Responding to Climate change, Overview (PDF). http://www.nap.edu/catalog.php?record_id=11676

Tyndall Centre for Climate Change Research, Norwich, UK at http://www.tyndall.ac.uk/

CLIMATE FORCING AND FEEDBACKS

Climate forcing and feedbacks include processes among atmospheric constituents that sway the climate in given directions from its equilibrium condition. The combined effects of climate forcing lead to alteration of the Earth's radiation budget (Fig. 74). Changes in radiation are not only affected by changing concentrations of gases and particles in the atmosphere, but also by external sources such as Earth–Sun orbital factors. Any change that causes enhanced changes in the same direction is a positive feedback. Conversely, if a change in the environment leads to a process that dampens any change, it is a negative feedback. In most climate change research, the focus is on the atmospheric radiation forcing the climate system. Major research has focused on radiative forcing associated with the steadily increasing concentrations of different gases in the atmosphere – the so-called greenhouse gases: CO_2, CH_4, N_2O, CFC-gases etc, and particles. Other changes in the environment can also lead to changes in the radiative budget, such as desertification, forestation, and other changes in land use and air pollution (ozone, SO_4-aerosols, and contrails). Natural changes are important in perturbing the radiative balance, such as fluctuations in the solar output and volcanic activity. Important positive feedback mechanisms include the ice-albedo mechanism, lower tropospheric water vapour content changes, and ocean warming.

A negative feedback mechanism might include cloud development and aerosol cooling. Clouds are known to have a negative impact on the surface temperatures in the present climate system. However, a changing climate may involve changes in the types of clouds with both positive and negative effects on the radiative balance. It is unclear whether the total effect of these changes will be a negative or positive feedback. Sulphates in the atmosphere filter the

Radiative Forcing Components

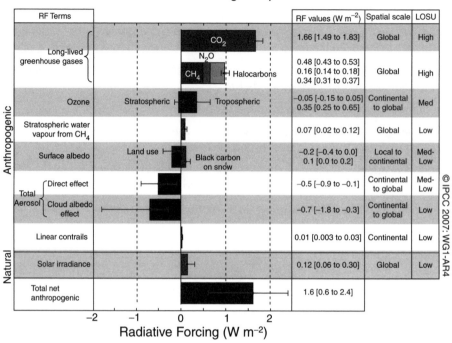

RF Terms		RF values (W m^{-2})	Spatial scale	LOSU
Long-lived greenhouse gases	CO$_2$	1.66 [1.49 to 1.83]	Global	High
	N$_2$O CH$_4$ Halocarbons	0.48 [0.43 to 0.53] 0.16 [0.14 to 0.18] 0.34 [0.31 to 0.37]	Global	High
Ozone	Stratospheric Tropospheric	−0.05 [−0.15 to 0.05] 0.35 [0.25 to 0.65]	Continental to global	Med
Stratospheric water vapour from CH$_4$		0.07 [0.02 to 0.12]	Global	Low
Surface albedo	Land use Black carbon on snow	−0.2 [−0.4 to 0.0] 0.1 [0.0 to 0.2]	Local to continental	Med- Low
Total Aerosol Direct effect		−0.5 [−0.9 to −0.1]	Continental to global	Med- Low
Cloud albedo effect		−0.7 [−1.8 to −0.3]	Continental to global	Low
Linear contrails		0.01 [0.003 to 0.03]	Continental	Low
Solar irradiance		0.12 [0.06 to 0.30]	Global	Low
Total net anthropogenic		1.6 [0.6 to 2.4]		

Radiative Forcing (W m^{-2})

© IPCC 2007: WG1-AR4

Anthropogenic / Natural

Figure 74 *IPCC 4th Assessment Executive Summary radiative forcing sensitivity of the climate system*

Sun's light and can aid in stemming global warming and thus would be a negative feedback. Ocean warming provides a good example of a potential positive feedback mechanism. The oceans are an important sink for CO$_2$ through absorption of the gas into the water surface. As CO$_2$ increases, it increases the warming potential of the atmosphere. If air temperatures warm, it should in turn warm the oceans. The ability of the ocean to remove CO$_2$ from the atmosphere decreases with increasing temperatures. Hence, increasing CO$_2$ in the atmosphere could have effects that exacerbate the increase in CO$_2$. As ice melts, it changes the surface characteristics and yields a lower albedo, and hence creates an enhanced ability to absorb solar radiation. However, the increase in temperature may cause more water vapour to be stored in the atmosphere. The increased water vapour, as a greenhouse gas, enhances the greenhouse effect and could lead to further warming, as long as this positive feedback does not cause an increase in cloud cover that would lead to a negative feedback. The increased cloud thickness or extent could reduce incoming solar radiation and curtail warming. Thick low clouds would have a stronger ability to block sunlight than extensive high (cirrus) type clouds. Low clouds tend to cool; high clouds tend to warm. High clouds tend to have lower albedo and reflect less sunlight back to space than low clouds. Clouds are generally good absorbers of infrared radiation, but high clouds have colder tops than low clouds, so they emit less infrared radiation spaceward, thus reducing the overall loss of radiation from the Earth–atmosphere system. Cloud properties may change with a changing climate, and emitted aerosols may confound the effect of greenhouse gas forcing on clouds. With fixed clouds and sea ice, models suggest climate sensitivities between 2 to 3°C for a CO_2 doubling. Depending on whether and how cloud cover changes, the cloud feedback could almost halve or almost double the warming. Feedback mechanisms exist that can potentially exacerbate or negate temperature responses. Temperature increases may enhance processes that either increase CO_2 concentrations or the absorption of incoming solar radiation. Many of the feedback mechanisms are poorly understood, so there is ample uncertainty in ongoing scientific estimates of future climates.

See also: CLIMATE CHANGE, GLOBAL CHANGE, GLOBAL DIMMING, GLOBAL WARMING

Further reading

Intergovernmental Panel on Climate Change (IPCC), 2007. Climate Change 2007: The Physical Science Basis, Summary for Policymakers, WMO, UNEP, follow from http://www.ipcc.ch/

IPCC, 2001. Climate Change, UNEP/WMO at http://www.ipcc.ch/pub/un/syreng/spm.pdf

Ackerman, S.A. and Chung, H., 1992. Radiative effects of airborne dust on regional energy budgets at the top of the atmosphere. *Journal of Applied Meteorology* 31, 223–233

Andronova, N.G., Rozanov, E.V., Yang, F., Schlesinger, M.E. and Stenchikov, G.L., 1999. Radiative forcing by volcanic aerosols from 1850 through 1994. *Journal of Geophysical Research* 104, 16807–16826.

Kirkevåg, K.R. *et al.*, 1998. Intercomparison of models representing direct shortwave radiative forcing by sulphate aerosols. *Journal of Geophysical Research* 103, 16979–16998.

Cess, R.D. and Potter, G.L., 1988. A methodology for understanding and intercomparing atmospheric climate feedback processes in GCMs. *Journal of Geophysical Research* 93, 8305–8314.

Charlson, R.J., Anderson, T.L. and Rodhe, H., 1999: Direct climate forcing by anthropogenic aerosols: quantifying the link between sulphate and radiation. *Contributions in Atmospheric Physics* 72, 79–94.

Christiansen, B., 1999. Radiative forcing and climate sensitivity: the ozone experience. *Quarterly Journal of the Royal Meteorological Society* 125, 3011–3035.

Wetherald, R. and Manabe, S., 1988. Cloud feedback processes in a general circulation model. *Journal of Atmospheric Science* 45, 1397–1415.

CLIMATE SENSITIVITY

C limate sensitivity refers to the change in surface air temperature for a given unit change in radiative forcing, expressed in units of, for example, °C or K /(W/m²). The evaluation of climate sensitivity from models requires detailed simulations with local climate models. It can be achieved with observational analysis. For example, Shaviv in 2005 did an analysis for six different timescales, ranging from an 11-year solar cycle to the climate variations over geological timescales. What was found was a typical sensitivity of from 1.3 °C to 2.0 °C. Using simple climate models, Andronoma and Schlesinger in 2006 found that the sensitivity could be between 1 and 10 °C. The exact range depends on which factors are most important during the instrumental period. Forest *et al.* in 2002 estimated a 95% confidence interval of 1.4–7.7° C for the climate sensitivity, and a 30% probability that sensitivity was outside the 1.5–4.5 °C range. Frame *et al.* in 2005 noted that the size of the confidence limits are dependent on the nature of prior assumptions made. Climate sensitivity is not the same thing as expected climate change by some year into the future. The Transient Climate Response (TCR) – a term first used in the IPCC 2001 report – is the temperature change at the time of CO_2 doubling in a run with CO_2 increasing at 1% per year (Fig. 75). The effective climate

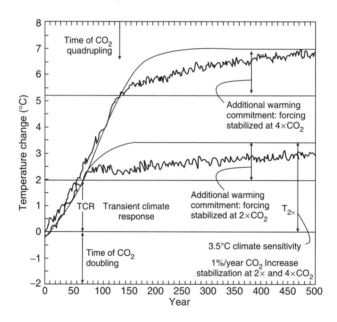

Figure 75 *Global mean temperature change for 1%/yr CO2 increase with subsequent stabilisation at 2xCO2 and 4cCO2. From Climate Change 2001. Chp 9 Projections of Future Climate Change. IPCC. Figure 9.1. The jagged curves are from a coupled AOGCM simulation (GFDL_R15_a) while the solid curves are from a simple illustrative model with no exchange of energy with the deep ocean. The "transient climate response", TCR, is the temperature change at the time of CO2 doubling and the "equilibrium climate sensitivity", T2x, is the temperature change after the system has reached a new equilibrium for doubled CO2, i.e., after the "additional warming commitment" has been realized.*

sensitivity is a related measure that is different than this. It is evaluated from model output for evolving non-equilibrium conditions and is a measure of the strengths of the feedbacks at a particular time and may vary with forcing history and climate state. Considerable research has been accomplished since the TAR (Third Assessment Report, IPCC, 2001) to estimate climate sensitivity and to provide a better quantification of relative probabilities, including a most likely value, rather than just a subjective range of uncertainty. the global mean equilibrium warming for doubling CO2, or 'equilibrium climate sensitivity', is likely to lie in the range 2–4.5 °C, with a most likely value of about 3 °C. Equilibrium climate sensitivity is very likely larger than 1.5 °C.

See also: CLIMATE CHANGE, CLIMATE FORCING AND FEEDBACKS

Further reading

Andronova, N. and Schlesinger, M.E., 2001. Objective estimation of the probability distribution for climate sensitivity. *Journal of Geophysical Research* 106, D19, 22605–22612.

Annan, J.D. and Hargreaves, J.C., 2006. Using multiple observationally-based constraints to estimate climate sensitivity. *Geophysical Research Letters* 33, L06704. doi:10.1029/2005GL025259.

Forest, C.E. *et al.*, 2002. Quantifying uncertainties in climate system properties with the use of recent observations. *Science* 295, no. 5552, 113–117.

Frame, D.J., Booth, B.B.B., Kettleborough, J.A., Stainforth, D.A., Gregory, J.M., Collins, M. and Allen, M. R., 2005. Constraining climate forecasts: the role of prior assumptions. *Geophysical Research Letters* 32, L09702 doi:10.1029/2004GL022241.

Gregory J.M. *et al.*, 2002. An observationally based estimate of the climate sensitivity. *Journal of Climate* 15, 3117–3121.

Shaviv, N.J., 2005. On climate response to changes in the cosmic ray flux and radiative budget. *Journal of Geophysical Research* 110, A08105, doi:10.1029/2004JA010866.

COASTAL MANAGEMENT

Coastal Management, or Coastal Zone Management as it is sometimes termed, is an important topic in environmental science because the coastlines of Earth are experiencing rapid rise in sea level as a result of global warming. As glacial ice melts and the oceans of the world warm and expand, they are rising at rates that are changing coastal ecosystems at low-lying areas around the globe. Coastal erosion is threatening low-lying islands and in some places tens of millions of people may be displaced due to rising water during the next 100 years. At present, sea level is rising world-wide at the rate of a millimetre or so per year and various mathematical models suggest that during the next century, the rise is likely to be about 20–40 cm with as much as 15 cm by the year 2050. Rise in sea level puts pressure on human structures in the coastal zone and threatens ground water supplies for coastal communities through intrusion of salt water into wells.

At the very least, it appears that projected rise in sea level will have serious consequences and will lead to further investment to protect cities in the coastal zone. It is likely that there will be a necessity for construction of sea walls and dikes, as well as other erosion control structures that threaten urban property. Increasingly there will arise hard choices of how to manage the coastal zone. The choice will be to either spend vast amounts of money in an attempt to control rising sea levels and coastal erosion or to allow loss of coastal property. In rural areas or wild-land-coasts, the most likely response and management to rising sea level will be to adjust to the rising levels and allow change to occur. A choice to defend our coastlines against erosion will be made only where it is absolutely necessary.

Nevertheless, societies are at a crossroads with respect to coastal zone management. One management path leads to ever-increasing coastal defence in attempts to control the processes of erosion. The second management path involves learning to live with rising sea level and coastal erosion through flexible environmental planning linked to wise land use in the coastal zone. The first path will leads to spending large sums of money to control the coastal environment and protect development from erosion. The latter path considers that land in some parts of the coastal zone is temporary and expendable. Whichever path is chosen is likely to be a mixture of the two approaches. Sometimes this is termed managed retreat. That is, a recognition that the coastal zone is subject to change and that change, as a result of rising sea levels, is likely to accelerate. In some areas, people can live with that change and maintain a more natural coastline, but in others where considerable urban development and in particularly urban regions, the shoreline will have to be protected through construction of engineering structures that protect the development. The path taken will ultimately reflect social and economic values concerning the coastal zone and its ecosystem. Management of many of the coastal zones of the world is becoming an ecosystem management issue. Many of the coasts of the world are bordered by wetlands such as fresh and saltwater marshes and mangrove forest. These ecosystems provide a buffer to inland areas from storm waves and tsunami. These wetlands and coastal forest also serve as important habitat for fish and wildlife. When wetlands and coastal forest are removed, ecosystems and species are threatened and the land is rendered more vulnerable to rising sea level.

In summary, a flexible approach to coastal zone management will likely yield the most favourable results. In some areas, through necessity, our coastlines will be defended while in others they will be managed to maintain natural ecosystems with their service functions to inland development. In other more mixed areas of development, there is likely to be a combination of flexible approaches such as managed retreat to nourishing beach with sand to provide for natural ecosystems of beaches and maintaining recreational beaches for humans.

See also: INTEGRATED COASTAL AND MARINE MANAGEMENT

Further reading

Komar, P. D. 1998. *Beach processes and sedimentation*, Prentice Hall, Upper Saddle River, NJ, 2nd edn.

McDonald, K. A. 1993. A geology professor's fervent battle with coastal developers and residents. *Chronicle of Higher Education*, 40(7), A8–89, A12.

National Research Council. 1990. *Managing coastal erosion*. National Academy Press, Washington, DC.

Neal, W.J., Blakeney, W.C., Jr., Pilkey, O.H., Jr., and Pilkey, O.H., 1984. *Living with the South Carolina shore*. Duke University Press, Durham, NC.

Pennsylvania: Cause and repair. *Environmental Geology and Water Science* 12(2), 89–98.

Pilkey, O.H., and Dixon, K.L., 1996. *The Corps and the shore*. Island Press, Washington, DC.

CONSERVATION

U sually thought of as environmental conservation whereby environmental resources and processes are sustainably utlized but in the context of the human environment, this term concerns perpetuation, is the desire that some entity shall not become extinct or unreachable to human access. The entities to which the term is most applied are (a) resources and (b) habitats and species. Resource conservation is driven by the fear that some vital under-pinning of the human economy will run out, either on a world-wide or a more regional scale. At various times in human history, such fears have been expressed about food, land, coal and most recently about oil. So far, technological advances have made these worries redundant, although in the early 21st century there has been a persistent thought that the years of peak oil production have been passed and that there will be progressively less to extract. It is not the case, of course, that the oil will have been totally taken out of the Earth's crust but that it will become too expensive to reach, process and distribute. Conservation policies are aimed at stretching the resource by getting consumers to use less and by increasing the efficiency of use. In the case of oil, this means getting more work per joule of energy. The economist's attitude is that the resource will be stretched in availability just so long as the price is allowed to reflect the balance of supply and demand; an environmentalist is more likely to demand government action to reduce the consumption by limiting individuals' access via the tax system or even by rationing.

The reasons for wanting to secure the future of habitats and species are more diverse. In some cases they are instrumental in the sense that the habitat provides something useful for human societies: forests for example are the sources of wood for construction (for which substitutes could be found, at a price) and for pulp for paper, for which there is at present no obvious replacement material. Coastal ecosystems such as sand dunes and estuarine mudflats protect against storm damage or act as nurseries for commercial fish species. By contrast, many habitats and species are valued for their intrinsic qualities: that is, people want to share the planet with them and do not want to see a world bereft of wild places and untamed animals. In these cases, the economists' focus on the market as a determinant of availability usually fails, for the relation between value (as a feature of culture) and price (as a marker of cost to the consumer though not necessarily to the wider environment) is rarely agreed upon. One oak tree can have a price as timber, but what value can be put upon an oak wood as an element of landscape? There may be a price for an elephant's tusks, but what is the value of an intact herd that attracts thousands of tourists? Thus the environmentally-oriented rely mostly on regulation by governmental agencies to allow the wild to thrive. Conservation has a very long history, was a major theme of the World Conservation Strategy proposed in 1980, and is now believed to be an integral part of sustainable development.

Both strands can be put in a wider context; broadly, it seems that the more fossil fuel that is used, the less 'space' there is for the wild at the expense of the tame. Where plentiful supplies of oil are used, for example, then the whole land is likely to be made over with a consequent high use of materials with large amounts of embedded energy. Hence, there is less chance of maintaining wild habitats and species. Many environmentalists will take that a stage further and argue that the root problem is population growth.

See also: CULTURES, ENERGY, ENVIRONMENTALISM, ENVIRONMENTAL VALUES, NATURE, RESOURCE DEPLETION, SUSTAINABILITY, ZOOS BOTANICAL GARDENS AND MUSEUMS

Further reading

This area is well served by authoritative websites except for some of the polemic over oil reserves, their exploitation and their geopolitical context.

For habitats and wildlife:

Sinclair, A.R.E., Fryxell, J.M., Caughley, G. *et al* 2006. *Wildlife ecology, conservation, and management.* Blackwell, Oxford and Malden, MA.

http://www.iucn.org/ – the World Conservation Union

http://www.panda.org/ – the Worldwide Fund for Wildlife

On oil:

http://www.eia.doe.gov/pub/oil_gas/petroleum/feature_articles/2004/worldoilsupply/oilsupply04.html – concise and not apparently tied to any vested interest.

Books from both sides:

Clarke, D., 2007. *The battle of barrels: peak oil myths and world oil futures*. Profile Books, London.

Deffreyes, K.S., 2005. *Beyond oil*. Hill and Wang, New York.

DEFORESTATION

Ever since hunter-gatherer times, humans have interacted with forests in ways which have resulted in the disappearance of trees, either temporarily or permanently. Using fire, some hunting groups maintained openings in woodlands which then attracted game animals. Shifting agriculture had been present for thousands of years in both tropical and temperate zones and can be seen as a way of growing crops in forests without clearing them entirely. Since trees are the dominants in the local ecosystem, their removal has considerable ecological consequences. They form the primary energy-fixing role and so any replacement vegetation is likely to have a lower capacity to turn solar radiation into chemical energy. They are also the physical habitat for many other species; further the fragmentation of a formerly continuous forest may make reproduction and evolution of many species more problematic and lead to extinction. The water relations between soils and runoff are also changed by the loss of trees. The clearance of forests on a large scale is a result of the tension between their role as providers of resources (timber, underwood, bush meat and medicinal plants, for example) and their capacity to be a land bank which has stored fertility pre-adapted as it were for agriculture. Thus most of the major removals of forest have been in the cause of extending either cropland or grazing areas. The 19th and 20th centuries have far exceeded any previous periods, with an especially heavy impact on the tropics in the period after 1950. In 1700 there were about 53 million km^2 of forest and woodland and 4.0 million km^2 of cropland; in 1950 the figures were 46 and 15 respectively and in 1990, 44 and 18.0. There have been some large reforestations as well: in the tropic plantations of eucalypts and conifers are found and in North America, the withdrawal of farming from poor soils and harsh climates has allowed the growth of much secondary forest.

See also: ANTHROPOGENIC IMPACT

Further reading

Williams, M., 2006. *Deforesting the earth. From prehistory to global crisis*. Chicago University Press, Chicago. Abridged edn.

Ramankutty, N. and Foley, J.A., 1999. Estimating historic changes in global land cover: croplands from 1700 to 1992. *Global Biogeochemical Cycles* 13, 997–1027.

DESERTIFICATION

Desertification may result from increased population pressures on land resources that stress crops and animals and may cause biodiversity loss, reduction in productive capacity, and loss of soil fertility (Fig. 76). In some areas, deserts are distinct from their surrounding less arid environments (e.g., mountains and other contrasting landforms). The margins of the desert represent a gradual transition from a dry to a more humid environment. These transition zones are heavily used by populations and may consist of sensitive ecosystems.

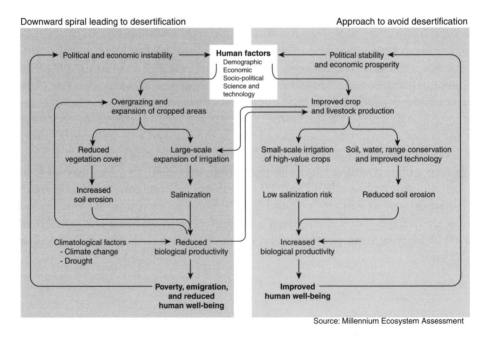

Source: Millennium Ecosystem Assessment

Figure 76   *Processes leading to desertification and ways to abate it. http://www.maweb.org/en/index.aspx*

A common misconception is that drought – common in arid and semi-arid lands – causes desertification. It is the pressure on marginal lands, primarily due to increased populations and grazing of animals which has been shown to promote a speeding up of desertification. Some arid and semi-arid lands can support crops, but additional pressure from increased populations or decreases in rainfall have led to the demise of sustainable vegetation for animals. Desertification may produce a positive feedback to a climatic trend toward greater lack of moisture. It could aid in initiating a change in local climate.

   Although the word 'Desertification' was not a household term until the 1950s, the effect was dramatically felt, for example, in the 1930s in the United States Great Plains which came to be known as the 'Dust Bowl' primarily as a result of overgrazing, drought, settlements, and poor farming practices. Several solutions have been tried in the effort to slow the rate of desertification and restoring affected areas. Plants, which extract nitrogen from the air by bringing it to the soil, can be planted to restore soil fertility. Stones stacked around the base of trees help retain soil moisture by collecting morning dew. Artificial grooves dug in the ground can help retain rainfall and trap windblown seeds. Windbreaks designed with trees and bushes can reduce soil erosion. Water conservation measures in the form of efficient agricultural irrigation practices and tillage of soil to reduce wind effects have aided in restoring land. At the local level, individuals and local governments should collaborate in an effort to reclaim and protect afflicted areas. On a larger scale, a 'Green Wall,' which will stretch more than 5700 kilometres in length (nearly as long as the Great Wall of China) is being implemented in northeastern China to protect 'sandy lands' – deserts believed to have been created by human activities. Novel water enhancement technology has been pursued such as harvesting rainwater and irrigating with seasonal runoffs from adjacent highlands. Additional ways are sought to tap ground water resources and to develop more effective ways of irrigating arid and semi-arid lands. Research on the reclamation of deserts also is focusing on discovering proper crop rotation to protect fragile soil, on understanding how sand-fixing plants can be adapted to local environments, and on how grazing lands and water resources can be developed effectively without overuse.

See also: ARID ZONES/DESERTS, CLIMATE CHANGE

Further reading

In www.oasisglobal.net the following references are useful:

Abahussain, A.A., Abdu, A.S., Al-Zubari, W.K., El-Deen, N.A. and Abdul-Raheem. M., 2002. Desertification in the Arab region: analysis of current status and trends. *Journal of Arid Environments* 51, 521–545.

Batterbury, S.P.J. and Warren, A., 2001. Desertification. in Smelser, N. and Baltes, P. (eds) *International encyclopædia of the social and behavioural sciences.* Elsevier Press, Amsterdam, 3526–3529.

Garcia Latorre, J., Sanchez Picon, A. and Garcia Latorre, J. 2001. The man-made desert: effects of economic and demographic growth on the ecosystems of arid southeastern Spain. *Environmental History* 6, 75–94.

Geist, H., 2005. *The causes and progression of desertification.* Ashgate, London.

Holzel, N., Haub, C., Ingelfinger, M.P., Otte, A. and Pilipenko, V.N., 2002. The return of the steppe – large-scale restoration of degraded land in southern Russia during the post-Soviet era. Journal for Nature Conservation 10, 75–85.

Millennium Ecosystem Assessment (2005): Desertification Synthesis Report, follow from http://ma.caudillweb.com/en/Products.Synthesis.aspx

Reynolds, J.F. and Stafford Smith, D.M. (eds) 2002. *Global desertification – do humans cause deserts?* Dahlem University, Berlin.

Saiko, T.A. and Zonn, I.S. 2000. Irrigation expansion and dynamics of desertification in the Circum-Aral region of central Asia. *Applied Geography* 20, 349–367.

UNCCD, 1995. *Down to Earth: A simplified guide to the Convention to Combat Desertification, why it is necessary and what is important and different about it.* Bonn, Germany: Secretariat for the United Nations Convention to Combat Desertification. (see also Facts Sheet, FAQ and Explanatory Leaflet at http://www.unccd.int/knowledge/menu.php

UNEP (2006): Global Deserts Outlook at http://new.unep.org/geo/news_centre/pdfs/Chapter6.pdf

Yassoglou, N.J. 2000. History of desertification in the European Mediterranean. In Enne, G., D'Angelo, M. and Zanolla, C. (eds). Indicators for assessing desertification in the Mediterranean. Proceedings of the International Seminar held in Porto Torres, Italy, 18-20 September, 1998. University of Sassari Nucleo Ricerca Desertificazion Sassari, Italy, 9-15.

DISASTERS

A disaster from the perspective of environmental science may be defined as an event that results from interaction between humans and natural processes resulting in injuries or loss of life accompanied by significant damage to property. The event generally occurs over a limited time span and within a defined geographic area. Disasters that are more massive in extent and require significant expenditure in time and money for recovery to take place are catastrophes. Examples of environmental disasters include:

» Natural hazard events such as earthquakes, volcanic eruptions, tsunamis, landslides, and floods.

» Release of toxic gas into the environment as for example, the 1984 release of toxic liquid that formed a deadly cloud of gas in Bhopal, India. The chemical was methyl isocyanate, which vaporized and although the leak only lasted about one hour, more than 2000 people were killed and 15,000 were injured.

» Nuclear power plant accidents that release radiation into the environment as at Chernobyl in 1986.

» Natural release of poisonous gases into the environment as, for example, the 1986 release of carbon dioxide from a dormant volcanic lake in Africa that spilled over the rim of a crater lake to suffocate over 1500 people as well as killing several thousand cattle and other animals.

» Weather events such as heat waves, blizzards and hurricanes.

» Air pollution events such as the 1952 London smog event and particulate air pollution from 1997 fires in Indonesia.

Important topics related to disasters in the environment include disaster preparedness, education of the public about natural disasters, and plans for recovery following disasters. These three adjustments to potential disasters are proactive and anticipatory rather than reactive, which refers to actions taken only after an event has occurred. Advanced planning that includes components of land use planning to avoid hazardous area, constructing buildings and homes to withstand the forces of nature along with education are three adjustments that may reduce our vulnerability to future natural hazards.

See also: ENVIRONMENTAL PERCEPTION, HAZARDS, PERCEPTION

Further reading

Advisory Committee on the International Decade for Natural Hazard Reduction. 1989. *Reducing disaster's toll*. National Academy Press, Washington, DC.

Abramovitz, J.N. and Dunn, S., 1998. *Record year for weather-related disasters*. World Watch Institute, Vital Signs Brief 98–5.

Abramovitz, J.N. 2001. Averting unnatural disasters. In Brown, L.R. *et al.*, *State of the World 2001*. World Watch Institute. W.W. Norton, New York; 123–142.

del Moral, R. and Walker, L.R., 2007. *Environmental disasters, natural recovery and human responses*. Cambridge University Press, Cambridge.

International Federation of Red Cross and Red Crescent Societies. 1993– [annual] *World disaster report*. Nijhoff, Dordrecht.

ECOLOGICAL ENGINEERING

This term was first defined by H.T. Odum in 1962 as 'environmental manipulation by man using small amounts of supplementary energy to control systems in which the main energy drives are still coming from natural sources'. This was because although the energy supplied by man is small relative to natural sources it can be sufficient to produce large effects. Hence there was a perceived need to provide more ecologically sound methods of environmental management so that in China since 1978 the term 'ecological engineering' has been used. Ecological engineering employs biological species, communities, ecosystems together with self-design through ecotechnology to achieve a more sustainable and less harsh management solution. It employs natural energy sources and self-regulating processes to design, construct, operate and manage sustainable landscape and aquatic systems, consistent with ecological principles, to benefit both humanity and nature. The aims of ecological engineering and ecotechnology are to achieve the restoration of ecosystems that have been substantially disturbed by human activities such as environmental pollution, climate change or land disturbance, the development of new sustainable ecosystems that have human and ecological value, and the identification of the life-support value of ecosystems ultimately to lead to their conservation. Ecological engineering depends on the self-designing capability of ecosystems and nature. The discipline demands knowledge both of engineering theory and core elements of quantitative ecology, systems ecology, restoration ecology, ecological engineering, ecological modelling and ecological engineering and ecological economics.

Applications have concentrated on the restoration of disturbed ecosystems and on the creation of new ecosystems, with a goal of better performance, less cost, multiple benefits and acceptance by the public and regulators, and underpinned by concepts of self-design and self-organization, a systems approach to biology, and the goal of sustainable ecosystems. Recent applications have been investigated in experimental ecosystems and have included the construction of shallow ponds and wetlands for ecological value such as water pollution control or enhancement of biodiversity. Although ecological engineering can be applied in the course of environmental management, it is particularly valuable in restoration which includes all those activities which aim to restore an environment to a former state or to an unimpaired condition. Hence ecological restoration has been used when restoring lakes and reservoirs that have suffered because of their high eutrophic status; when restoring rivers using soft rather than hard engineering approaches and for which a variety of terms are available; for recreating wetlands in areas which have been drained; for restoring coastal environments including salt marshes, mangroves and deltas; for bioremediation of contaminated soils; or for restoration of mineland and degraded land.

See also: ECOTECHNOLOGY, SELF-DESIGN, SUSTAINABILITY

Further reading

Ecological Engineering. International journal published by Elsevier since 1992.

Mitsch, W.J. and Jorgensen, S.E. (eds), 1989. *Ecological engineering: an introduction to ecotechnology.* Wiley, New York.

Mitsch, W.J. and Jorgensen, S.E., 2004. *Ecological engineering and ecosystem restoration.* Wiley, New York.

Odum, H.T., 1962. Man in the ecosystem. *Proceedings of the Lockwood Conference on the Suburban Forest and Ecology.* Bulletin 652 Connecticut Agricultural Station, Storrs, CT, 57–75.

ECOLOGICAL INTEGRITY

As ecosystems worldwide are put increasingly under stress, the need to preserve the ecological processes, on which all life depends, has been widely recognized. The United Nation's World Commission on Environment and Economy recognized that human society was dependent on uses of the biosphere that were sustainable. Human values are an integral part of decisions to protect or rehabilitate, but the goals and objectives for such actions are often implicit or unclear; ecosystems are often valued because of certain features they contain or functions they serve.

There are several published definitions of ecosystem integrity in the literature, most of them subjective, general approaches, and with ethical basis for understanding integrity. More quantitatively terms and including a quantitative index of biological integrity, as well as a definition that recognizes both ethical judgement and quantitative elements provided by ecosystem science, we propose a combined definition, which is shown in the box below.

> Combined Definition Biological integrity is the capability to support and maintain a balanced, integrated, adaptive community of organisms having a species composition and functional organization comparable to that of the natural habitat on the region as suggested by Karr and Dudley in 1981; so that the ecological integrity is the state of the ecosystem development that is optimized for its geographic location, including energy input, available water, nutrients and colonization history, implying that ecosystem structures and functions are unimpaired by human-caused stresses on environmental units and that native species are present at variable population levels.

The state of ecosystems is commonly defined in terms of ecosystem integrity, ecosystem health or even biological integrity, having a basis of biodiversity. Such terms are used in a normative sense and cannot be defined easily or fully by coastal ecosystem science. Thus, resource management agencies have been struggling to operationalize these terms. This is a complicated issue, particularly under the perspective of sustainable economic development in the Gulf of Mexico as linked to the ecological integrity of coastal ecosystems, at present under severe risk to healthy and resilient conditions.

See also: ECOLOGICAL ENGINEERING, ECOSYSTEM-BASED MANAGEMENT, ENVIRONMENTAL ETHICS, ENVIRONMENTAL VALUES, SELF-DESIGN

Further reading

Bruntland, G., 1987. *Our common future.* World Commission on Environment and Development. Oxford University Press, New York.

Costanza, R., d'Arge, R., de Groot, R., Farber, S., Grasso, M., Hannon, B., Naeem, S., Limburg, K., Paruelo, J., O'Neill, R.V., Raskin, R., Sutton, P. and van den Belt, M., 1997. The value of the world's ecosystem services and natural capital. *Nature* 387, 253.

Day, J.W., Martin, J.F. Cardoch, L. and Templet, P.H., 1997. System functioning as a basis for sustainable management of deltaic ecosystems. *Coastal Management* 25 (2), 115–154.

Karr, J.R. and Dudley, D.R., 1981. Ecological perspective on water quality goals. *Environmental Management* 5 (1), 55–68.

Woodley, S. 1993. Monitoring and measuring ecosystem integrity in Canadian National Parks. *In*: Woodley, S., Kay, J. and Francis, G. (eds.), *Ecological Integrity and the Management of Ecosystems*, Chapter 9. University of Waterloo and Canadian Parks Service, Ottawa.

Yáñez-Arancibia, A, Ramírez-Gordillo, J.J., Day, J.W. and Yoskowitz, D., 2008. Environmental sustainability of economic trends in the Gulf of Mexico: what is the limit for Mexican coastal development?. In Cato, J. (ed.) *The changing ocean and coastal economy of the Gulf of Mexico*, The Harte Research Institute for Gulf of Mexico Studies, Texas A&M University Press, College Station, TX, Chapter 5.

ECOSYSTEM-BASED MANAGEMENT

Progressive concepts of 'ecosystem-based management' emphasize four common principles, namely that effective management must: (1) be integrated among components of the ecosystem and resource uses and users, (2) lead sustainable outcomes, (3) take precaution in avoiding deleterious actions, and (4) be adaptive in seeking more effective approaches based on experiences. These principles have important implications for addressing the coastal environmental crises world-wide. Although frameworks exist for integration of management objectives in a number of regions, the technical ability for the quantitatively assessment of multiple stressors and strategies is still in an early stage of development, particularly concerning planning and restoring major coastal ecosystems in the Gulf of Mexico and in Latin America as well.

Science is also being challenged to identify sustainable futures, but emerging concepts of ecosystem resilience offer some promising approaches. Precautionary management is best conceived with regards to fisheries, but also should become a more explicit consideration for managing risk and avoiding unanticipated consequences of restoration activities. Adaptive management is embraced as a central process in coastal Gulf of Mexico restoration, but has not formally been implemented in the more mature public policy restoration. Based on these experiences, ecosystem-based management could be advanced by: (1) orienting more scientific activity to providing the solutions needed for ecosystem restoration, (2) building bridges crossing scientific management barriers to more effectively integrate science and management, (3) directing more attention to understanding and predicting achievable restoration outcomes that consider possible state changes and ecosystem resilience, (4) improving the capacity of science to characterize and effectively communicate uncertainty, and (5) fully integrating modelling, observations, and research to facilitate more adaptive management.

See also: ECOSYSTEM, ENVIRONMENTAL MANAGEMENT, RESTORATION OF ECOSYSTEMS

Further reading

Boesch, D.F., 2006. Scientific requirements for ecosystem-based management in the restoration of Chesapeake bay and Coastal Louisiana. *Ecological Engineering* 26 (1), 6–26.

Boesch, D. F., Burreson, E., Dennison, W. *et al.*, 2001. Factors in the decline of coastal ecosystems. *Science* 293, 1589–1590.

Day, J.W., Jae-Young Ko, Rybczyk, J., Sabins, D., Bean, R., Berthelot, G., Brantley, C., Cardoch, L., Conner, W., Day, J.N., Englande, A.J., Feagley, S., Hyfield, E., Lane, R., Lindsey, J., Mitsch, J., Reyes, E., and Twilley, R., 2004. The use of wetlands in the Mississippi Delta for wastewater assimilation: A review. *Ocean and Coastal Management* 47, 671–692.

Day, J.W. and Yáñez-Arancibia, A., 2008. *The Gulf of Mexico: Ecosystem-based management*. The Harte Research Institute for Gulf of Mexico Studies. Texas A and M University Press, College Station TX (in press).

Yáñez-Arancibia, A. and J. W. Day, 2004. Environmental sub-regions in the Gulf of Mexico coastal zone: The ecosystem approach as an integrated management tool. *Ocean and Coastal Management* 47, 727–757.

Yáñez-Arancibia, A., Day, J.W., Boesch, D.F. and Mitsch, W.J., 2006. Following the ecosystem approach for developing projects on coastal habitat restoration in the Gulf of Mexico. IUCN Commission on Ecosystem Management Newsletter, Gland Switzerland, December 2006, 2 pp.

ENERGY SCARCITY

E nergy cost and availability will likely become an important factor affecting future natural resource management. Recent information suggests that world oil production will peak within a decade or two implying that demand will outstrip supply and energy costs will increase significantly and become more scarce in the coming decades. Some have augured that additional discoveries will provide abundant oil well into the future. But most estimates of ultimately recoverable oil (URO) have remained relatively constant since 1965 at about 2 trillion barrels. Oil discoveries peaked in the 1950s and 1960s and have declined since. The world now consumes about three barrels of oil for each one discovered. An important factor that affects consideration of energy use is energy return on investment (EROI); the ratio of the energy used to all the energy used to discover and produce that energy source. During the period of exponential growth in conventional oil production, EROI was between 100:1 and 50:1. Over the last two decades, EROI for world oil production has fallen to about 25:1. Thus, it is costing more and more to find and produce oil; and other energy sources. EROI for non-conventional sources of oil (oil shale and sands) and most renewables are generally less than 15:1 and most are less than 10:1. This implies that oil production, and natural gas as well, will not be able to meet demand and the cost of energy will increase substantially. Humans will have to become much more efficient in using energy, and turn more to the use of renewable energy sources. Humans will also have to enhance and restore the natural environment since natural systems provide ecosystem services that are important to the human economy.

See also: ENERGY, FOSSIL FUELS, NATURAL RESOURCES

Further reading

Campbell, C.J. and Laherrère, J.H., 1998. The end of cheap oil. *Scientific American* March, 78–83.

Cleveland, C.J., 2005. Net energy from the extraction of oil and gas in the United States. *Energy* 30, 769–782.

Deffeyes, K.S., 2001. *Hubbert's Peak – The impending world oil shortage.* Princeton University Press, Princeton, NJ.

Deffeyes, K.S., 2002. World's oil production peak reckoned in near future. *Oil and Gas Journal* 100(46), 46–48.

Hall, C.A.S., Tharakan, P., Hallock, J., Cleveland C. and Jefferson, M., 2003. Hydrocarbons and the evolution of human culture. *Nature* 426, 318–322.

Heinberg, R., 2003. The party's over – oil, war and the fate of industrial societies. Clairview Books, Sussex, UK.

ENVIRONMENTAL ETHICS

That ethics, as a set of moral codes or principles, should apply to the environment, arose with increasing awareness of environmental problems, for example consequent upon increased human population, conservation of resources, nuclear power developments, or pollution. Environmental ethics should foster a closer relationship between humans and nature, countering the historical antipathy of Western philosophy to conservation, environmentalism and their values, which had led Europeans to encourage human alienation from the natural environment and an exploitative practical relationship with it. A 'land ethic' was proposed by Aldo Leopold (1949), to change the role of *Homo sapiens* from conqueror of the land-community to plain member and citizen of it, but not until the 'green movement' of the 1980s did environmental issues become the concern of millions of Americans and of millions of others. Environmental ethics, concerned with the moral relations that hold between humans and the natural world, was seen initially as a rearguard action against those who saw values in society dominated by self-interest, thus establishing a trend away from the NIMBY (not in my backyard) approach.

Exploring the ethical relationship of humans and nature prompted new concepts such as 'deep ecology', contending that all species have an intrinsic right to exist in the natural environment, in contrast to 'shallow ecology' whereby nature is valued only on the basis of its value to humans, with the interests of non human species subordinate to those of human beings. Alternative to the sharp contrast between the two extremes of deep and shallow ecology is the distinction of four types of environmental ethics:

» shallow, anthropocentric;
» intermediate, which denies that only humans are of value and therefore includes the Aldo Leopold land ethic;
» deep environmental movements which embrace deep ecology;
» deep green theory as a philosophical approach to environmental problems and issues.

The management implications of deep ecology include viewing environments such as rivers from a biocentric perspective whereby all species have an intrinsic right to exist in the natural environment, encouraging preservation of unspoilt wilderness and the restoration of degraded areas. Ethical considerations must now underpin policies applied both to environmental management and to human development, involving the precautionary principle (avoiding harm that is either irreversible or serious and reversible with great difficulty and great effort) and the polluter pays principle and leading towards more participatory, community-led development strategies being adopted by government institutions such as bodies for environmental regulation and management.

Although there is no general agreement as to which stance, shallow or deep ecology, is appropriate it is generally agreed that environmental ethics should always be considered although their significance varies with the environment and the place. The journal *Environmental Ethics* has been published since 1979.

See also: ECOLOGY, ENVIRONMENTALISM, NATURE

Further reading

Cahn, R., 1978. *Footprints on the planet: the search for an environmental ethic.* Universe Books, New York.

Guha, R. and Martinez-Alier, J., 1997. *Varieties of environmentalism: essays north and south.* Earthscan, London.

Leopold, A. 1949. The land ethic. In *A sand county almanac and sketches here and there.* Oxford University Press, New York.

Light, A. and Rolston, H. III., (eds), 2003. *Environmental ethics. An anthology.* Blackwell, Malden, MA.

Naess, A., 1973. The shallow and the deep, long range ecology movement: a summary. *Inquiry* 16, 95–100.

Richards, K., 2003. Ethical grounds for an integrated geography. In Trudgill, S. and Roy, A. (eds), *Contemporary meanings in physical geography. From what to why?* Arnold, London, 233–258.

Sylvan, R. and Bennett, D., 1994. *The greening of ethics: from human chauvinism to deep-green theory.* The Whitehorse Press, Cambridge, MA.

Taylor, P., 1986. *Respect for nature.* Princeton University Press, Princeton.

Environmental Ethics. 1979–. Published by The Center for Environmental Philosophy, University of North Texas.

ENVIRONMENTAL IMPACT ASSESSMENT (EIA)

An agency intending a major development is required by EIA to file an EI Statement with the appropriate authority. Its purpose is to estimate the effect on land, water, biota and atmosphere that the development is likely to have and to offer mitigations. Informal statements and the production of analogous evidence at inquiries and hearings long preceded the introduction of EIA into legislation during the 1970s. Because it relies on a high degree of scientific input and a competence to deliver compliance, it has been most effective in developed countries; court challenges to the validity of statements are relatively common. Most statements include a description of the project, its site. The main components of the development throughout its lifetime, and the sources of environmental disturbance and the production of wastes. A major section of the statement must examine alternatives that have been considered. All aspects that might receive an impact from the development must be listed, if necessary with the help of NGOs and non-specialists. Statements which use the word 'significant' at this stage are likely to have a difficult time defining some of the less obvious and minor impacts and this carries through into mitigation, which is an obvious location of contention since the developer is likely to have to meet extra costs required by such means. Lastly, it is essential that a wide public has access to the whole process, which means among other things writing non-technical summaries of the scientific and engineering dimensions of the environment, the development and any proposed mitigation. First introduced into the USA, the process is now found in e.g., Canada, New Zealand, China and the European Union.

See also: ENVIRONMENTAL LAW, ENVIRONMENTAL POLICY

Further reading

The EU website: http://ec.europa.eu/environment/eia/home.htm

Petts, J. (ed), 1999. *Handbook of environmental impact assessment.* 2 vols. Blackwell, Oxford.

ENVIRONMENTAL LAW

As with other forms of law, the spatial scale of environmental law can run all the way from local bye-laws (dealing with matters like litter or chewing gum) through national law (perhaps the commonest scale) to international law which is binding on states. The laws may carry the title of treaties, protocols, and conventions but there are also bodies of common law, implanted by time and precedent, which carry weight in judicial decisions. The subject matter of environmental laws is extremely diverse but most statutes are likely to be either regulatory (i.e., govern the conditions under which an environmental resource may be used, as in delimiting a nature reserve or game park), precautionary (i.e., act against a possible future conditions, as with preventing housing developments on a flood plain) or safety-related (i.e., prevent actual harm to humans or other environmental components, as with emission control from refineries or of sewage).

The precautionary principle may be carried out with the help of an environmental impact statement. International law is also keen on the precautionary principle, which is, for example, the basis of action on climatic changes, as in the Montreal Protocol of 1989 on substances that deplete the ozone layer. It also incorporates the principle of 'the polluter pays', though it is fair to say that many of them get away with it. The ideas behind sustainable development have also appeared in the early paragraphs of much recent international agreement. Major pieces of international law include the 1992 Framework Convention on Climatic Change, the 1992 Convention on Biodiversity (both stemming from the 1992 Rio 'Earth Summit' and the 1963 Convention on International Trade in Endangered Species of Wild Fauna and Flora (CITES) which protects some 33,000 species. The Convention on the Prohibition of Military or Any Other Hostile Use of Environmental Modification Techniques (abbreviated as the ENMOD Convention) is a 1976 international treaty prohibiting the military or other hostile use of environmental modification techniques. It entered into force on October 5, 1978.

See also: ENVIRONMENTAL IMPACT ASSESSMENT, SUSTAINABLE DEVELOPMENT

Further reading

Menell, P.S. (ed), 2003. *Environmental law*. Ashgate Publishing, Burlington VT.

Richardson, B.J. and Wood, S. (eds), 2006. *Environmental law for sustainability*. Hart Publishing, Oxford.

Westing, A.H., 1990. *Environmental hazards of war: releasing dangerous forces in an industrialized world*. IPPR/UNEP, Stockholm.

ENVIRONMENTAL MANAGEMENT

Environmental management, which involves the management of all components of the biophysical environment, both living (biotic) and non-living (abiotic), is now so universal that it can produce 80 million responses on a search engine. Whereas management focuses on resources and has always been necessary because of the impact of *Homo sapiens* upon the Earth, planning can be undertaken with a cultural or demand bias. Early attempts at environmental management were direct responses to problems or hazards posed by the environment, such as flooding, but subsequently conservation movements developed as it was appreciated that certain environments or species were at risk and should be preserved. With greater population pressure it became appreciated that long-term environmental management must be based upon both scientific understanding and a system of morals and values, and this has required development of approaches, policies and legislation.

Although cultures in different parts of the world show differing attitudes to environmental management according to national priorities, varying attitudes to wealth and population growth and other factors, several approaches to management have been evolved. Whereas an early approach was to preserve environment but in other cases a 'technology can fix it' approach was employed, but it was later appreciated that to protect environment, conservation requires the sustainable utilization of species together with the conservation of essential ecological species and the preservation of genetic diversity. Sustainable development as defined by the United Nations in 1987 advanced seven strategic imperatives and seven preconditions for sustainability to be achieved. In addition specific approaches to environmental management including those necessary to control environmental hazards, required working with the environment employing softer rather than hard approaches. Thus in river channel management wherever possible harsh engineering techniques including channelization were replaced by softer techniques collectively described as 'working with the river rather than against it'. However although one approach of Gaia envisaged the planet as a self-regulating, living entity which was resilient to human impact, more recently there are fears that the resilience is under threat as a result of global climate change.

Whatever approaches are adopted it has been necessary to develop policies for environmental management. Whereas for specific types of impact these can be expressed in the form of protocols or guidelines for example, they are needed to achieve the management of specific hazards and to reduce environmental risk. The inception of environmental impact assessment (EIA) was important as a process determining and evaluating all the effects that a proposed action could have on the environment before the decision was taken as to whether to proceed. EIA originated in the USA from the National Environmental Policy Act (NEPA) of 1969 requiring environmental impact statements (EIS) to be undertaken by all agencies of the US Federal Government. Most countries have introduced policies requiring EIA and EIS. A recent development is that when undertaking such assessments it is necessary to obtain involvement of stakeholders/communities taking account of their views as appropriate.

If policies are formulated then legislation is necessary to ensure that they are effective and implemented. Environmental laws are required and although some are long established they are now required at international, national, state and local levels to deal with conservation of natural resources and with human health issues arising from contact with pollutants in the air, water, or land. A number of environmental protection agencies have been required to implement and monitor environmental legislation and their work can often include an element of environmental ethics.

Environmental management is wide-ranging and has to embrace the facts that there is often an incomplete communication between scientists and managers, so that managers may not be aware of the implications of the most recent environmental research; that some aspects of environment may be considered more than others; and that there are environmental risks which are difficult to manage. Thus in environmental management of the flood hazard the original idea of designing a structure for a particular recurrence interval, such as the 100 year event, promoted the perception that areas would be protected for 100 years – whereas the 100 year event could occur immediately. Therefore a more holistic approach has been advocated to environmental management, taking account of all aspects of the environment together with an acceptance of the need to accommodate risk and uncertainty.

See also: CONSERVATION, ENVIRONMENTAL IMPACT ASSESSMENT, ENVIRONMENTAL LAW, ENVIRONMENTAL POLICY, HOLISM

Further reading

Downs, P.W. and Gregory, K.J., 2004. *River channel management.* Arnold, London.

National Research Council, 1999: *New strategies for America's watersheds.* National Academy Press, Washington, DC.

Simmons, I.G., 1993. *Interpreting nature: Cultural constructions of the environment.* Routledge, London and New York.

UNWCED, 1987. *Our common future.* World Commission on Environment and Development, Oxford University Press, Oxford.

Wathern, P. (ed), 1988. *Environment impact assessment: theory and practice.* Unwin Hyman, London.

ENVIRONMENTAL PERCEPTION

erception is our sensory experience of the environment, involving both the recognition of environmental stimuli and actions in response to these stimuli. It is through neural networks that individuals know about the environment, make plans, and decide upon activities. However, in addition to direct experience of the environment, through the senses of taste, touch, sight, hearing and smell, information is also obtained indirectly from other people, from science and other educational sources, and from the mass media. Information received can be influenced by our own personalities, values, roles and attitudes and environmental perception will also vary with culture – hence environment will not be perceived in the same way in all places. Psychologists investigate the processes of perception but the study of environmental perception has developed to position the individual psychological processes of prediction, evaluation and explanation into a relevant social and political framework. Its importance for environmental sciences arises because the environment as perceived may affect the human response – including how the environment is managed or planned.

Ways in which environmental perception has been investigated include:

» How people image the natural and built environment. The physical structure of cities was the basis for some of the earliest perception studies – thus establishing what type of urban environments are preferred by population. Studies of the perception of riverine flooding in the USA were initiated by Gilbert White in the 1960s, leading to the study of other natural hazards and a world wide programme of studies of perceptions of hazards and of response to the hazards, helping to develop recommendations for hazard mitigation strategies. Such a behavioural approach of human response to natural hazards was influenced by methods used in psychology and social sciences. Although perception of, and response to, hazards depend on societal pressures and goals, the main objective of a perception approach to environmental management is to analyze how decision-making and choice of particular adjustments are affected by perception of individuals. It was shown that flood damage can increase with greater expenditure on flood control simply because the perception prevails that the flood hazard has been reduced. Mental maps or cognitive maps provide a graphical way of representing public or an individual's perception of environment and its spatial structure.

» How people react to environment. The book *Silent Spring* by Rachel Carson is often heralded as initiating environmental awareness with its repercussions for politics, especially the creation of Green Parties, education, and lifestyles. Although it is generally thought that people prefer 'natural' environments to urban environments and believe them to be natural and restorative, nevertheless studies have shown that many individuals who express preference for 'natural' landscapes or environments in fact often prefer 'tidy' environments that have been managed. Specifically in studies of landscape aesthetics this has included establishing what is perceived to be aesthetically valuable or of high quality and what should be preserved when developments occur. Since the 1960s and 1970s in addition to aesthetics and wilderness, there have been more public objections to natural resource management, to pollution, or to developments such as roads, often reflecting the way in which environment is perceived. This has occurred at a

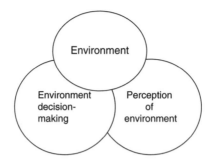

Figure 77 *Framework for environmental perception*

time when there has been increased demand for public consultation in relation to resource development.

» How perception of environment conditions management decisions. Thus oral tradition, that a river is aggrading, although not supported by scientific data, may provide the basis for river management. Similarly in ecological restoration the objective may be to create a perceived environment rather than an actual one. Decision-making about consequences of global climate change is also affected by perception in that change is perceived or not despite the scientific evidence presented.

A simple framework (Fig. 77) demonstrates the interaction between environment (characteristics, variability), decision-making (collective at different scales, individual) and perception of environment because it is this interaction that determines how courses of action will be determined.

See also: AESTHETICS, ENVIRONMENTAL ETHICS, HAZARDS, PERCEPTION

Further reading

Carson, R., 1962. *Silent spring*. Houghton Mifflin, Boston.

Finlayson, B.L. and Brizga, S.O., 2000. The oral tradition; environmental change and river basin management: case studies from Queensland and Victoria. *Australian Geographical Studies* 33, 180–192.

Tuan, Yi Fu, 1990. *Topophilia: a study of environmental perception, attitudes and values*. Prentice Hall, Englewood Cliffs, NJ. Revised edn.

Urban, M. and Rhoads, B., 2003. Conceptions of nature: implications for an integrated geography. In Trudgill, S. and Roy, A. (eds), *Contemporary meanings in physical geography. From what to why?* Arnold, London, 211–231.

White, G.F. (ed), 1974. *Natural hazards. Local, national, global*. Oxford University Press, New York.

ENVIRONMENTAL POLICY

In 2008, a well-known Internet search engine yielded about 43 million hits for 'Environmental Policy', about the same as for 'Football'. It is a term very widely applied and concise guides difficult to formulate. Firstly, there is a very wide variety of scale. It is possible for an individual to have a goal-directed, value-infused behaviour which uses the latest information to minimize his/her environmental impact: using public transport or cycling, refusing unnecessary packaging, offering wastes for recycling are all examples. Firms and other institutions like universities and colleges will also have such policies, aimed at such things as reducing energy consumption (switching off lights) or combating the disamenities of litter or discarded chewing gum. Cities will do similar things on larger scale, and so on to the formulation of international policies through environmental law and formulations like Agenda 21.

There are perhaps two main contextual considerations. The first is that many members of any given society do not have such as policy and there is nothing by way of environmental law to compel them to have one. Indeed, the mores of their society may shower praise of those with the heaviest impact: the rich and the ruling class, for example, whose energy use through vehicles alone will wildly exceed that of poorer people. The second is the meshing of policies at different scales both within a society and internationally. The attitudes to the generation of carbon dioxide are different between the State of California and the (2006) Federal Government of the USA, for instance; in north-east England, one county collects only paper and glass for recycling but the adjacent county processes many other domestic wastes as well: should people use their cars to cross the boundary with their disjecta? International co-ordination is a major task of the agencies of the United Nations Organization. It may be that this term 'policy' needs some kind of space-time qualifier in order to be really useful.

See also: AGENDA 21, RECYCLING, ENVIRONMENTAL LAW, UNITED NATIONS PROGRAMMES

Further reading

Cohen, S. 2006 *Understanding environmental policy*. Columbia University Press, New York.

Many types of environmental policy are dealt with on the websites of:

Resources for the Future Inc (www.rff.org)

World Resources Institute (www.wri.org)

ENVIRONMENTAL VALUES

The very concept of value stems from human cultures. There is no totally objective measure of it and so in matters of the environment there is always a tension between those who think that they may be absolute values and those who think that value is necessarily relative to culture. Values teach about the difference between right and wrong and are often derived from experience and history rather than a more disinterested examination of the evidence. In western societies the tendency to adopt binary classifications like 'right' and 'wrong' (also as in 'economic' and 'uneconomic' as a result of cost-benefit analysis) make well make for unwise decisions in a world with many more intermediate shades. Values lead to all kinds of other manifestations of culture with environmental relevance. Ethics, for example, is concerned with whether people ought to use a particular resource or alter a particular environment: a decision on possession of a large gas-guzzler is an ethical matter. Aesthetics also stems from ideas of value: in western Europe some groups consider wind-turbines to be ugly defacements of the landscape whereas others see them as logical heirs of other industrial installations with a vertical dimension. Some have seen that women's values might be especially relevant to caring for the Earth (ecofeminism). Doctrines are also infused with value: followers of the message of the Book of Genesis are in no doubt that humans are to have control of the Earth and its non-human inhabitants even to the limit of being able to name them. A concern for the future is also an appearance of values: a decision not to use a resource now so as to leave it for the next generation but one is giving value to potential rather than present actuality. All these may be part of an individual's value system, find their place in a communal system, or try to alter those values. In neo-classical economics, value is measured in money by the price of something in an open and competitive market: many commentators suggest this has poor applicability to many environmental matters.

See also: AESTHETICS, COST-BENEFIT ANALYSIS, CULTURES, LANDSCAPE

Further reading

Kempton, W., Boster, J.S. and Hartley, J.A., 1996. *Environmental values in American culture*. MIT Press, Cambridge, MA.

Lockwood, M., 1999. Humans valuing nature: synthesising insights from philosophy, psychology and economics. *Environmental Values* 8, 381–401.

Munda, G., 1997. Environmental economics, ecological economics, and the concept of sustainable development. *Environmental Values* 6, 213–233.

FIRE

From time to time, the news media highlight major 'wildfires' in the world: recent examples have been in eastern Australia, coastal California, Indonesia, southern France, and Greece. These can imply that the burning of wild and semi-natural vegetation is something new, which is true and not true at the same time. Fire occurs naturally on the planet: the expulsion of volcanic materials will set light to almost anything, and a lightning strike onto dry fuel will also start a fire. In both cases, any spread depends on the availability of fuel, usually dry plant material. Perhaps the most interesting thing about fire, though, is adoption by humans from a very early stage in their Palaeolithic evolution, probably in Africa. This was not only for cooking and as a domestic focus but applied at landscape scale to manipulate vegetation. This produced favoured plants for enhanced food supply and was also useful as a hunting tool since wild animals all flee fire. So even before agriculture, fire landscapes evolved, for example in savanna, steppe and prairie environments. Any forest with a dry period in the year might also be subject to firing by local people: charcoal layers are found in soils beneath tropical rain forests and those of the California Coast redwood.

Today, large-scale fires occur in dry periods in forest, savanna and grassland zones; these dry periods may be connected to global climate change and to the ENSO cycle. Open biomass burning is now a significant component in the carbon cycle, accounting for a significant proportion of that from fossil fuel burning, thus constituting a major factor in controlling spatial and temporal variations in atmospheric composition. Paradoxically, many protected forest areas are at great risk because effective fire suppression allows the build-up of combustible material on the woodland floor and when fire does break out (whether from lightning or from human activity) then it has a major fuel store to keep it going, especially if fanned by a strong wind. Control of such fires is expensive and success can be patchy, resulting in the loss of property and human life. Recent extensive fires, in Greece and the Canary Islands in 2007 for example, may have been fostered by extremely dry conditions and very high summer temperatures but sometimes triggered by deliberate arson – a case of environmental terrorism?

See also: BIOGEOGRAPHY, CARBON CYCLE, ECOLOGY, HAZARDS

Further reading

Pyne, S., 1995. *World fire*. University of Washington Press, Seattle, WA.

Pyne, S., 1997. *Fire in America*. University of Washington Press, Seattle, WA.

Whelan, R.J., 1995. *The ecology of fire*. Cambridge University Press, Cambridge.

http://www.pacificbio.org/Projects/Fire2001/fire_ecology.htm – basic information on fire in western North America.

http://www.talltimbers.org/ – a Florida-based research organization with a leading reputation in fire ecology and management.

FISHERY MANAGEMENT

For decades, the objective of fishery management was to maintain the resource at the level of maximum sustainable yield (MSY). It now is clear that MSY is no longer acceptable. The principal reason is that marine fishery resources fluctuate widely in abundance from natural causes such as environmental variability and biotic and ecological interactions, so that MSY cannot be represented by a simple number. A MSY that can safely be taken at one time may be much too high at another, and the resources would be in jeopardy if this were not recognized. For this reason and others, the comprehensive understanding of modern fishery management concepts has a threshold as defined by the Food and Agricultural Organization of the United Nations.

Fisheries are an excuse for doing the right thing. In the public mind, fishes are good and adequate justification for resource management policies and actions that also serve other aspects of the Public Trust. To quote Aldo Leopold writing in 1949, 'A thing is right when it tends to preserve the integrity, stability, and beauty of the biotic community. It is wrong when it tends otherwise'. Thus identifying fish needs and managing for sustainable fisheries advances many cultural, economic, and environmental matters are required. Long-term stewardship of fisheries is more cost-effective than short-term profitability. With a position near the top of the trophic pyramid, the health of fish populations reflects the health of lower trophic levels and influences the health of higher levels. To be healthy, sustainable fishery populations need more than clean water, they need to be part of a productive and healthy ecosystem. Proper management to maintain the health of fish populations also has indirect effects on the resilience of ecosystems upon which they and human populations depend. Therefore, developing knowledge that leads to the wise and sustainable management of fishes through studies that advance our understanding of bioenergetics, habitat requirements, predator-prey interactions, community structure, responses to perturbations, population dynamics, recruitment potential, and fisheries yields, will have far reaching benefits for any region, now and in the future.

There are consistent influences of ecological factors on the production of fish and shellfish in coastal ecosystems, as for example in tropical and sub-tropical areas (e.g., Gulf of Mexico), such as: river runoff onto the continental shelf, terrigenous sediment discharge (and organic detritus) onto coastal soft bottom shelf communities, primary production, coastal wetland vegetated areas, and climatic events and coastal pulsing (for example, local oceanographic features). Estuaries and coastal lagoons are, overall, more productive than other ecosystems in terms of fisheries yield, for which three reasons are suggested, for example, nutrient and organic matter inputs via rivers; shallowness, conducive to rapid remobilization of nutrients; velocity and volume of water exchanges between the sea and the estuarine system, which also directly affects fish production via recruitment. In the Gulf of Mexico, the high productivity of the Mississippi delta in Louisiana and adjacent states (the so-called fertile fisheries crescent) and the Grijalva–Usumacinta delta in Mexico is enhanced by a combination of factors: shallow, turbid waters and expansive marshes with extensive marsh-edge ecotones, and a microtidal (i.e., a tidal range of <2 m) system that is easily dominated by meteorological events. Destroying intertidal vegetation will affect shrimp fisheries, especially in those areas that have little of such vegetation. Whatever the reasons, it is clear that the interactions among habitats, river discharge, primary production, and fisheries are complex and more study is needed to fully understand these ecological interactions.

To understand recruitment in tropical coastal demersal (bottom) communities, early life history, and hence recruitment studies cannot be based on egg and larval surveys. This may imply that methodologies for the quantitative study of recruitment in tropical demersal communities can begin, with early juveniles at best. Thus, estimating number of newly metamorphosed juveniles can be measured by direct sampling along beaches, marshes, swamps, mangroves, and lagoons and estuaries. The fundamental problems related to such recruitment studies are that most species have a clear separation of habitat between the egg, juvenile and adult stages. In addition, eggs and larvae are often taxonomically indistinguishable. There is strong biotic interactions between different species, especially trophic competition and predation in the juvenile stages. Lifecycles are short, often less than one year, and recruitment is often continuous for different species with interspecific programming between them. Ecological implications of recruitment in functional structures of estuarine and coastal fish communities, and implications to fisheries in coastal lagoons and estuaries are key concerns towards ecosystem-based management of coastal fisheries.

The concept of Essential Fish Habitat (EFH) is aimed at enhancing the sustainability of coastal fisheries through legislation that established four levels of data quality in defining EFH: (I) Presence/absence. (II) Density patterns (e.g., population responses to gradients, suitability). (III) Growth/condition/health (e.g., growth, parasite loads, pollution loads, RNA/DNA ratios). (IV) Production (e.g., secondary production, reproductive output). So what is EFH? Essential is a qualifier that carries a notion of quality. The question is not just asking where a species lives (Level I), but where it lives well (Levels II–IV). Thus, essential habitat protection is a major concern in the sustainable management of fish stocks in tropical and other coastal zones. The responses of the ecosystem as a whole are modulated by 'natural production mechanisms' strongly influenced by such factors as rain, river discharge, tidal cycles and sea-level pulses, areas of coastal vegetation, and subsequent responses of ecosystem in aquatic primary productivity and fish biomass. Management decisions must take into consideration the integrated seasonal pulsing of such 'production mechanisms', towards a more comprehensive environmental planning. Coastal fish resources are an expression of ecosystem functioning and to assure the persistence of fish resources, the protection and conservation of essential habitats is the keystone for sustainable management of coastal fisheries.

See also: ECOSYSTEM-BASED MANAGEMENT

Further reading

Adger, W.N. *et al.*, 2005. Social-ecological resilience to coastal disasters. *Science* 309: 1036–1039.

Baltz, D. and Yáñez-Arancibia, A., 2008. Ecosystem-based management of coastal fisheries in the Gulf of Mexico: Environmental and anthropogenic impacts and essential habitat protection. In Day, J.W. and Yáñez-Arancibia, A. (eds), *The Gulf of Mexico: ecosystem-based management*, The Harte Research Institute for Gulf of Mexico Studies, Texas A and M University Press, College Station TX, Chapter 18 (in press).

Caddy, J.F. and Sharp, G.P., 1986. *An ecological framework for marine fishery investigation*. FAO Fisheries Technical Paper No. 283, FAO, Rome.

Caddy, J.F. and Garibaldi, L., 2001. Apparent changes in the tropic composition of World marine harvest: the perspective from FAO capture database. *Ocean and Coastal Management* 43: 611–655.

Leopold, A., 1949. *A sand county almanac*. Oxford University Press, New York.

McHugh, J.L., 1984. *Fishery management. lecture notes on coastal and estuarine studies*. Springer Verlag, Berlin.

Pauly, D., 1994. *On the sex of fish and the gender of scientists: a collection of essays in fisheries science*. Chapman and Hall, London. Fish and Fisheries Series 14.

Pauly, D. and Yáñez-Arancibia, A., 1994. Fisheries in coastal lagoons. In Kjerfve, B. (ed), *Coastal Lagoon Processes*. Elsevier, Amsterdam, 377–399.

Pauly, D. and Christensen, V., 1995. Primary production required to sustain global fisheries. *Nature* 374: 255–257.

Pauly, D., Christensen, Y., Dalsgaard, J. *et al.*, 1998. Fishing down marine food webs. *Science* 279, 860–863.

Seijo, J.C. and Caddy, J.F., 2000. Uncertainty in bio-economic reference points and indicators of marine fisheries. *Marine and Freshwater Research*, CSIRO Publishing, 51, 477–483.

Yáñez-Arancibia, A. and Pauly, D. (eds), 1986. *IOC/FAO workshop on recruitment in tropical coastal demersal communities*. Intergovernmental Oceanographic Commission, UNESCO Paris. Workshop Report 44.

Yáñez-Arancibia. A., Sánchez-Gil, P. and Soberón, G., 1985. Ecology of control mechanisms of natural fish production in the coastal zone. In A. Yáñez-Arancibia (ed.), *Fish community ecology in estuaries and coastal lagoons: towards an ecosystem integration*. UNAM Press Mexico, Mexico DF, chap. 27, 571–594.

GENETIC MODIFICATION (GM)

Even before the origins of agriculture, humans were modifying the genetics of plants and animals: consider the relations of modern breeds of dogs to ancestral wolves in the light of hunter-gatherers' domestication of the wild species. Agriculture itself is founded on tailoring plants and animals in a cultural image to yield maximal amounts of food (wheat or beef) or drugs (opium poppy) or pleasure (roses). The discoveries of Gregor Mendel (1822–1884) of the laws governing the inheritance of plant characteristics led to the science of genetics and many targeted developments (such as the high-yielding hybrid strains of rice, maize and wheat that underlay the Green Revolution of the 1960s). The term GM is now reserved almost entirely for the practice of the isolation, manipulation and reintroduction of DNA into cells, usually to express a protein. The aim is to introduce new characteristics or attributes physiologically or physically, such as making a crop resistant to a herbicide, introducing a novel trait, or producing a new protein or enzyme. Thus its protagonists promise that crops resistance to pests, or saline soils (for example) will magnify crop yields while lessening the role of harmful pesticides. Or, organisms can be modified to produce medicine such as insulin or drugs. The sceptics are concerned about the leakage of modified DNA into the wild and either killing or modifying wild species. They also point to the way in which sterile seeds may be a result and thus farmers are locked into buying seeds from a monopoly supplier. The experience of the Green Revolution of the 1960s was that sudden increases in crop yield made a few farmers much richer and displaced the rest from the land. In the EU, food products made with GM crops have to be labelled; elsewhere in the world this is rarely the case.

See also: ENVIRONMENTAL POLICY, ENVIRONMENTAL LAW

Further reading

Conway, G.R. and Barbier, E.B., 1990. *After the Green Revolution: sustainable agriculture for development*. Earthscan, London.

Evenson, R.E. and Gollin, D., 2003. Assessing the impact of the Green Revolution, 1960–2000. *Science* 300, 758–762.

http://ideas.repec.org/p/fao/wpaper/0509.html

GEOTHERMAL ENERGY

Geothermal energy is energy obtained from the natural heat from the interior of Earth. Geothermal energy was used as early as 1904 in Italy and today is being used in over twenty countries around the world. Today, tens of millions of people receive their electricity from geothermal sources at costs that are competitive with fossil fuels. Most geothermal energy projects involve utilization of high-heat sources such as very hot water and steam, but low temperature geothermal energy is also being used. For example, ground water at a normal ambient temperature of Earth can be considered a source of geothermal energy. The temperature of ground water at depths of about 100 metres is about 13 °C, which admittedly is cold for a bath. However, compared with cold winter temperatures in much of the Northern Hemisphere, it is warm. Furthermore, in the summer, water at a temperature of 13°C is cool relative to outside temperatures in many places. Thus, groundwater can help heat homes and buildings and assist with cooling in the summer.

World-wide, geothermal energy is relatively unexploited. Although energy from the heat of the Earth is not likely to be a major source of electricity, in some areas such as California and the United States in general, it has the potential to supply about 10% of the country's total electric capacity. This is equivalent to the electricity produced from waterpower today. Geothermal energy does not come without an environmental price. The energy is generally developed at a particular site and there are several potential environmental problems such as emission of gases, disturbance of the land, on site noise, and disposal of mineralized water after energy has been extracted from steam or hot water. Finally, geothermal energy is a form of nonrenewable energy when rates of extraction are greater than the natural replenishment. Thus, geothermal fields may be depleted, at least in the short term. What is apparent is that geothermal energy is one of those alternatives to fossil fuels that may help reduce dependency on oil, gas and coal.

See also: ENERGY, NATURAL RESOURCES

Further reading

Duff, W.A., 2003. *Geothermal energy from the Earth's heat*. USGS Geology Circular 1249, available at http://geopubs.wr.usgs.gov/circular/c1249/C1249.pdf

Duffield, W.A., Sass, J.H. and Sorey, M.L., 1994. *Tapping the Earth's natural heat*. U.S Geological Survey Circular 1125.

Gupta, H.K. and Roy, H., 2007. *Geothermal energy: an alternative resource for the 21st century*. Elsevier, Amsterdam.

Lund, J.W. and Freeston, D.H., 2001. Worldwide direct uses of geothermal energy 2000. *Geothermics* 30, 29–68.

Muffler, L.J.P. 1973. Geothermal resources. In D. A. Brobst and W.P. Pratt, (eds.), *United States Mineral Resources*, US Geological Survey Professional Paper 820, 251–61.

Tenenbaum, D., 1994. Deep heat. *Earth* 3(1), 58–63.

Wright, P., 2000. Geothermal energy. *Geotimes* 45(7), 16–18.

GLOBAL CHANGE

Global change is the term used to generally define multiple environmental and ecological changes taking place or projected to take place at a global level (with regional implications). Global change includes the study of climate change, species extinction, land-use change, as well as other areas. A major issue is to reduce human impact on the globe, particularly since recent studies have shown global warming, for example, is now significantly attributed to human societal activities. Some scientists think that if we applied 'fundamental scientific,

Table 51 *Strategies to reduce carbon emission rate in 2054 by 1 GtC/year or to reduce carbon emissions from 2004 to 2054 by 25 GtC (all the below have specific targets – see Pacala and Socolow, 2004)*

1. Efficient vehicles
2. Reduced use of vehicles
3. Efficient buildings
4. Efficient baseload coal plants
5. Gas baseload power for coal baseload power
6. capture CO_2 at baseload power plant
7. Capture CO_2 at H_2 plant
8. Capture CO_2 at coal-to-synfuel plant geologic storage
9. Nuclear power for coal power
10. Wind power for coal power
11. PV power for coal power
12. Wind H_2 in fuel-cell car for gasoline in hybrid car
13. Biomass fuel for fossil fuel
14. Reduced deforestation, plus reforestation, afforestation, and new plantations
15. Conservation tillage

technical, and industrial know-how' that we already possess, we could solve, for example, the carbon and climate problem and possibly avoid catastrophic climate change and its global change consequences (e.g., Pacala and Socolow, 2004). Table 51 list those ideas in general terms. Global change addresses the most important feedbacks between ecological systems and global change (especially climate), and their quantitative relationships; the potential consequences of global change for ecological systems; the options for sustaining and improving ecological systems and related goods and services, given projected global changes; the environmental, social, economic, and human health consequences of current and potential land-use and land-cover change over the next 5 to 50 years; the current and potential future impacts of global environmental variability and change on human welfare; factors that influence the capacity of human societies to respond to change; and how resilience can be increased and vulnerability reduced. These are a few of the objectives among scientists and policy-makers. Global forecasting includes assessing vital trends which affect corporate survival, globalization, economic instability, market changes, production and distribution, technology, computers, networking, virtual offices, socio-demographic changes, biotechnology, science, medicine, financial services, tribalism, political changes, single issues, lifestyle changes and global ethics.

A series of international **Stratospheric Ozone Assessments** were initiated in the mid 1980s to examine ozone-depleting chemicals and the current and projected state of the stratosphere. These assessments benefited from excellent leadership, succeeded at meeting the needs of decision-makers, and proved effective in mobilizing participants to render scientific and techinic judgements. More recently, however, the frequency of these assessments (every four years) has become has become somewhat burdensome; periodic update highlighting new findings would likely suffice. In addition, participation from industry has waned as the scale of economic implications declined.

The **Intergovernmental Panel on Climate Change (IPCC)** conducts periodic assessments (1990, 1995, 2001, 2007), mandated by the UN Framework Convention on Climate Change, on the scientific basis of climate change, impacts of Climate Change, on the scientific basis of climate change, impacts of climate change on natural and human systems, and options for mitigation and adaptation. With a well-developed organizational structure and strong ties to scientists and government, IPCC assessments are highly credible and effectively communicate to multiple audiences. The process could be improved by strengthening the coordination among individual working groups and rethinking the assessment strategy to take into consideration the rate at which new knowledge becomes available and the burden on the scientific community.

The **Global Biodiversity Assessment (GBA)**, published in 1995, provided a synthesis of available science to support the work of the United National Convection on Biological Diversity. Covering the many dimensions of biological diversity, the GBA achieved high scientific credibility due to involvement of the world's leading scientists. However, the lack of an authorizing environment limited its acceptance by governments. Further, efforts at outreach and interaction among working groups were hindered by a limited budget.

A **National Assessment of Climate Change Impacts** on the United States released in 2000 by US Global Change Research Program, as mandated by the 1990 Global Change Research Act (GCRA). Benefiting from a well-defined mandate and clearly articulate questions, the National Assessment succeeded in involving a broad range of stakeholders, in part through its well-planned communication strategy. However, it was the subject of considerable criticism and had limited impact on US policy or in funding new directions in research. Specific shortcoming included problems with the phasing of different assessment steps and uneven funding availability.

The **Arctic Climate Impact Assessment** was conducted in response to growing concern about how global warming and other associated changes could affect the Arctic environment. The assessment, completed in 2004, had a clear and strong mandate, with support from decision-makers, a well-planned communication strategy, and a transparent model for the science-policy interface. However, it could have been stronger if economic impacts had been considered and if follow-up activities had been better defined.

The **Millenium Ecosystem Assessment** was designed to answer a question fundamental to various UN conventions: what are the consequences of environmental change on the functioning of ecosystems and their continuing capacity to deliver services that are essential to human well-being? Published in 2005, strengths included broad participation from business, industry, academia, non-governmental organizations, UN agencies, and indigenous groups, and a conceptual model that was well designed to answer the central question. It could have been improved with more direct government interaction and plans for follow-up activities.

The **German Enquete Kommission on 'Preventative Measures to Protect the Earth's Atmosphere'** was set up by the German Parliament (Bundestag) to assess the importance and consequences to the country of stratospheric ozone depletion and of climate change. Strengths included good support and participation from political decision-makers, broad participation by stakeholders and a wide range of experts, and a good communication strategy. In some cases, the involvement of parliamentarians hindered the assessment process, for example, because they had little expertise on the subject or political differences made it hard to agree on specific resolutions.

A set of 21 **US Climate Change Science Program (CCSP) Synthesis and Assessment Products**, addressing various aspects of climate change, are currently being conducted to meet the requirements of the 1990 GCRA. The first product, released in May 2006, appears to have authoritatively resolved a long-standing discrepancy in the scientific community regarding global temperature trends in the lower atmosphere. Although individual products may be effective, it is not clear that the collection of assessment products will provide an integrated view of climate change impacts and possible response options.

See also: CLIMATE CHANGE, GLOBAL WARMING

Further reading

IPCC 4th Assessment – Summary for Policymakers. http://www.ipcc.ch/pdf/assessment-report/ar4/syr/ar4_syr_spm.pdf

Munn, T. (ed), 2002. *Encyclopedia of global environmental change*. 5 vols. Wiley, Chichester.

Pacala, S. and Socolow, R. 2004. Stabilization wedges: solving the climate problem for the next 50 years with current technologies, Science, 305, 968–972.

http://gcmd.nasa.gov/– a portal to many data sources.

GLOBAL CONVENTIONS AND TREATIES

A number of global conventions and treaties have taken place over the last several decades (Table 52). A recent example is the Kyoto Protocol to the United Nations Framework Convention on Climate Change, which is an amendment to the international treaty on climate change. The protocol assigns mandatory targets for the reduction of greenhouse gas emissions to nations which have signed on. The Kyoto Protocol is an agreement made under the United Nations Framework Convention on Climate Change. Countries that ratify this protocol commit to reducing their emissions of carbon dioxide and five other greenhouse gases, or engaging in emissions trading if they maintain or increase emissions of these gases. The Kyoto Protocol now covers more than 160 countries globally and over 55% of global greenhouse gas (GHG) emissions. The Kyoto linking mechanisms are in place for two reasons: the cost of complying with Kyoto is prohibitive for many so-called Annex 1 countries (countries such as Japan or the Netherlands that have highly efficient, low GHG polluting industries, and high prevailing environmental standards). Kyoto, therefore, allows these countries to purchase Carbon Credits instead of reducing GHG emissions domestically. This is seen as a means of encouraging Non-Annex 1 developing economies to reduce GHG emissions since by doing so, it is economically viable for them because of the capability to sell Carbon Credits. Many global conventions and treaties have been developed (Table 52).

Table 52   *Examples of world conventions and treaties*

CBD Convention on Biological Diversity www.biodiv.org

UNFCCC United Nations Framework Convention on Climate Change (1992/1994) www.unfccc.int

UNCCD The United Nations Convention to Combat Desertification (1994/1996) www.unccd.int

Ramsar Convention on Wetlands (1971/1975) www.ramsar.org

Convention concerning the Protection of World Cultural and Natural Heritage (1972/1975) whc.unesco.org

UNEP Convention on International Trade in Endangered Species of Wild Fauna and Flora (1973/1975) www.cites.org

Convention on Conservation of Migratory Species of wild Animals (1979/1983) www.cms.int

Convention on the Protection and Use of Transboundary watercourses and International Lakes (1992/1996) www.unece.org/env/water

UNEP Basel Convention on the Control of Transboundary Movements of Hazardous Wastes and their Disposal (1989/1992) www.basel.int

Rotterdam Convention on the Prior Informed Consent Procedure for Certain Hazardous Chemicals and Pesticides in International Trade (1998/2004) www.pic.int

Stockholm Convention on Persistent Organic Pollutants POPs (2001/2004) www.pops.int

Framework Convention on Climate Change (UNFCCC) in 1992 – Kyoto Protocol was adopted at the third Conference of the Parties to the UNFCCC (COP 3) in Kyoto, Japan, on 11 December 1997 http://unfccc.int/2860.php.

Based upon InfoResources Focus No. 3/2005 InfoResources
Länggasse 85 3052 Zollikofen Info@inforesources.ch
www.inforesources.ch

See also: UN PROGRAMMES

Further reading

Available in the websites given in Table 52.

GLOBAL DIMMING

Global dimming is the gradual reduction in the amount of global hemispherical solar radiation at the Earth's surface, observed since the beginning of systematic measurements in the 1950s. The effect varies by location, but worldwide it is of the order of a 5% reduction over three decades (1960 to 1990). This trend may have reversed during the past decade. The temporary disappearance of contrails due to plane groundings after the September

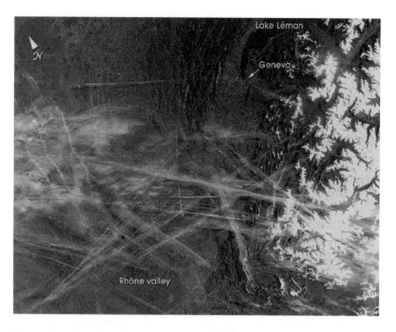

Figure 78 *Contrails can reduce the Sun's light and have been implicated in the global dimming idea. (http://visibleearth.nasa.gov/view_rec.php?id=6143)*

11, 2001 attacks in New York City, USA and the resulting increase in diurnal temperature range provided empirical evidence of the effect of thin ice clouds on the Earth's surface energy budget (Fig. 78). Global dimming creates a cooling effect that may have partially masked the effect of greenhouse gases on global warming over decades. The idea is that the effect of global dimming is probably due to the increased presence of aerosol particles in the atmosphere. Aerosol particles and other particulate pollutants absorb solar radiation and reflect sunlight back into space. Pollutants can also become nuclei for cloud droplets. Increased pollution, resulting in more particulates, creates clouds consisting of a greater number of smaller droplets, which in turn makes them more reflective, thereby reflecting more sunlight back into space. Clouds intercept both heat from the Sun and heat radiated from the Earth. Their effects are complex and vary in time, location, and altitude. During the day, the interception of sunlight predominates, giving a cooling effect. However, in the evenings, the re-radiation of heat to the Earth slows the Earth's heat loss.

See also: CLIMATE CHANGE, GLOBAL WARMING

Further reading

Abakumova, G.M., Feigelson, E.M., Russak, V. *et al.*, 1996. Evaluation of long-term changes in radiation, cloudiness and surface temperature on the territory of the former Soviet Union. *Journal of Climate* 9, 1319–1327.

Budyko, M.I., 1968. The effect of solar radiation variations on the climate of the Earth, *Tellus*, 21, 5.

Liepert, B.G., 2002. Observed reductions in surface solar radiation in the United States and worldwide from 1961 to 1990. *Geophysics Research Letters*, 29/12, 10.1029/2002GL014910.

Ohmura, A. and Lang, H., 1989. Secular variation of global radiation in Europe. In Lenoble, J. and Geleyn, J.-F. (eds), IRS'88: *Current problems in atmospheric radiation*, A. Deepak Publ., Hampton, VA, 298–301.

Srinivasan X.X. *et al.*, 2002. Asian Brown Cloud – fact and fantasy. *Current Science*, 83, 586–592.

Stanhill, G. and Cohen, S., 2001. Global Dimming: a review of the evidence for a widespread and significant reduction in global radiation with discussion of its probable causes and possible agricultural consequences. *Agricultural and Forest Meteorology* 107, 255–278.

Travis, D.J., Carleton, A.M., Lauritsen, R.G. *et al.*, 2002. Contrails reduce daily temperature range. *Nature*, 418, 601.

Wild, M., Gilgen, H., Roesch, A., Ohmura, A., Long, C.N., Dutton, E.G., Forgan, B., Kallis, A., Russak, V. and Tsvetkov, A., 2005. Changes in Earth's reflectance over the past two decades. *Science*, 308, 847–850.

http://www.pbs.org/wgbh/nova/sun/

GLOBAL WARMING

In the 20th century, the Earth's average near-surface atmospheric temperature rose $0.6 \pm 0.2\,°C$ ($1.1 \pm 0.4°$ F). The general scientific consensus on climate change is that most of the warming observed over the last 50 years is attributable to human activities (Fig. 79, Table 53). The increased amounts of carbon dioxide (CO_2) and other greenhouse gases (GHGs) are the primary causes of the human-induced component of warming. CO_2 and GHGs are released by the burning of fossil fuels, land clearing, and agriculture that have led to an increase in the greenhouse effect. The measure of climate response to increased GHGs, climate sensitivity, is found by observational studies and climate models. This sensitivity is usually expressed in terms of the temperature response expected from a doubling of CO_2 in the atmosphere. The current literature estimates sensitivity in the range of 1.5–$4.5\,°C$ (2.7–$8.1\,°F$). Models referenced by the Intergovernmental Panel on Climate Change (IPCC) predict that global temperatures may increase by between 1.4 and 5.8 °C (2.5 – $10.5\,°F$) between 1990 and 2100. The uncertainty

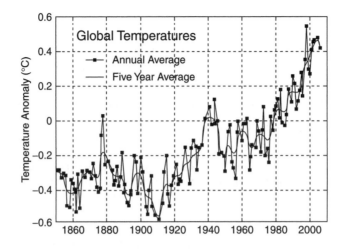

Figure 79 *Famous global warming curve of the 20th century and into the 21st century. From http://data.giss.nasa.gov/gistemp/graphs/. The dotted black line is the annual mean; solid red line is the five-year mean; green bars show uncertainty estimates*

Table 53 *These data can be used for graphing your own graphs and doing analysis: Global Land and Ocean Surface Temperature Anomaly (C) (Base: 1951–1980)*

Year	Annual_Mean	5-year_Mean
1880	−.25	*
1881	−.20	*
1882	−.23	−.24
1883	−.24	−.25
1884	−.30	−.26
1885	−.31	−.29
1886	−.25	−.30
1887	−.35	−.27
1888	−.27	−.28
1889	−.15	−.28
1890	−.37	−.28
1891	−.28	−.29
1892	−.32	−.32
1893	−.32	−.30
1894	−.33	−.28
1895	−.27	−.24
1896	−.17	−.23
1897	−.12	−.20
1898	−.25	−.16

1899	−.17	−.16
1900	−.10	−.19
1901	−.16	−.20
1902	−.27	−.24
1903	−.31	−.27
1904	−.34	−.27
1905	−.25	−.30
1906	−.20	−.30
1907	−.39	−.30
1908	−.34	−.32
1909	−.35	−.35
1910	−.33	−.34
1911	−.34	−.34
1912	−.34	−.30
1913	−.32	−.25
1914	−.15	−.24
1915	−.09	−.25
1916	−.30	−.25
1917	−.40	−.26
1918	−.32	−.28
1919	−.20	−.25
1920	−.19	−.22
1921	−.13	−.19
1922	−.24	−.20
1923	−.21	−.19
1924	−.21	−.17
1925	−.16	−.14
1926	−.01	−.13
1927	−.13	−.13
1928	−.11	−.11
1929	−.25	−.11
1930	−.07	−.10
1931	−.01	−.11
1932	−.06	−.07
1933	−.17	−.08
1934	−.05	−.08

(Cont'd)

Year	Annual_Mean	5-year_Mean
1935	−.10	−.06
1936	−.04	.00
1937	.08	.02
1938	.11	.05
1939	.03	.08
1940	.05	.07
1941	.11	.06
1942	.03	.10
1943	.10	.10
1944	.20	.07
1945	.07	.07
1946	−.04	.04
1947	.01	−.01
1948	−.04	−.06
1949	−.06	−.05
1950	−.15	−.05
1951	−.04	−.02
1952	.03	−.03
1953	.11	−.02
1954	−.10	−.05
1955	−.10	−.04
1956	−.17	−.04
1957	.08	−.01
1958	.08	.01
1959	.06	.06
1960	−.01	.05
1961	.08	.05
1962	.04	.00
1963	.08	−.02
1964	−.21	−.05
1965	−.11	−.05
1966	−.03	−.08
1967	.00	−.02
1968	−.04	.01
1969	.08	−.01
1970	.03	−.01

1971	−.10	.03
1972	.00	.00
1973	.14	−.02
1974	−.08	−.03
1975	−.05	.00
1976	−.16	−.03
1977	.13	.00
1978	.02	.05
1979	.09	.13
1980	.18	.12
1981	.27	.17
1982	.05	.17
1983	.26	.14
1984	.09	.12
1985	.05	.16
1986	.13	.17
1987	.27	.19
1988	.31	.25
1989	.19	.30
1990	.38	.27
1991	.35	.24
1992	.12	.25
1993	.14	.25
1994	.24	.24
1995	.38	.29
1996	.30	.38
1997	.40	.39
1998	.57	.38
1999	.33	.42
2000	.33	.45
2001	.48	.45
2 002	.56	.48
2003	.55	.54
2004	.49	.55
2005	.62	.55
2006	.54	*
2007	.57	*

From http://data.giss.nasa.gov/gistemp/graphs/

in this range results from both the difficulty of predicting the volume of future greenhouse gas emissions and the uncertainty about climate sensitivity.

An increase in global temperatures can cause changes such as a rising sea level and differences in the amount and pattern of precipitation. These changes may increase the frequency and intensity of extreme weather events – floods, droughts, heat waves, hurricanes, and tornados. Other consequences are reduced summer stream-flows, higher or lower agricultural yields, glacier retreat, species extinctions, and increases in diseases. On the one hand, warming is expected to affect the number and severity of these events. On the other hand, it is virtually impossible to connect specific events singularly to global warming. Although most studies focus on the period up to 2100, global warming (and sea-level rise due to thermal expansion) is expected to continue past 2100, since CO_2 has a long atmospheric lifetime. Few scientists dispute humanity's role in recent global warming. There is uncertainty regarding how much climate change is anticipated in the future, but scenarios are for substantial warming and updates come from the IPCC periodically.

See also: ANTHROPOGENIC IMPACT, CARBON CYCLE, CLIMATE CHANGE, GLOBAL DIMMING, HUMAN IMPACTS ON THE ENVIRONMENT

Further reading

Barnett, T.P., Adam, J.C. and Lettenmaier, D.P., 2005. Potential impacts of a warming climate on water availability in snow-dominated regions. *Nature,* 438, 303–309.

Choi, O. and Fisher, A., 2003. The impacts of socioeconomic development and climate change on severe weather catastrophe losses: mid-Atlantic region (MAR) and the U.S. *Climate Change,* 58, 149–170.

Emanuel, K.A., 2005. Increasing destructiveness of tropical cyclones over the past 30 years. *Nature,* 436, 686–688.

Oerlemans, J., 2005. Extracting a Climate Signal from 169 Glacier Records. *Science,* 308, 675–677.

Ruddiman, W.F., 2005. *Plows, plagues, and petroleum: how humans took control of climate.* Princeton University Press, Princeton NJ.

Shaviv, N. and Veizer, J., 2004. Comment, *Eos Trans. AGU,* 85 (48), 510–511.

Smith, T.M. and Reynolds, R.W., 2005. A global merged land and sea surface temperature reconstruction based on historical observations (1880–1997). *Journal of Climate,* 18, 2021–2036.

UNEP Summary (2002) Climate risk to global economy, Climate Change and the Financial Services Industry, United Nations Environment Programme Finance Initiatives Executive Briefing Paper; available at http://www.unepfi.org/fileadmin/documents/CEO_briefing_climate_change_2002_en.pdf

Intergovernmental Panel on Climate Change (IPCC) web site www.ipcc.ch

HUMAN IMPACTS ON ENVIRONMENT

Human impacts on Earth include by-products of industrialization, pollution, global climate change, acid precipitation, ozone depletion, water, soil erosion, desertification, salinization, and deforestation. With intelligence and technology to exploit energy resources, human beings have advanced into new habitats and extended the carrying capacity of the Earth. Sometime in the future, however, population growth will far exceed the carrying capacity of the Earth. In the past 25 years, the world's population has increased 3.7 fold to over 6 billion people. Continued growth will consume more and more natural resources.

Infrastructure developments such as highways, dams, and utilities, are key components for accessing and delivering products and services from nature and are essential to economic networks. Some of the consequences are increased emissions of greenhouse gases, pollutants and depletion of fisheries, and destruction of plants, wildlife, and fish habitats. Highways, power lines, airports, harbours, and dams are all a part of the infrastructure, connecting cities and propagating human expansion. Expansion of roads into previously undeveloped areas have opened up areas for industrialization, increased activities in explorations of energy resources and minerals, as well as logging, hunting, and poaching, have all exacerbated the environmental sustainability of the Earth. As a cycle, these activities in turn bring about deforestation and land and water degradation. While dams are intended to supply water for irrigation, they have also caused decreased water supply further down stream, increased methane emissions from reservoirs, and affected wetlands by their impacts on biodiversity. Growing land-based pressures on food, water, land, and fibre supplies are directly related to the rate of human expansion as infrastructure develops.

'Ecological footprint' is a phrase used to depict the amount of land and water area a human population would need to provide the necessary resources to support itself and to absorb its wastes. The term 'human footprint' was first coined in 1992 by Canadian ecologist William Rees, and is now widely used as an indicator of environmental sustainability. It is commonly used to explore the sustainability of individual lifestyles, goods and services, organizations, industrial sectors, and cities and nations by measuring demands on nature and comparing human consumption of natural resources with the Earth's regeneration capabilities. Human footprint has exceeded the biocapacity of the planet by 25%. Ecological footprint analysis approximates the amount of ecologically productive land and water mass areas necessary to manufacture a product or to sustain a population, all of which requires the use of energy, food, water, building material, and other consumables. The calculations used typically convert this into a measure of land area used in 'global hectares' (gha) per person. It is human use of renewable resources, not of non-renewable ones, at a rate beyond their carrying capacity, that poses the real sustainability crisis. Nature can restore renewable resources at a certain rate. Humans increasingly consume renewable resources faster than wilderness and ecosystems can replenish them. This state of ecological burden threatens ecosystems by not allowing them sufficient time to be replenished. Tables 54 and 55 below suggest varying impacts depending on how society engages in approaching environmental issues of the future.

Table 54 *Percentage land area impacted under different scenarios (world figures). Impacts on ecosystems, infrastructure, natural environments*

	Year 2000	Year 2050
Great Transitions	42	58
Policy Reform	44	70
Fortress World	44	75
Market Forces	48	80

Table 55

Great transitions – visionary solutions to sustainability, preservation of natural systems, high levels of welfare, material sufficiency, equitable distribution, population stabilized, green technologies, lower consumerism.

Policy reform – strong coordinated government action, policy discussions on sustainability, strengthen management systems, environmentally-friendly technology, sustainability as a strategic priority.

Fortress world – authoritarian response to threat of breakdown, protective enclaves develop, safeguarding of elite. Outside the fortress is repression, environmental distruction and misery.

Market forces – mid-range population and development projections. The resolving of social/environmental stresses left to competitive markets.

Information in Tables 54 and 55 is obtained from UNEP/RIVM (2004). José Potting and Jan Bakkes (eds.). The GEO-3 Scenarios 2002-2032: Quantification and analysis of environmental impacts. UNEP/DEWA/RS.03-4 and RIVM 402001022.

See also: ANTHROPOGENIC IMPACT, NATURE, SUSTAINABILITY

Further reading

Chambers, N., Simmons, C. and Wackernagel, M., 2000. *Sharing nature's interest: ecological footprints as an indicator of sustainability.* Earthscan, London.

Chambers, N. *et al.,* 2004. *Scotland's footprint. Best foot forward,* Oxford. Available at http://www.scotlands-footprint.com/downloads/Full%20Report.pdf

Lenzen, M. and Murray, S.A., 2003. The ecological footprint – issues and trends. *ISA Research Paper* 01-03.

Rees, W., 1992. Ecological footprints and appropriated carrying capacity: what urban economics leaves out. *Environment and Urbanisation,* 4, 121–130.

Tinsley, S. and George, H. 2006. *Ecological footprint of the Findhorn foundation and community.* Sustainable Development Research Centre, UHI Millennium Institute, Moray, Scotland.

van den Bergh, J.C.J.M. and Verbruggen, H., 1999. Spatial sustainability, trade and indicators: an evaluation of the 'ecological footprint'. *Ecological Economics,* 29, 63–74.

Wackernagel, M. and Rees, W., 1996. *Our ecological footprint: reducing human impact on the earth.* New Society Publishers, Gabriola Island BC.

http://www.ecologicalfootprint.com – allows individuals to calculate their own footprint.

INTEGRATED BASIN MANAGEMENT

The process of managing water resources within a drainage basin or watershed in a way which optimizes water use throughout the basin and minimizes deleterious effects for water, river channels and land use. Management methods were often developed separately as necessary for water supply, flood control, water quality control, and water as a power source, concentrating on the solution of a single problem. Such an approach, undertaken with little consideration of the effects that the 'solution' would have upstream or downstream of the project and elsewhere in the basin, was succeeded by an integrated approach, intended to consider all known uses and feedbacks together, described by a range of terms including river basin management, integrated river basin planning, and total catchment management.

Early integrated multipurpose approaches to water resources management are exemplified by the TVA (Tennessee Valley Authority) established in 1933, to improve navigability, provide for flood control, for land use and industrial needs, and later to include provision of hydroelectric power (HEP). River basins are ideal planning units for an integrated approach because they are clearly bounded, physically functional, hierarchical in scale, and culturally meaningful. The four essential stages recommended by the United Nations for integrated river basin planning are:

1. A preliminary stage, usually responding to an immediate problem such as erosion or flooding.
2. The existing physical, socioeconomic and administrative system is studied in detail and a development strategy formulated.
3. Small scale pilot projects are implemented to investigate the likely success of the large strategies proposed in Stage 2.
4. Completion of the project by physical structures and management of the overall scheme.

Recent developments have been achieved by more 'holistic' fully comprehensive approaches where 'the whole is more than the sum of the parts', incorporating energetics of the fluvial system, consideration of potential impacts throughout the drainage basin, of the longer-term history, and adopting a dynamic approach to management and planning by ecosystem-based approaches. Such an integrated watershed view has been advocated in the USA, and has been required by the European Union Water Framework Directive which obliges European countries to establish integrated river basin management, employing more holistic approaches which balance sustainability with environmental change. Differences continue to occur between areas and countries for reasons indicated in Table 56.

Table 56 *Reasons for differences in integrated basin management (Developed from Downs and Gregory, 2004; see also Diplas, 2002)*

Aspect	Categories	Reasons for differences from one area to another
Subjects	Water, Land	One particular subject may be emphasized, e.g., water resources, land conservation
Approach	Discipline, Techniques	Particular academic training may provide a dominant influence (e.g., engineering, economics). Techniques adopted (e.g., GIS) may be influential
Involvement	Stakeholders	Range of stakeholders consulted (e.g., public participation and involvement) and involved
Strategy	Single or multipurpose, integrated, holistic, sustainable	Variations according to character of area and history of watershed management
Framework	Legal, structural, political	Administrative structure may affect implementation
Decision-making process	Integrated decision-making	Extent to which decision making is truly integrated and considers all aspects
Subsequent action	Adaptive management	Degree to which adaptive management implemented and results are heeded

See also: DRAINAGE BASINS, ENERGETICS, HOLISM, INTEGRATED COASTAL AND MARINE MANAGEMENT, LAND EVALUATION

Further reading

Diplas, P., 2002. Integrated decision making for watershed management. *Journal of the American Water Resources Association* 38, 337–340.

Downs, P.W. and Gregory, K.J., 2004. *River channel management*. Arnold, London.

Downs, P.W., Gregory, K.J. and Brookes, A., 1991. How integrated is river basin management? *Environmental Management*. 15, 299–309.

European Community, 2000. Directive 2000/60/EC of the European Parliament and of the Council of 23 October 2000 establishing a framework for Community action in the field of water policy. *Official Journal of the European Communities*, L327: 1–72.

Newson, M.D., 1997. *Land, water and development: sustainable management of river basin systems.* Routledge, London, 2nd edn.

Newson, M.D., 2002. Geomorphological concepts and tools for sustainable river ecosystem management. *Aquatic Conservation: Marine and Freshwater Ecosystems* 12, 365–379.

National Research Council, 1999. *New strategies for America's watersheds.* National Academy Press, Washington, DC.

United Nations, 1970. *Integrated river basin development: report of a panel of experts.* UN Department of Economic and Social Affairs, New York.

INTEGRATED COASTAL AND MARINE MANAGEMENT

Coastal zone management represents different things to different people. To the conservation minded, the concept represents either a panacea for every excess of the private sector or governmental agencies, or the solution to every unsolved coastal-related problem. A successful programme is based on a comprehensive and integrated planning process, which aims at harmonizing cultural, economic, social, and environmental values and regulations (CEP/UNEP, 1995). Management without an appropriate planning process tends to be neither integrated nor comprehensive, but rather a sectorial activity. A consensus emerged after UNCED (1992) to work towards *Integrated Coastal Area Management* (ICAM). The new acronym ICAM and the old *Integrated Coastal Zone Management* (ICZM) are used interchangeably in the literature.

There is no shortage of definitions of Integrated Coastal Zone Management. A slightly more activist, interventionist definition suggested by Knecht and Archer in 1993 is:

> Integrated coastal management is a dynamic process by which decisions are taken for the use, development and protection of coastal areas and resources to achieve goals established in cooperation with user groups and national, regional and local authorities. Integrated coastal management recognizes the distinctive character of the coastal zone – itself a valuable resource – for current and future generations. Integrated coastal management is multiple purpose oriented, it analyzes implications of development, conflicting uses, and interrelationships between physical processes and human activities, and it promotes linkages and harmonization between sectorial coastal and ocean activities.

The dimensional aspects of ICZM are a function of the kinds of integration required, which sets the pattern of outreach, peripheral involvement, and the nature of partnership, participation and negotiation with other coastal resources users and institutions. According to CEP/

UNEP in 1995 there are at least seven different kinds of integration, each of which has its own dimensional limits: (1) intergovernmental, (2) land–water interface, (3) inter-sectorial, (4) interdisciplinary, (5) interinstitutional, (6) inter-temporal, and (7) managerial. It is to be noted that the inclusion of NGOs, the local public, the ecosystem approach, and the multi sectorial approach into the management process is of vital importance for a successful ICZM programme.

See also: COASTAL MANAGEMENT, INTEGRATED BASIN MANAGEMENT

Further reading

Cicin-Sain, B., 1993. Sustainable development and integrated coastal management. *Ocean and Coastal Management* 21, 11–43.

Cicin-Sain, B, and Knecht, R. W., 1998. *Integrated coastal and ocean management.* UNESCO and University of Delaware CSMP. Island Pres, Inc., Washington DC.

CEP/UNEP, 1995. *Guidelines for integrated planning and management of coastal and marine areas in the wider Caribbean region.* UNEP (OCA)/CAR W.G. 17/3, 28-30/06/95, Kingston, Jamaica.

Chua, T. E. and Pauly, D. (eds), 1989. *Coastal area management in Southeast Asia: policies, management strategies and case studies.* ICLARM Conference Proceedings 19, Manila, Philippines.

Clark, J.R., 1996. *Coastal zone management handbook.* CRC. Lewis Publishers, Ltd., Boca Raton, FL.

Knecht, R. and Archer, J., 1993. 'Integration' in the United States coastal zone management programme. *Ocean and Coastal Management* 21, 183–199.

UNCED, 1992. *United Nations conference on environment and development.* Agenda 21, Rio de Janeiro, Brazil, chap. 17.

Vallega, A., 1992. *Sea management. A theoretical approach.* Elsevier Applied Science, London and New York.

Yáñez-Arancibia, A., 1999. Terms of reference towards coastal management and sustainable development in Latin America: introduction to special issue on progress and experiences. *Ocean and Coastal Management* 42, 77–104.

Yáñez-Arancibia, A., 2000. Coastal management in Latin America. In C. Sheppard (ed), *The seas at the millennium: an environmental evaluation.* Elsevier, Amsterdam. 3 vols, volume 1, 447–456.

LIMITS TO GROWTH

The thrust of the major work of Thomas Malthus (1776–1834) was that human population growth would always exceed the capacity to produce adequate resources; the *Limits* study, published as a book in 1972, was an attempt to use the newly developed powers of digital computers to model the interaction between exponentially growing populations and their resource uses. The conclusions were that some resources (notably those of the fossil fuels) would run out in relatively short periods: for petroleum, some 50 years was forecast. The object of the authors was to issue a warning to the world that it was on a track that would not only produce resource insufficiencies but considerable environmental damage as well. There was a considerable furore about the book, with sharp criticism from economists and growth-oriented international agencies: in particular, the details of the models used were not published at the same time, and also that although population and resource use grew exponentially, technology did not. So the favourite mantra of neo-classical economics, that rising prices bring about shifts in demand and innovation was not addressable in these models. In 1992, an updated version was published, with modified conclusions resulting from the 20 years of experience in both the real world and the models of it. The slowing-down of population growth in the interim may well have been partially the result of the concerns raised by the *Limits* study, and recent (2008) statements that the peak of oil production has been passed seem to suggest that although Malthus (as so often) has been disapproved, he has not yet been disproved.

See also: NATURAL RESOURCES, POPULATION, RESOURCE DEPLETION

Further reading

Meadows, D.H., Randers, J. and Behrens, W.W., 1972. *The limits to growth.* Universe Books, New York.

Meadows, D. H., Randers, J. and Meadows, D.L., 1994. *Limits to growth. The 30 year update.* Earthscan, London.

Lomborg, B., 2001. *The sceptical environmentalist: measuring the real state of the world.* Cambridge University Press, Cambridge.

http://www.clubofrome.org/archive/reports.php is an archive from the Club of Rome which commissioned the original and later *Limits* studies.

OZONE DEPLETION

Ozone depletion describes two distinct but related observations: a slow, steady decline of about 3% per decade in the total amount of ozone in Earth's stratosphere since around 1980; and a much larger, but seasonal, decrease in stratospheric ozone over Earth's polar regions during the same period. The latter phenomenon is commonly referred to as the ozone hole (Fig. 80). The most important process in both trends is catalytic destruction of ozone by atomic chlorine and bromine. The main source of these halogen atoms in the stratosphere is photodissociation of chlorofluorocarbon (CFC) compounds, commonly called freons, and of bromofluorocarbon compounds known as halons. These compounds are transported into the stratosphere after being emitted at the surface. Both ozone depletion mechanisms strengthened as emissions of CFCs and halons increased. CFCs, halons and other contributory substances are commonly referred to as ozone-depleting substances (ODS). Since the ozone layer prevents most harmful UVB wavelengths (270–315 nm) of ultraviolet light (UV light) from passing through the Earth's atmosphere, observed and projected decreases in ozone have generated worldwide concern leading to adoption of the Montreal Protocol banning the production of CFCs and halons as well as related ozone depleting chemicals such as carbon tetrachloride and trichloroethane (also known as methyl chloroform). It is suspected that a variety of biological consequences such as increases in skin cancer, damage to plants, and reduction of plankton populations in the ocean's photic zone may result from the increased UV exposure due to ozone depletion. However, the ozone hole is most usually measured not in terms of ozone concentrations at these levels (which are typically of a few parts per million) but by reduction in the total column ozone, above a point on the Earth's surface, which is normally

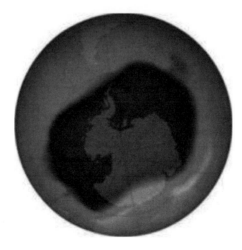

Figure 80   *Image of the largest Antarctic ozone hole ever recorded in September 2006. Image from http://www.nasa.gov/vision/earth/environment/ozone_resource_page.html.*

expressed in Dobson units, abbreviated as 'DU'. Marked decreases in column ozone in the Antarctic spring and early summer compared to the early 1970s and before have been observed using instruments such as the Total Ozone Mapping Spectrometer (TOMS). Reductions of up to 70% in the ozone column observed in the austral (southern hemispheric) spring over Antarctica and first reported in 1985 are continuing (Fig. 80). Throughout the 1990s, total column ozone in September and October continued to be 40–50% lower than pre-ozone-hole values. In the Arctic the amount lost is more variable year-to-year than in the Antarctic. The greatest declines, up to 30%, are in the winter and spring, when the stratosphere is colder. It is calculated that a CFC molecule takes an average of 15 years to go from the ground level up to the upper atmosphere, and it can stay there for about a century.

See also: GLOBAL CHANGE

Further reading

Abarca, J.F. and Casiccia, C.C., 2002. Skin cancer and ultraviolet-B radiation under the Antarctic ozone hole: southern Chile, 1987–2000. *Photodermatology, Photoimmunology and Photomedicine* 18, 294–302.

Benedick, R.E. 1991. *Ozone diplomacy*. Harvard University Press, Cambridge, MA. (Ambassador Benedick was the Chief US Negotiator at the meetings that resulted in the Montreal Protocol.)

Cagin, S. and Dray, P., 1993. *Between earth and sky: how CFCs changed our world and endangered the ozone layer*. Pantheon, New York.

Hoffman, M.J., 2005. *Ozone depletion and climate change: constructing a global response*. State University of New York Press, Albany, NY.

Newman, P.A. et al., 2004. On the size of the Antarctic ozone hole. *Geophysical Research Letters 31*, L12814.

Newman, P.A., et al., 2006. When will the Antarctic ozone hole recover? *Geophysical Research Letters 33*, L12814.

Roan, S., 1990. *Ozone crisis, the 15 year evolution of a sudden global emergency*. Wiley, New York.

Wayne, R.P., 2000. *Chemistry of atmospheres*. Oxford University Press, New York.

Weatherhead, E.C. and Andersen, S. B., 2006. The search for signs of recovery of the ozone layer. *Nature* 441, 39–45.

POLLUTION OF AIR AND WATER

Throughout human history, there has always been some level of local air and water pollution, but with the coming of the industrial age, and especially in the 20th century, human activity is causing air and water pollution on a global scale.

Air pollution can be categorized in several ways. There are stationary and mobile sources of air pollution. Stationary sources include point, fugitive, and area sources. Point sources come from one or more fixed sources such as industrial smoke stacks. Fugitive sources come from open areas exposed to winds such as burning of agricultural fields and dust from construction sites. Area sources are areas where there are several air pollution emitters such as a small town or an industrial complex. Mobile sources move while emitting pollution; such as cars, planes, or ships. Air pollution has a number of effects on many aspects of the environment including human health, aesthetics, soils, water quality, structures, and vegetation. In humans, air pollution can cause poisoning, cancer, birth defects, and irritation of eyes and lungs. Air pollution can also affect water quality, in some watersheds the major source of nitrogen is in precipitation. Major air pollutants include ozone, sulphur dioxide, nitrogen oxides, carbon monoxide and carbon dioxide, particulate matter, and lead. One of the most dramatic affects of air pollution is global warming caused by excessive levels of carbon dioxide and other green house gases such as nitrous oxide and methane. The gases are predicted to have major impacts on climate in the 21st century. Acid rain is another well-known air pollutant. Acid rain is precipitation that is highly acidic, with pH values ranging from 5.6 to less than 2.0. It includes both wet (rain, fog, and snow) and dry (particulate matter) deposition. The primary causes of acid rain are due to sulphur dioxide and nitrogen oxides. These form sulphuric acid and nitric acid. Sulphur oxide comes mainly from stationary sources such as power plants and industry and nitrogen oxides come mainly from mobile sources such as automobiles. Acid rain has caused death of trees in forest ecosystems, acidification of lakes (sometimes with loss of all fish populations), corrosion of metal and stone (including historical buildings in Europe and elsewhere), and human health problems. Indoor air pollution is also a widespread problem.

The most common indoor air pollutant is second-hand tobacco smoke. Other sources of indoor pollutants are organisms (i.e., *Legionella pnuemophila*, a bacterium that causes Legionnaires disease, a variety of molds or fungi, and dust mites), radon gas, pesticides applied in buildings, asbestos, and a variety of chemicals used in construction, furniture, and other items found in buildings. Formaldehyde is used in foam insulation, plywood, ceiling tiles, etc. and can cause skin irritation and cancer. A special kind of air pollution is ozone depletion in the atmosphere. Near the ground, ozone is a pollutant, but ozone at high altitudes in the stratosphere helps shield the Earth from ultraviolet radiation. The release of CFCs or chloro-fluorocarbons (used in refrigeration systems) is related to ozone depletion. The depletion of the ozone is thought to be a factor in the high rates of skin cancer. Many nations have agreed to limit CFC production to halt loss of the ozone layer.

Pollution of surface and ground water resources is also a world-wide problem due to a variety of causes. Water pollution is the degradation of water quality. As with air pollution, water pollution is caused by point and non-point sources. Point sources include sewage treatment plants, industrial discharges, leaks from pipelines, spills from ships, and leakage from storage tanks. Non-point sources include agricultural runoff, sediment runoff from a variety of sources, salt-water intrusion into coastal aquifers and coastal surface waters, and acid runoff from mines. Types of water pollution include organic matter as in sewage, pathogens such as coliform bacteria in faeces, organic chemicals such as pesticides and industrial wastes (i.e., PCPs), nutrients such as nitrogen and phosphorus in agricultural runoff, heavy metals such as lead and mercury, acids, sediments, heat such as from power plants, and radioactivity. Many of these pollutants can cause acute and chronic human health problems resulting from pathogens (bacteria and viruses) and toxins (pesticides, industrial chemical, heavy metals).

The introduction of excessive levels of organic matter (as in sewage) and nutrients leads to water quality deterioration. High levels of nutrients lead to eutrophication, which is characterized by algal blooms, low oxygen levels, and fish kills. Runoff from agricultural fields and feed lots are major causes of eutrophication. Sometimes, harmful algal bloom species develop. These can cause health effects in aquatic organisms and humans. Hypoxia, where bottom waters are 2 mg/l dissolved oxygen or less, is a widespread problem in eutrophic waters, especially in coastal waters. The large hypoxic zone in the Gulf of Mexico in front of the Mississippi delta is one of the best known examples. The low oxygen is caused mainly by agricultural runoff from the Midwestern states which produce large quantities of corn and soy beans. Corn is heavily fertilized and a portion of the nitrogen in the fertilizer, mainly in the form of nitrate, is washed off the land and enters the Mississippi River and its tributaries. When this nitrogen reaches the Gulf of Mexico, it causes large algal blooms, some of which dies and sinks to the bottom. As the cells decompose, oxygen is used. Because the Gulf is stratified, oxygen in surface waters cannot reach the bottom. A solution to this problem is the construction of wetlands throughout the agricultural landscape to remove oxygen from agricultural runoff. Pollution of ground water is a serious problem. Most classes of pollutants can enter ground water as surface water sinks into the ground. If high levels of pollutants enter ground water, it can threaten drinking water supplies. Because it is below the surface, ground water pollution is very difficult to clean up. There are widespread efforts to clean up water pollution using both advanced wastewater technology as well as ecotechnology, where humans work with nature. The use of wetlands for water quality improvement is an example of ecotechnology.

See also: HUMAN IMPACTS ON ENVIRONMENT

Further reading

Botkin, D.B. and Keller, E.A., 2007. *Environmental science – earth as a living planet*. Wiley, New York.

Mitsch, W.J., Day, J.W., Wendell, J. *et al.*, 2001. Reducing nitrogen loading to the Gulf of Mexico from the Mississippi river basin: strategies to counter a persistent problem. *BioScience* 51, 373–388.

Journal: *Water, Air and Soil Pollution,* published by Springer, Berlin-Heidelberg.

http://earthtrends.wri.org/ *is* a gateway to material on pollution.

POPULATION GROWTH

The course of the growth of human numbers is well known. This world-wide, long-timescale graphic conceals a number of variations which need explanations. For example, there have been times of slow growth or even negative growth: hunter-gatherer populations grow very slowly; world pandemics like the Black Death of the 14th century, or the introduction of European diseases into Latin America with the Spanish and Portuguese conquests decimated native populations; some European nations are in the early 21st century declining a little. The effect of HIV/AIDS is to slow growth rates but not as yet put them into reverse. But the overall story is one of inexorable rise. There seems to have been a take-off in the 16th–17th centuries, possibly due to better nutrition as crops like maize and the potato improved calorie production in the Old World; maybe as intercontinental connections increased then immunity to diseases improved; tea and coffee (it is said, more or less seriously) involved boiling water and thus raised hygiene levels. Above all, the ability to cope with famines by transporting relief, the improvement of longevity and the reduction in infant mortality by the application of modern medicine in the 19th and 20th centuries have been critical. In the 1950s, world growth rates were as high as 2.6% per annum. but they have fallen back to below 2% now. Most societies now have the means to control their fertility levels but some choose not to use them since family size has a large cultural element. The significance of population growth however is not simply that of numbers: the environmental impact depends upon many other factors, of which access to technology is one. But a total world population of nine billion seems inevitable unless some disaster intervenes.

See also: RESOURCE DEPLETION

Further reading

Livi-Bacci, M., 2006. *A concise history of world population.* Blackwell, Oxford. 4th edn.

McEvedy, C. and Jones, R. (eds), 1978. *Atlas of world population history*. Penguin Books, London. Revised edn.

http://popindex.princeton.edu/

RECYCLING

Recycling is the most visible of the resource-conserving quartet of Re-use, Repair, Refuse and Recycle, where 'refuse' means to decline to acquire goods and services rather than to create refuse. It is not a new idea: most pre-industrial societies practice recycling and re-use as a normal way of using materials. In many such groups, iron for example was a precious metal which had to be re-used as many times as possible and so the blacksmith was a key member of the community. It is still common to see the waste heaps of the cities of developing nations being picked over for every scrap of saleable and re-usable material. Only in western industrial countries since the 19th century has the availability of material goods been so lavish that a throwaway society could develop. As a reaction to this rapid throughput of materials and energy, a set of environmental values has developed which tries to reduce the quantity of materials for which an environmental sink (e.g., as landfill or in the sea) is sought, and by short-circuiting the manufacturing process reduce the amount of energy being consumed, with its consequent emissions. The EU produces about 2 billion tonnes of solid wastes per year: 29% from mining and quarrying, 26% from manufacturing and 14% is municipal wastes, which includes domestic households. Most household bins contain about 550 kg per person per year. These data are the target for intensive attempts to change consumer behaviour, which is very variable between countries: the USA recycles about 28% of such wastes and the UK about 18%, both out of a theoretical maximum of 60%. Industry has a greater interest since it is a way of reducing costs: for every kilo of copper extracted, some 420 kg of unused material has to be processed; equally some industries have to pay for disposal of such wastes and also buy permits to pollute the atmosphere with the emissions form the energy used.

See also: RESOURCE DEPLETION, ENERGY, ENVIRONMENTAL VALUES

Further reading

Gandy, M., 1994. *Recycling and the politics of urban waste*. Earthscan, London.

Lund, F., 2000. *The McGraw-Hill recycling book*. McGraw-Hill, New York, 2nd edn.

RENEWABLE ENERGY

Renewable energy means garnering useful energy from existing flows, from ongoing natural processes, such as sunshine, wind, flowing water (hydropower), biological processes (such as anaerobic digestion), and geothermal heat flow. A common definition is that renewable energy is from an energy resource that is replaced by a natural process at a rate that is equal to or faster than the rate at which that resource is being consumed. For example, solar photovoltaic technology harvests energy from the Sun, but only a fraction of the total amount of solar energy is harvested. It is a subset of sustainable energy. Most renewable forms of energy, other than geothermal and tidal power, generally come from the Sun. Some forms are stored solar energy such as rainfall and wind power, which are considered short-term solar-energy storage, while the energy in biomass is accumulated over a period of months, as in straw, or through a period of years as in wood. By capturing renewable energy from plants, animals and humans do not permanently deplete the resource. Fossil fuels, while theoretically renewable on a long timescale, are exploited at rates that could deplete these resources in the near future. Renewable energy resources may be used directly, or used to create other more convenient forms of energy. Examples of direct use are solar ovens, geothermal heating, and watermills and windmills. A few examples of indirect use that require energy harvesting are electricity generation through wind turbines or photovoltaic cells (PV cells), and production of fuels such as biogas from anaerobic digestion or ethanol from biomass (using alcohol as a fuel). In a sense, renewable energy may be categorized as free energy, although most renewable energy sources are not normally called 'free energy'. In engineering, free energy means an energy source available directly from the environment, which is not expected to be depletable by humans. Renewable energy development is concerned with the use of renewable energy sources by humans. Today's interest in renewable energy development is linked to concerns about exhaustion of fossil fuels and environmental, social, and political risks of over-use of fossil fuels and nuclear energy.

See also: ENERGY, ENERGY SCARCITY, RESOURCE DEPLETION

Further reading

Bird, L and Swezey, B., 2005. *Estimates of New Renewable Energy Capacity Serving U.S. Green Power Markets in 2004.* National Renewable Energy Laboratory, Golden, CO. Available at www. eere.energy.gov/

EUROPA/European Commission. 2006. *EU Energy Fact Sheet* available at http://ec.europa.eu/ energy/green-paper-energy/index_en.htm

IEA, 2006. *Global Renewable Energy Policies and Measures Database*, Paris, available at www.iea. org/textbase/pamsdb/grindex.aspx

World Bank, 2005. *World Bank group progress on renewable energy and energy efficiency, Fiscal Year 2005*. Washington, DC. Available at http://siteresources.worldbank.org/INTENERGY/Resources/Annual_Report_Final.pdf

Worldwatch Institute, 2006. *Vital signs 2006–2007*. Norton, New York.

RESOURCE DEPLETION

All societies have had worries from time to time about the scarcity or even the non-existence of key resources such as food and water. These have been local in impact and many traditional societies had developed cultural ways of coping eg., by migration, by mutual dependence on less affected kin groups, or by going to war. Now that the resource pool is virtually world-wide due to trade and there are so many more human beings around since the 19th century, the dimensions of the anxieties are much magnified. Examples of concerns over resource depletion include for example food. The small proportion of solar radiation which is fixed by photosynthesis suggests that in a carbon dioxide-rich atmosphere, there should be no problem in growing more plant material for food, especially of genetically modified (GM) organisms can be more effective at utilizing the sunlight. The depletive aspect comes from the use of much plant material to feed animals that provide meat. A great deal of the garnered solar energy is wasted since so many animals give it off as heat and also need a great deal of processing and transport. If *Homo sapiens* were totally vegetarian, this anxiety would virtually disappear.

Another worry is about energy, since it is possible that the peak of oil production has passed. Here again, profligacy of use is a main element: attention to energy conservation and the determination not to be 'hooked on oil' would mean changes in western lifestyles but also decreased political dependencies. Water may turn out to be a less tractable resource. It is already scarce in some countries, which have had to turn to desalination (islands such as Malta are good examples), and seasonally so in heavily urban regions that have an unpredictable drought year, such as southeast England, or have a dense development for tourism, such as southern Spain's coast. Space is required to produce all types of resources: energy (e.g., refineries, wind power installations, agriculture, forestry, the conservation of biodiversity). These functions have to be combined with human activities that are resource-using, such as housing, industry and transport. For some fortunate nations, space is not a problem except in limited areas (the USA would be a good example), or can be created if the demand exists (as with the recreational islands built off Abu Dhabi); for others, there are difficulties in reconciling all the legitimate demands for space, as in the Netherlands. Here, intertidal reclamation was the safety valve but this is becoming impossible in an era of rising sea levels.

See also: BIODIVERSITY, ENERGY, GENETIC MODIFICATION

Further reading

Boyle, G., 1996. *Renewable energy: power for a sustainable future.* Open University/Oxford University Press, Oxford.

Hilborn, R., Walters, C.J., and Ludwig, D., 1995. Sustainable exploitation of renewable resources. *Annual Review of Ecology and Systematics* 26, 45–67.

Just, R.E., Netangahu S., Olson, L.J. *et al.,* 2005. Depletion of natural resources, technological uncertainty, and the adoption of technological substitutes. *Resource and Energy Economics* 27, 91–108.

World Resources Institute: http://www.wri.org

RESTORATION OF ECOSYSTEMS (NEW FOCUS FOR WETLANDS)

Living on the coast of the Gulf of Mexico for centuries has been a story of coping with the spring floods (in United States) and summer-autumn floods (in Mexico), as well as fall major storms and hurricanes in both coasts. On one hand, floods and hurricanes are driving forces shape coastal landscapes and controls on what and where people can base their livelihoods. On the other hand, both driving forces control the natural production mechanisms in the inner sea shelf. A number of fish resources couple their lifecycles with these environmental processes and utilize the continuous landscape from the river basin to coastal lagoons and estuaries, deltas, and the adjacent ocean. These two forces also drive the diverse and productive array of coastal wetlands found in the Gulf of Mexico, but in the 21st century it is also necessary to consider how these coastal habitats both support and are affected by the coastal communities. Attempts to restore and sustain these habitats in the future must be set in a social, economic, and ecological context that incorporates an understanding the ecosystem approach, or ecosystem-based management. In the 21st century new components in the equation for restoring coastal habitats will become important including global climatic change, sea-level rise, energy crisis, self-sustaining ecosystems, and environmental units for dealing with such as problems.

Ecosystem-based management requires integration of multiple system components and uses, identifying and striving for sustainable outcomes, precaution in avoiding deleterious actions, and adaptation based on experience to achieve effective solutions. For example, an ecological and hydrologic restoration of the Mississippi–Ohio–Missouri (MOM) basin in the United States has been proposed as the solution to the recurring hypoxic conditions in the Gulf of Mexico, because of nitrate-nitrogen is the cause of this eutrophication in the Gulf and its source is mainly due to increased fertilizer use in the United States Midwest. Water quality problems have been greatly increased by the loss of wetlands, 80–90% in some states. In the entire Gulf of Mexico region, about 250 km² of wetlands are lost per year. Contributing to this loss is the isolation of rivers from deltaic systems, mostly due to the construction of flood control levees, and pervasive hydrological disruption of the deltaic plain. Coastal restoration efforts will have to be more intensive to offset the impacts of climate change including accelerated sea-level rise (i.e., 40 to 100 cm, or more, for the end of the 21st century), and changes in precipitation patterns. Future coastal restoration efforts should also focus on less energy-intensive, ecologically engineered management techniques that use the energy of nature as much as possible, because energy intensive restoration such as the pumping of dredged sediments will likely become much more expensive in the future. With increasing restoration initiatives for coastal wetlands, the self-sustaining ecosystem processes must be enhanced in restoration planning, design, and implementation. All of these approaches present commonalities and differences in the contrasting environmental sub-regions throughout the over 6,000 km of length in the Gulf of Mexico coastal zone form Florida to Yucatan.

Adaptive management is embraced as a central process in coastal ecosystem restoration, as for example in the Gulf of Mexico. The ecosystem approach-based management should be characterized by: (1) orienting more scientific activity to providing the solutions needed for ecosystem restoration; (2) building bridges across scientific and management barriers to more effectively integrate science and management; (3) directing more attention to understanding and predicting achievable restoration outcomes that consider possible state changes and ecosystem resilience; (4) improving the capacity of science to characterize and effectively communicate uncertainty; and (5) fully integrating modelling, observations, and research to facilitate more adaptive management. Benefits of this ecosystem approach for restoration include: (1) water quality improvement; (2) reduction of public health threats; (3) habitat creation and enhanced landscapes; (4) flood mitigation; and (5) significant economic savings. There is a need for formal and rigorous large-scale ecosystem research to reduce uncertainties.

Wetlands restoration, in addition to traditional approaches, should also be viewed as the restoration of natural capital from a functional ecosystem perspective. This perspective include consideration of ecological integrity and ecosystem services. As an example, Day *et al.* (2007) emphasized science must guide reestablishment of dynamic interactions in the restoration of the Mississippi delta. This included reconnecting the river to the deltaic plain. Costanza *et al.* (2006) suggested sustainable Mississippi delta restoration and storm protection be based on seven principles, which also applied to coasts in general.

1   Let the water decide.
2   Avoid abrupt boundaries between deepwater systems and uplands.
3   Restore natural capital.
4   Use the resources of the river-delta system to rebuild the coast, changing the current system that constrains the river between levees.

5   Restore the built capital of New Orleans to the highest standards of high-performance green building and a car-limited urban environment, with high mobility for everyone.

6   Rebuild the social capital to 21st century standards of diversity, tolerance, fairness, and justice.

7   Restore the river basin-deltaic system to minimize coastal pollution and the threats of river flooding.

See also: ECOLOGICAL ENGINEERING, ECOSYSTEM-BASED MANAGEMENT

Further reading

Boesch, D.F., 2006. Scientific requirements for ecosystem-based management in the restoration of Chesapeake Bay and Coastal Louisiana. *Ecological Engineering* 26, 6–26.

Costanza, R., d'Arge, R., de Groot, R. *et al.*, 1997. The value of the world's ecosystem services and natural capital. *Nature* 387, 253–260.

Costanza, R., Mitsch, W.J. and Day, J.W., 2006. A new vision for New Orleans and the Mississippi delta: applying ecological economics and ecological engineering. *Frontiers in Ecology* 4 (9), 465–472.

Day, J.W. and Yáñez-Arancibia A. (eds), 2008. The Gulf of Mexico: Ecosystem-Based Management. The Harte Research Institute for Gulf of Mexico Studies. Texas A and M University Press, College Station, Texas (in press).

Day, J.W., Barras, J., Clairain, E., Johnston, J., Justix, D., Kemp, P., Jae-Young Ko, Lane, R., Mitsch, W., Steyer, G., Templet, P., and Yáñez-Arancibia, A., 2005. Implications of global climatic change and energy cost and availability for the restoration of the Mississippi delta. *Ecological Engineering* 24, 253–266.

Day, J.W., Boesch, D., Clairain, E., Kemp, P., Laska, S., Mitsch, W., Orth, K., Mashriqui, H., Reed, D., Shabman, L., Simenstad, C., Streever, B., Twilley, R., Watson, C., Wells, J., and Whigham, D., 2007. Restoration of the Mississippi delta: lessons from hurricanes Katrina and Rita. *Science* 315, 1679–1684.

Mitsch, J.W. and Day, J.W., 2006. Restoration of wetlands in the Mississippi–Ohio–Missouri (MOM) river basin: experience and needed research. *Ecological Engineering* 26, 55–69.

Yáñez-Arancibia, A. and Day, J.W., 2004. Environmental sub-regions in the Gulf of Mexico coastal zone: the ecosystem approach as an integrated management tool. Ocean and Coastal Management, 47, 727–757.

Yáñez-Arancibia, A., Day, J.W., Mitsch, W. J., and Boesch, D. F., 2006. Following the ecosystem approach for developing projects on coastal habitat restoration in the Gulf of Mexico. Commission on Ecosystem Management Newsletter 5, 2006, Highlights News, CEM-IUCN, Gland Switzerland. www.iucn.org/themes/cem/documents/cem/members_2006/restoration_esa_a.yanez_arancibia_nov2006.pdf

Yáñez-Arancibia, A., Day, J.W., Twilley, R.R., Mitsch, W.J., 2007. Ecosystem approach for restoring wetlands because of global changes. *Ambientico* 165, 35–38.

RISK-BASED MANAGEMENT

Risk-based management is a process of measuring or assessing risk, and then developing strategies to manage it. The strategies include avoiding the risk, transferring the risk to another party, reducing the negative effect of the risk, and accepting some or all of the consequences of a particular risk. Traditional risk management focuses on risks from physical or legal causes (e.g. natural disasters or fires, accidents, death, and lawsuits). Financial risk management focuses on risks that can be managed using traded financial instruments. Regardless of the type of risk management, most corporations have risk management strategies. For ideal risk management, a prioritization process is followed so that the risks with the greatest loss and the greatest probability of occurring are addressed first, while risks with lower probability of occurrence and lower loss are approached later. Balancing between risks with a high probability of occurrence but lower loss vs. risks with high loss but lower probability of occurrence often can be challenging.

Intangible risk management identifies a new type of risk – a risk that has a 100% probability of occurring but is ignored by the organization due to a lack of identification ability. Risk management faces a difficulty in allocating resources properly – an opportunity cost. Resources spent on risk management could be spent instead on more profitable opportunities. Ideal risk management spends the least amount of resources in the process while reducing the negative effects of risks as much as possible.

See also: ENVIRONMENTAL MANAGEMENT, RISK ASSESSMENT

Further reading

Alijoyo, A., 2004. *Focused enterprise risk management*. PT Ray, Jakarta, Indonesia.

Damen, K., Faaij, A. and Turkenburg, W., 2006. Health, safety and environmental risks of underground CO_2 storage – overview of mechanisms and current knowledge. *Climatic Change* 74, 289–318.

Dorfman, M.S., 2008. *Introduction to risk management and insurance*. Prentice Hall, New York, 9th edn.

Jones, R.B., 1995. *Risk-based management*. Gulf Pub Co (Elsevier), Amsterdam.

Stulz, R., 2003. *Risk management and derivatives*. Thomson South-Western, Mason OH.

Thomsett, R., 2002. *Radical project management*. Prentice Hall, Upper Saddle River, NJ.

RISK ASSESSMENT

R isk assessment is an evaluation of risk of that hazard before action is pursued to address it. Risk assessment, for example, has been at the root of studying cancer risks associated with pesticides and some toxic chemicals. Most assessment has been created to study risks to human health. There are several stages in the process of Risk Assessment – assessment of the hazard, study of the exposure of humans to various levels of a hazard, and a risk characterization and probability of fatal levels of the hazard. Often, animal tests are used to determine the dose-response and exposure assessment of a hazard that involves pollutants and chemicals. The study of epidemiology is the analysis of how a sickness, for example, multiplies through a population. As a result many chemicals can be determined to be human carcinogens. Regulatory agencies rely on scientific analysis of this type to determine regulatory legislation for enhancing quality of life of the citizenry. Table 57 represents an example of results of a risk assessment of many situations.

Table 57

Hazard	Annual Risk (probability of dying)
Cigarette smokers	10 per 1,000
All cancers	2.0 per 1,000
Firefighters	4.0 per 10,000
Police on duty	2.9 per 10,000
Hang gliding	2.6 per 10,000
Air pollution	2.5 per 10,000
Vehicle accident	1.6 per 10,000
Snowmobiling	1.3 per 10,000
Home accidents	1.1 per 10,000
Firearms	1.1 per 10,000
Airline pilot	10 per 100,000
Mountain hiking	6.4 per 100,000
Alcohol consumption	6.4 per 100,000
Boating	5 per 100,000
Swimming	3 per 100,000

See also: HAZARDS

Further reading

Wilson, R. and Crouch, E.A.C., 2001. *Risk–benefit analysis*. Harvard Center for Risk Analysis, Cambridge, MA, 2nd edn.

SENSITIVITY

A term referring to the likelihood of responding to slight changes. In environmental sciences it has been used to connote the propensity of a system to respond to a minor external change. It therefore embraces the proximity of a system to a threshold; if it is near and sensitive it will readily respond to an external influence. Sensitivity has been employed in relation to individual landforms where it has been described as the likelihood, extent and rapidity with which a given landform will change in response to a single unusual process event or to a reinforcing sequence of such events. Landscape stability can be regarded as a function of the temporal and spatial distributions of the resisting and disturbing forces, being described by the landscape change safety factor which is considered to be ratio of the magnitude of barriers of change to the magnitude of the disturbing forces. Sensitivity of river channels can be an integral component in successfully evaluating the risk of geomorphological hazards in river channel management. Hypersensitivity denotes an apparently disproportionately large channel adjustment resulting from a small hydrological change, whereas undersensitivity applies to a disproportionately small response. Sensitivity can also be used when disturbance regimes in ecology are disrupted by anthropogenic impacts, and sensitivity has been described more generally (e.g., by UNEP) as the degree to which a system will change in response to climate conditions including changes in ecosystem composition, structure, and functioning. This has also been accompanied by concepts of adaptability and vulnerability. Environmental sensitivity mapping is a technique developed to integrate numerous datasets into a single composite layer in a GIS, thus giving a rapid, objective and straightforward method of identifying areas which may be particularly sensitive to development. This is somewhat analogous to sensitivity analysis which is the study of how the variation in the output of a model (numerical or otherwise) can be apportioned, qualitatively or quantitatively, to different sources of variation.

See also: EARTH SURFACE PROCESSES, PULSING, THRESHOLDS

Further reading

Brunsden, D. and Thornes, J.B., 1979. Landscape sensitivity and change. *Transactions of the Institute of British Geographers* 4, 463–484.

Downs, P.W. and Gregory, K.J., 2004. *River channel management*. Arnold, London.

Ward, J.V., 1998. Riverine landscapes: biodiversity patterns, disturbance regimes, and aquatic conservation. *Biological Conservation* 83, 269–278.

SUSTAINABILITY

'The emergence of sustainability science builds toward an understanding of the human-environment condition with the dual objectives of meeting the needs of society while sustaining the life support systems of the planet' (National Academy of Science, 2003). Sustainability is a concept to provide the best outcomes for the human and natural environments both now and into the future. The concept relates to social, economic, environmental, and institutional aspects of human society and the physical environment. Sustainability is accomodating human activity so economies are able to achieve their potential in the present, while preserving biodiversity and natural ecosystems, and acting and planning for the ability to maintain these ideals for the future. It affects every level of the organization, from local communities to the entire Earth. The word sustainability (German: Nachhaltigkeit) was used for the first time in 1712 by the German forester and scientist Hans Carl von Carlowitz) in his book *Sylvicultura Oeconomica*. French and English scientists adopted the concept of planting trees and used the term 'sustained yield forestry'. The 1987 Brundtland Report defined sustainable development as development that 'meets the needs of the present generation without compromising the ability of future generations to meet their own needs'. The term 'sustainable development' was adopted by the Agenda 21 program of the United Nations. The 1995 World Summit on Social Development further defined this term as 'the framework for our efforts to achieve a higher quality of life for all people', in which 'economic development, social development and environmental protection are interdependent and mutually reinforcing components'. The 2002 World Summit on Sustainable Development expanded this definition identifying the "three overarching objectives of sustainable development" to be (1) eradicating poverty, (2) protecting natural resources, and (3) changing unsustainable production and consumption patterns. With burgeoning world populations, pressures on land and ocean resources, tension on the geopolitcal spheres, and the specter of global climate change in the next century, sustainability will attain increased attention in world summits to ensure future generation survival on Earth.

See also: SUSTAINABLE DEVELOPMENT, UN PROGRAMMES

Further reading

Allen, P. (ed), 1993. *Food for the future: conditions and contradictions of sustainability.* Wiley, New York.

Bartlett, A., 1994. Reflections on sustainability, population growth, and the environment. *Population and Environment,* 16, 5–35.

Brown, M.T. and Ulgiati, S., 1999. Energy evaluation of natural capital and biosphere services. *Ambio,* 28, 486–493.

Brundtland, G.H. (ed.), 1987. Our common future: The World Commission on Environment and Development. Oxford University Press, Oxford.

Dalal-Clayton, B. (1993) Modified EIA and indicators of sustainability: first steps towards sustainability analysis. Environmental Planning Issues No. 1, International Institute For Environment and Development, Environmental Planning Group, Washington DC. Available at http://www.iied.org/pubs/pdf/full/7766IIED.pdf

Daly H., 1996. *Beyond growth: the economics of sustainable development.* Beacon Press, Boston.

Gallopin, G.A., Hammond, Raskin, P. and Swart, R., 1997. Branch points: global scenarios and human choice. Stockholm Environment Institute. PoleStar Series Report No. 7. http://www.gsg.org.

Hargroves, K. and Smith, M. (eds.), 2005. *The natural advantage of nations: business opportunities, innovation and governance in the 21st Century.* Earthscan, London. (See the book's online companion at www.thenaturaladvantage.info)

Hawken, P., Lovins, A., and Lovins, L.H., 1999. *Natural capitalism: creating the next industrial revolution.,* Earthscan, London (Available at from www.natcap.org)

Nelson, E.H. 1986. *New values and attitudes throughout Europe.* Taylor-Nelson, Epsom.

SUSTAINABLE DEVELOPMENT

Sustainable development creates ways to increase relief from poverty, make for equitable standards of living, satisfy the basic needs of all peoples, and establish sustainable political practices all while taking the steps necessary to avoid irreversible damages to natural capital in the long term. The field of sustainable development can be conceptually broken into four constituent parts: environmental sustainability, economic sustainability, social sustainability and political sustainability. Sustainable development does not focus solely on environmental issues. More broadly, sustainable development policies encompass three general policy areas: economic, environmental and social. In support of this, several United Nations texts, most recently the 2005 World Summit Outcome Document, refer to the 'interdependent and mutually reinforcing pillars' of sustainable development as economic development, social development, and environmental protection. The Universal Declaration on Cultural Diversity of 2001 elaborates further the concept by stating that '...cultural diversity is as necessary for humankind as biodiversity is for nature'; it becomes 'one of the roots of development understood not simply in terms of economic growth, but also as a means to achieve a more satisfactory intellectual, emotional, moral and spiritual existence'. In this vision, cultural diversity is the fourth policy area of sustainable development. Sustainable development can also be explained as a process rather than an end goal.

Green development is generally differentiated from sustainable development in that Green development prioritizes environmental sustainability over economic and cultural sustainability. Green development is not necessarily practical in all applications, so sustainable development provides a context in which to improve overall sustainability when cutting edge Green development is unattainable. Sustainable development is a notoriously ambiguous concept, as a wide array of views have fallen under its umbrella. The concept has included notions of weak sustainability, strong sustainability and deep ecology. The concept remains weakly defined and contains a large amount of debate as to its precise definition. Environmental degradation occurs when nature's resources (such as trees, habitat, earth, water and air) are being consumed faster than nature can replenish them, when pollution results in irreparable damage done to the environment or when human beings destroy or damage ecosystems in the process of development. Environmental degradation can take many forms including desertification, deforestation, extinction and radioactivity. Some of the major causes of such degradation include: overpopulation, urban sprawl, industrial pollution, waste dumping, intensive farming, over fishing, industrialization, introduction of invasive species and a lack of environmental regulations. The goal of environmental sustainability is to minimize or bring to halt these degradations and, ideally, reverse the processes. An unsustainable situation occurs when natural capital (the sum total of nature's resources) is used up faster than it can be replenished. Sustainability requires that human activity, at a minimum, only uses nature's resources at a rate at which they can be replenished naturally. Theoretically, the long term final result of environmental degradation would result in local environments that are no longer able to sustain human populations. Such degradation on a global scale would lead to extinction for humanity. In the short-term, environmental degradation leads to declining standards of living, the extinctions of large numbers of species, health problems in the human population, conflicts, sometimes violent, between groups fighting for a dwindling resource, water scarcity and many other major problems.

See also: SUSTAINABILITY

Further reading

Green Building Council – http://www.cascadiagbc.org

Tellus Institute – Organization that uses scenario analysis to inform Sustainable Development practices: http://www.tellus.org/

http://www.sustainable-development.gov.uk/ – a national government's attempt to formulate a pathway.

International Journal of Ecological Economics and Statistics (IJEES) – http://www.ceser.res. in/ijees.html

THERMOHALINE CIRCULATION

Much of the circulation of the oceans, especially in water deeper than a few hundred metres is driven by density differences between water masses. Density of ocean water is mostly related to temperature and salinity. Circulation caused by these density differences is called Thermohaline Circulation. This type of circulation is responsible for the circulation of water in the ocean. Thermohaline circulation is responsible for vertical water movements in the ocean and for circulation of the oceans a whole. Ocean water is stratified with the densest water at the bottom (very cold, high salinity water) and the least dense water at the surface (warmer, lower salinity water). The differences in density give rise to a number of different water masses which do not mix much.

At temperate and lower latitudes, five water masses are generally identified. Surface water is warmer, low salinity surface water to a depth of about 200 metres (this is the maximum depth that surface water can be mixed). Central water is located below the surface water to bottom of the thermocline (the zone where temperature changes most rapidly). Intermediate water is below the central water to a depth of about 1500 metres. Deep water is below the intermediate water to a depth of about 4000 m, but it does not come into contact with the bottom. Bottom water is the coldest, most dense water mass and is in contact with the bottom. The surface currents along the western boundaries of the ocean carry warm water from the tropics towards the poles. At higher latitudes, this water cools and sinks where it becomes deep water and bottom water. This sinking occurs to the greatest extent in the North Atlantic. The sinking water forms a water mass called North Atlantic deep water. This water then moves southerly towards the Southern Hemisphere. There is also a great deal of water that cools during the Southern Hemisphere winter and sinks along Antarctica, mostly south of South America, forming Anarctic bottom water. This water is the densest water in the ocean with a temperature of –0.5°C, a salinity of 34.65 parts per thousand, and a density of 1.0279 grams per cubic metre. It then comes to the surface in the Indian and Pacific oceans. Water movement at depth is very slow, and for example, it may take Antarctic bottom water a thousand years to reach the equator. This transport of water is called the global conveyor belt. One concern of global climate change is that cooling will be less at higher latitudes and that the conveyor belt will slow and perhaps even stop, with great consequences for global climate. The global nature of this circulation is illustrated by the measurements that indicate that some of the water that warms the coasts of western Europe originates near Australia and Indonesia and flows across the Indian Ocean, around southern Africa and through the Atlantic where it ultimately joins the Gulf Stream.

See also: OCEAN CIRCULATION

Further reading

Aken, H.M., 2007. *The oceanic thermohaline circulation: an introduction.* Springer, New York.

WASTE MANAGEMENT

For many years, the approach to waste management was to simply release the waste back into the environment and depend on dilution and dispersement to take care of the problem. In the past, this led to only localized problems because human populations were low. But dealing with wastes is a critical problem with large and dense populations and great economic activity. When dispersal and dilution did not work, the objective became to concentrate wastes, often at a single site such as municipal garbage dump. This has led to land and water pollution. Today, the main approach to managing waste is called integrated waste management. This includes an integrated approach that emphasizes re-use, source reduction, recycling, composting, landfill, and incineration. Integrated waste management can include both conventional industrial approaches and alternative, more environmentally friendly approaches such as ecological engineering and ecotechnology. Reduction of inputs to the waste stream, and re-using and recycling could reduce solid waste generation by 50% or more. The EPA reported that about half of solid waste was paper or yard trimmings. Much of this could be recycled or composed. Metals, plastics, wood, and glass are also important in solid waste and much could be recycled. A reduction in solid wastes would be good for the environment since the waste causes a number of problems. Pollutants from sanitary landfills can enter the environment via escape to the atmosphere, incorporation into the soil, into surface or groundwater, and incorporation into plant material. Hazardous wastes pose a serious problem because they must be kept separate from the environment. A number of approaches have been developed to deal with these wastes. Environmental justice is an aspect of waste management that looks at inequities related to the storage and processing of wastes. In general, waste facilities tend to be located near areas that have populations that are poor, minority, or disadvantaged. There is a growing movement to deal with these inequities.

See also: HAZARDS, POLLUTION OF AIR AND WATER

Further reading

Botkin, D.B. and Keller, E.A., 2007. *Environmental science – earth as a living planet.* Wiley, New York.

Brewer, R., 1994. *The science of ecology.* Saunders College Publications, London and Fort Worth TX, 2nd edn.

Environment Agency [of England and Wales], 2006. *What happened to waste – how to find out about waste production and management.* Environment Agency, Bristol.

Kangas, P.C., 2004. *Ecological engineering – principles and practice.* Lewis Publishers. Boca Raton, FL.

Mitsch, W.J. and Jørgensen, S.E., 2003. *Ecological engineering and ecosystem restoration.* Wiley, New York.

WATER RESOURCES

Water resources are those sources of water that are accessible for human use including for agriculture (more than 70% of water consumed with most for irrigation and livestock), industry (c. 20%), domestic (c. 6%) purposes. Just a small portion of the water in the world hydrological cycle (approximately 0.08%) is available for water resources. After use some water is returned to the environment whereas some is incorporated into products. Some parts of the world have more water available than others, and the demand for water is increasing (tripled during the second half of 20th Century, and agriculture could require 20% more by 2020), possibly by as much as 40% in the next 20 years. Therefore the potential water shortage, together with global warming, has been identified by the UNEP (United Nations Environment Programme) as one of two major problems for the 21st Century. International organizations have taken initiatives in relation to critical issues of water quantity, quality and availability such as *Vital Water Graphics*, produced by UNEP in 2002, including assessment of global water resources, the provision of early warnings on water issues; Agenda 21; The World Water Council (established in 1996 in response to increasing concern from the global community about world water issues); the World Commission on Water for the 21st century.

Water is obtained from surface supply from rivers, impoundments (large or small dams); or from ground water supply. Ground water represents about 90% of the world's readily available freshwater resources, and some 1.5 billion people depend upon ground water for their drinking water. In the USA, aquifers provide 50% of drinking water and 80% of rural domestic and livestock consumption. Increasingly, governments are seeking to solve their water problems by using subterranean supplies of ground water, despite the fact that the aquifers are being exploited by ground water mining at a rate more rapid than the rate of recharge.

As demand for water increases conflicts can arise, particularly with agricultural users. Water shortages are greatest in poorer countries where resources are limited and population growth is rapid including Africa, parts of Asia and the Middle East. Today, water scarcity already affects one third of the total world population, one person in five has no access to safe drinking water, one in two lacks safe sanitation, and polluted water is estimated to affect the health of 1.2 billion people, and contributes to the death of 15 million children annually.

Other problems include:

» more than one-half of the world's major rivers are being seriously depleted and polluted, thus threatening the health of people who depend upon them and including eutrophication, whereby accumulation of nutrients encourages growth of algae and vascular plants;
» drying out of rivers, wetlands and lakes includes the Amu Darya in Central Asia, the Colorado in the southwestern United States, and the Yellow River in China. Reduced river flows means that lakes decrease: the Dead Sea has dropped by 25 metres (82 feet) in the past 40 years. In Central Asia, the Aral Sea has lost >40% of its area and 60% of volume since 1960;
» lowering of ground water levels can lead to saline intrusions near the coast, and to surface subsidence (e.g. Bangkok, Mexico City and Venice);
» water wastage, with inefficient irrigation and domestic supply;
» global climate change could bring more precipitation to some regions and less to others, with major implications for water resources.

New sources of water and improvement in resource use include:

» desalination, expensive but already necessary in some parts of the world;
» new irrigation systems;
» planting less water-intensive crops;
» reduction of pollutants in agriculture;
» reduction of leakage in agricultural and domestic supply systems;
» refinement of water treatment facilities which do not have to produce drinking water quality for all purposes;
» greater re-use of water;
» multiple supply of water of different quality and used for different purposes.

In future water resource development it is increasingly appreciated that integrated schemes of water resources development and management together with responsible water-saving behaviour are necessary. National policies and laws are necessary for managing water quality and quantity, together with integrated basin management where appropriate, and international initiatives as required.

See also: HYDROLOGICAL CYCLE, INTEGRATED BASIN MANAGEMENT, RESOURCE DEPLETION

Further reading

Mygatt, E., 2006. World's water resources face mounting pressure. Earth Policy Institute, Washington, DC – www.Earth-Policy.org

Shiklomanov, I.A., 1993. World fresh water resources. In Gleick, P.H. (ed), *Water in crisis*. Oxford University Press, New York and Oxford, 13–24.

UNEP, 2002. Vital water graphics – www.unep.org/vitalwater

World Water Council World Commission on Water for the 21st century – www.worldwatercouncil.org

ZOOS, BOTANICAL GARDENS AND MUSEUMS

A zoo, zoological garden or zoological park is a facility where animals are confined in enclosures and displayed to the public; since ancient times rulers kept menageries of which the modern zoo is a descendant, with the first founded in Vienna, Austria in 1752, the *Ménagerie du Jardin des Plantes* in Paris in 1794 primarily for scientific and educational research, the London Zoological Garden in 1828; there are more than 1000 animal collections throughout the world, some of which concentrate on rare species with the hope of reintroducing them into the wild.

Museums gained enormous impetus during the 19th century and can be defined as 'institutions that collect, safeguard and make accessible artifacts and specimens, which they hold in trust for society'. Museums display archaeological artifacts, geological specimens or natural history collections, but their public displays are normally only a small part of their activity since they host primary scientific work (the Smithsonian Museum in Washington, DC is an outstanding example) as well as important reference collections, of rocks or animal specimens for example.

Botanical gardens which are now concerned with exhibiting ornamental plants were initially, in China and in Mediterranean countries, often used for growing plants for medicines and food and provided collections of living plants sometimes used to illustrate relationships within plant groups. The first modern botanical gardens were established in Italy in Pisa (1544), Padua (1545) and Florence (1545) and from the 17th Century onwards they sent plant-collecting expeditions to various parts of the world, thus establishing their research traditions on classification, propagation and adaptation of species. Many large cities now have botanical gardens and one of the most well known is Kew Gardens, London, UK, developed since the mid-18th century.

Why are environmental scientists interested in such facilities? Primarily because they provide ways in which aspects of environment were, and still are, introduced to the general public, but in addition they can be repositories for threatened species and plants, and the location for research especially in taxonomy. Kew, for example, plans to bank seeds from 10% of the world's species by 2010. Zoos, museums and botanical gardens have all evolved their missions, with conservation becoming more prominent from the 1970's, to be appropriate to the 21st Century. In addition other types of collections have developed including wildlife parks, the first established at Whipsnade in the UK in 1931, aquaria, marine mammal parks, safari parks, and larger enterprises which display whole ecosystems; the Eden Project (Fig. 81) opened in 2001 in Cornwall, UK, was constructed in a disused china clay pit, includes two giant biomes in conservatories exhibiting many world plant communities including tropical rainforest, and aims to tell the story of man's dependence on plants and to reconnect people with their environment both locally and globally. Biosphere 2 near Tucson, Arizona also includes large

Figure 81 *The Eden Project, Cornwall (Stock photo)*

conservatories, has been managed by the University of Arizona since 2007 and is now dedicated to research and the understanding of global scientific issues. All these facilities aim to provide an enjoyable recreational experience, but also have a more specialized role in research or collections. These two functions are becoming closer with greater public awareness of increasing human impact on the planet.

See also: CONSERVATION OF NATURAL RESOURCES

Further reading

Hanson, E., 2002. *Animal attractions*. Princeton University Press, Princeton NJ.

Kisling, V.N. (ed.) 2001. *Zoo and aquarium history*, CRC, Boca Raton, FL.

The world zoo and aquarium conservation strategy – http://www.waza.org/conservation/wzacs.php

Kew Gardens: http://www.kew.org/aboutus/annualreview07.pdf

INDEX

Headings for entries are shown in caps with their page numbers in bold; figures and tables are shown in italics.

The Qualitative Research Kit

Edited by Uwe Flick

Read sample chapters online now!

Doing Ethnographic and Observational Research — Michael Angrosino

Using Visual Data in Qualitative Research — Marcus Banks

Doing Focus Groups — Rosaline Barbour

Designing Qualitative Research — Uwe Flick

Managing Quality in Qualitative Research — Uwe Flick

Analyzing Qualitative Data — Graham Gibbs

Doing Interviews — Steinar Kvale

Doing Conversation, Discourse and Document Analysis — Tim Rapley

The SAGE Qualitative Research Kit
Edited by Uwe Flick

www.sagepub.co.uk

⑤SAGE